Monographs in Theoretical Computer Science
An EATCS Series

Springer

Berlin
Heidelberg
New York
Barcelona
Hong Kong
London
Milan
Paris
Singapore
Tokyo

Aldo de Luca · Stefano Varricchio

Finiteness and Regularity in Semigroups and Formal Languages

Springer

Authors

Prof. Dr. Aldo de Luca
Dipartimento di Matematica
Università di Roma „La Sapienza"
Piazzale Aldo Moro, 2
I-00185 Roma, Italy
deluca@mat.uniroma1.it

Prof. Dr. Stefano Varricchio
Dipartimento di Matematica
Università di Roma „Torvergata"
Viale della Ricerca Scientifica
I-00133 Roma, Italy
varricchio@mat.uniroma1.it

Series Editors

Prof. Dr. Wilfried Brauer
Department of Computer Science
Technical University of Munich
Arcisstrasse 21
D-80333 München, Germany
brauer@informatik.tu-muenchen.de

Prof. Dr. Grzegorz Rozenberg
Department of Computer Science
University of Leiden
Niels Bohrweg 1, P.O. Box 9512
NL-2300 RA Leiden, The Netherlands
rozenber@wi.leidenuniv.nl

Prof. Dr. Arto Salomaa
Data City
Turku Centre for Computer Science
FIN-20520 Turku, Finland
asalomaa@utu.fi

Library of Congress Cataloging-in-Publication Data

de Luca, A. (Aldo)
Finiteness and regularity in semigroups and formal languages /
Aldo de Luca, Stefano Varricchio.
p. cm. -- (Monographs in theoretical computer science)
Includes bibliographical references and index.
ISBN 978-3-642-64150-3 (hc.: alk. paper)
1. Semigroups. 2. Formal languages. I. Varricchio, Stefano, 1960- . II. Title.
III. Series.
QA182.D4 1998
512'. 2--dc21 98-42554 CIP

ACM Computing Classification (1998): F.4.2-3, G.2.1

ISBN-13: 978-3-642-64150-3 e-ISBN-13: 978-3-642-59849-4
DOI: 10.1007/978-3-642-59849-4

Softcover reprint of the hardcover 1st edition 1999

Typesetting: Camera ready by authors
Production: PRO EDIT GmbH, D-69126 Heidelberg
Cover Design: *design & production*, D-69121 Heidelberg
SPIN: 10552017 45/3142 5 4 3 2 1 0 – Printed on acid-free paper

Preface

The aim of this monograph is to present some recent research work on the combinatorial aspects of the theory of semigroups which are of great interest for both algebra and theoretical computer science. This research mainly concerns that part of combinatorics of finite and infinite words over a finite alphabet which is usually called the theory of "unavoidable" regularities.

The unavoidable regularities of sufficiently large words over a finite alphabet are very important in the study of finiteness conditions for semigroups. This problem consists in considering conditions which are satisfied by a finite semigroup and are such as to assure that a semigroup satisfying them is finite. The most natural requirement is that the semigroup is ***finitely generated.*** If one supposes that the semigroup is also ***periodic*** the study of finiteness conditions for these semigroups (or groups) is called the ***Burnside problem for semigroups*** (or groups). There exists an important relationship with the theory of ***finite automata*** because, as is well known, ***a language L over a finite alphabet is regular*** (that is, recognizable by a finite automaton) ***if and only if its syntactic monoid*** $S(L)$ ***is finite.*** Hence, in principle, any finiteness condition for semigroups can be translated into a regularity condition for languages. The study of finiteness conditions for periodic languages (i.e., such that the syntactic semigroup is periodic) has been called the ***Burnside problem for languages.***

Several finiteness conditions for finitely generated semigroups have been given in recent years based on different concepts such as: ***permutation properties, iteration conditions, minimal conditions on ideals, and repetitivity.***

These conditions are analyzed in some detail in Chap. 3. They are based, as we said before, on the existence of some different unavoidable regularities on very large words over a finite alphabet. As we shall see the permutation conditions are related to Shirshov's theorem and the iteration conditions to bi-ideal factorizations. A very recent result shows that these two regularities "appear" simultaneously in a suitable way in very large words. This fact gives rise to a new finiteness condition in which any sequence of $n > 1$ elements of the semigroup can be either permutable or iterable (on the right). Repetitivity is a concept related to the unavoidable regularities expressed in the theorems of van der Waerden and of Brown.

We present also finiteness conditions for semigroups based on *chain conditions*. In particular, we consider some remarkable generalizations of a theorem of Hotzel and of a theorem of Coudrain and Schützenberger. From these one derives an extension of the theorem of Green and Rees relating the bounded Burnside problem for semigroups with the corresponding problem for groups, and a new simple proof of the theorem of McNaughton and Zalcstein which gives a positive answer to the Burnside problem in the case of $n \times n$ matrices with elements in a field. The proof of these results requires also some structure theorems on semigroups based on the Green relations, as the $\mathcal{J}$-depth decomposition theorem, which are given in Chap. 3.

Chapter 4 concerns the following general problem: given a semigroup S, under what conditions can we say that the finite parts of S are recognizable sets? This is also equivalent to the following problem: let θ be a congruence in a finitely generated free semigroup A^+; when are the congruence classes of θ in A^+ *regular languages*? A semigroup whose finite parts are recognizable is called *finitely recognizable*. Some general results relating this problem to the $\mathcal{J}$-depth decomposition of S are shown. In particular we refer to the case when the semigroup S is the quotient semigroup $M_n = A^*/\theta_n$, where θ_n is the congruence generated by the relation $x^n = x^{n+1}$. The problem of the regularity of the congruence classes (noncounting classes) was posed for any $n > 0$ by Brzozowsky about thirty years ago. The authors have proved that this problem has a positive answer for $n > 4$. In this proof the finiteness condition for semigroups due to Hotzel and introduced in Chap. 3 has been used. The proof of this result allows one to show also that the *word problem* for the semigroup M_n when $n > 4$ is recursively solvable. This result was subsequently improved by other authors for the case $n > 2$ and extended to more general cases.

Chapter 5 deals with the Burnside problem for languages. From the finiteness conditions for semigroups one can easily find some *uniform* conditions which assure the regularity of a periodic language. However, the Burnside problem for languages is more complicated since the regularity conditions can be presented in a *non-uniform* way, that is, they depend on the contexts which complete the words in the language. The use of Ramsey's theorem is often a good tool to transform non-uniform conditions into uniform ones. Some important regularity conditions such as the *block-pumping* property of Ehrenfeucht, Parikh, and Rozenberg and the *permutative property* of Restivo and Reutenauer are proved. Moreover, the existence of a non-uniform and *positive* block pumping property is shown.

In Chap. 6 we present some further combinatorial aspects of semigroups related to the notion of *well quasi-order* which gives a new insight into the combinatorics of the free monoids.

Classical theorems of automata theory, such as the Myhill theorem, are extended by some regularity conditions based on well quasi-orders. For instance, one has that *a language is regular if and only if it is a closed part of a monotone well quasi-order*. Some applications of these notions and techniques for the regularity conditions are given.

In conclusion, this monograph, which greatly extends and completes a chapter written on the subject for a Handbook on formal languages [54], presents very recent research work, including new unpublished results, on those combinatorial aspects of the theory of semigroups which are intimately related to the fundamental property of finite automata, namely finiteness. This relation with automata gives rise to regularity conditions for formal languages. Of course there exist other regularity conditions, based on different techniques and concepts, which are not covered by this volume (see, e.g., [12, 35, 61, 79]).

The methods used are of a combinatorial and algebraic nature. However, the book presupposes no prior and deep knowledge of any particular mathematical topic, but only the background of an undergraduate student in mathematics or computer science, and the knowledge of some rudiments of the theory of automata and semigroups.

We thank deeply Dr. Flavio D'Alessandro, who patiently read the entire manuscript and helped us in improving the presentation.

Rome, February 1999 *A. de Luca, S. Varricchio*

Contents

1. Combinatorics on Words

The combinatorics of finite, as well as infinite, sequences of symbols (words) over a finite set can be considered at the present time as an independent mathematical topic which has arisen in such diverse fields as logic, algebra, physics, computer science, and, more recently, biology. The theorems and results of this theory are of wide interest and importance for the great number of applications in various fields. A first book, *Combinatorics on Words* by M. Lothaire [104], collects various basic results of the theory. Since there has been a great development of the theory during these last ten years, a second volume by M. Lothaire [105] called *Algebraic Combinatorics on Words*, presenting the most recent results of the research on the subject, will appear soon.

In this chapter we shall be concerned with some classical and basic notions of combinatorics on words such as ***periodicity, conjugation, Lyndon words,*** and ***subword complexity.*** We shall mainly deal with those aspects of the theory which will be used in the text and, in particular, in the next chapter which is devoted to the "unavoidable regularities" on free monoids.

1.1 Preliminaries

In the following A will denote a finite ***alphabet***, that is, a finite nonempty set whose elements are called ***letters.*** By A^+ we denote the set of all finite sequences of letters, or finite ***words.*** A finite word, or simply word, w can be uniquely represented by a juxtaposition of its letters:

$$w = a_1 \cdots a_n,$$

with $a_i \in A$, $1 \leq i \leq n$. The integer n is called the ***length*** of w and is denoted by $|w|$. The set A^+ is called the ***free semigroup*** on A, where the semigroup operation, called ***product***, is defined by concatenation or juxtaposition of the words. Thus if $u = a_1 \cdots a_n$ with $a_i \in A$, $1 \leq i \leq n$ and $v = b_1 \cdots b_m$ with $b_j \in A$, $1 \leq j \leq m$ are words, then their product uv is the word

$$uv = a_1 \cdots a_n b_1 \cdots b_m.$$

The product of words is associative, i.e., for all words u, v, w one has

$$u(vw) = (uv)w.$$

If one adds to A^+ the *identity* element ϵ, called the *empty word*, then one obtains the *free monoid* A^* over A. The length of ϵ is taken to be equal to 0. For any word $w \in A^*$ and $n \geq 0$, the *n-power* w^n is defined as

$$w^0 = \epsilon, \quad w^n = ww^{n-1}, \text{ for } n > 0.$$

A word $u \in A^*$ is a *factor* (or *subword* or *block*) of w if there exist words $p, q \in A^*$ such that $w = puq$. The factor u is called *proper* if $u \neq w$. If $p = \epsilon$ ($q = \epsilon$), then u is called *prefix* (*suffix*) of w. A prefix (suffix) is also called *left factor* (*right factor*).

Let $u \in A^*$. For any letter $a \in A$ we denote by $|u|_a$ the number of occurrences of the letter a in u. One has

$$|u| = \sum_{a \in A} |u|_a.$$

Two words $u, v \in A^*$ are *commutatively equivalent* if for all $a \in A$

$$|u|_a = |v|_a.$$

If $w = a_1 \cdots a_n$ is a word, $a_i \in A$, $i = 1, \ldots, n$, then the *reversal* $w^\sim$ of w is the word

$$w^\sim = a_n \cdots a_1.$$

Moreover, one sets $\epsilon^\sim = \epsilon$. A word w is called a *palindrome* if $w^\sim = w$.

In a free monoid A^* the following important property, usually referred as the *equidivisibility property* or *lemma of Levi*, holds. Let u, v, u', v' be words such that

$$uv = u'v'.$$

If $|u| \geq |u'|$ ($|u| \leq |u'|$), then there exists a word $\zeta \in A^*$ such that

$$u = u'\zeta \ (u' = u\zeta) \quad \text{and} \quad \zeta v = v' \ (v = \zeta v').$$

For each $n \geq 0$ we denote by A^n the set of all the words of length n. Moreover, $A^{[n]}$ will be the set of all the words of length $\leq n$. A word w of length n can be regarded as a map

$$w : [n] \to A,$$

where $[n]$ denotes the set $[n] = \{1, \ldots, n\}$. One can write

$$w = w_1 w_2 \cdots w_n,$$

having set $w_i = w(i)$ for all $i \in [n]$. For any pair (i, j) of integers such that $1 \leq i \leq j \leq n$ we denote by $w[i, j]$ the factor $w_i \cdots w_j$.

A *language* L over the alphabet A is any subset of A^*. For any language L, $F(L)$, $P(L)$, and $S(L)$ will denote the sets of the factors, prefixes, and suffixes of all the words of L, respectively.

For languages L_1 and L_2,

$$L_1 \cup L_2, \; L_1 \cap L_2, \text{ and } L_1 \setminus L_2$$

will denote their ***union***, ***intersection***, and ***difference***. For any $L \subseteq A^*$, L^c will denote its ***complement***, $L^c = A^* \setminus L$. The product operation of words can be extended to languages as follows. For $L_1, L_2 \subseteq A^*$ the product L_1L_2 of L_1 and L_2 is defined as

$$L_1L_2 = \{uv \in A^* \mid u \in L_1, v \in L_2\}.$$

For $u \in A^*$ and $L \subseteq A^*$ the products $\{u\}L$ and $L\{u\}$ will be simply denoted by uL and Lu, respectively. For any $n \geq 0$, the ***n-power*** of L is defined inductively as

$$L^0 = \{\epsilon\}, \quad L^n = LL^{n-1}, \quad \text{for } n > 0.$$

For any $L \subseteq A^*$, L^+ will denote the language

$$L^+ = L \cup L^2 \cup \cdots \cup L^n \cup \cdots.$$

Moreover, one sets

$$L^* = \bigcup_{n \geq 0} L^n = L^+ \cup \{\epsilon\}.$$

The languages L^+ and L^* are, respectively, the subsemigroup and the submonoid of A^* generated by L. For $u \in A^*$, $\{u\}^*$ will be simply denoted by u^*.

If L is a language over A and u is a word of A^*, then $u^{-1}L$ and Lu^{-1} will denote the sets:

$$u^{-1}L = \{w \in A^* \mid uw \in L\}, \quad Lu^{-1} = \{w \in A^* \mid wu \in L\}.$$

A nonempty subset R of A^* is a ***right ideal*** if $RA^* \subseteq R$. Similarly, a nonempty subset L of A^* is a ***left ideal*** if $A^*L \subseteq L$. A subset J of A^* is a ***two-sided ideal*** if it is both a right and a left ideal.

Given sets X, Y, Z, and maps $f : X \to Y$, $g : Y \to Z$, we denote by fg the ***composition map*** $fg : X \to Z$, defined for each $x \in X$ as

$$(fg)(x) = g(f(x)).$$

A ***morphism*** ϕ of the free monoid A^* in the free monoid B^* is a map $\phi : A^* \to B^*$ such that for all $u, v \in A^*$

$$\phi(uv) = \phi(u)\phi(v).$$

Moreover, the image of the empty word of A^* is the empty word of B^*. Hence, the morphism ϕ is completely determined by the images $\phi(a)$ with $a \in A$. The morphism ϕ is called ***nonerasing*** if $\phi(A^+) \subseteq B^+$.

Let $\phi : A^* \to A^*$ be a morphism. We set $\phi^0 = id$, where *id* is the identity map, and for $n \geq 0$

$$\phi^{n+1} = \phi^n \phi.$$

One easily verifies that for all $n \geq 0$, ϕ^n is a morphism of A^* in A^* called the n th *iterated* of ϕ.

In the following, $\mathbb{Z}$ will denote the set of the integers and $\mathbb{N}$, $\mathbb{N}_+$, and $\mathbb{N}_-$ the sets of non-negative, positive, and negative integers, respectively. Let x be a real number; $\lfloor x \rfloor$ denotes the greatest integer less than or equal to x; $\lceil x \rceil$ denotes the least integer greater than or equal to x. The integer $\lfloor x \rfloor$ is also called the *integer part* of x.

We shall often use in the following the very famous principle, called the *Dirichlet box principle* or *"pigeon-hole" principle*.

Box Principle. *For any $n > 0$, if $n + 1$ objects are distributed over n boxes, then some box must contain more than one of the objects.*

In the case of an infinite number of objects and a finite number of boxes the pigeon-hole principle becomes:

If infinitely many objects are distributed over a finite number of boxes, then some box must contain infinitely many objects.

Although intuitively obvious, the pigeon-hole principle is of great use in combinatorics on words, number theory, and all domains of mathematics.

1.2 Infinite words

An extension of the concept of finite word is obtained by considering infinite sequences of symbols over a finite set. One can have sequences which are infinite from left to right, from right to left, or in both directions. In the following we shall refer to *indexed* sequences of letters that are usually called also *infinite words*.

A *two-sided infinite* (or *bi-infinite*) word w over the alphabet A is any map

$$w : \mathbb{Z} \to A.$$

For each $n \in \mathbb{Z}$, we set $w_n = w(n)$ and denote w also as

$$w = \cdots w_{-3} w_{-2} w_{-1} w_0 w_1 w_2 w_3 \cdots .$$

The set of all two-sided infinite words over A will be denoted by $A^{\pm\omega}$.

Let us observe that the same *unindexed* sequence of letters can be described by different bi-infinite words. For instance, let $T : A^{\pm\omega} \to A^{\pm\omega}$ be the operator in $A^{\pm\omega}$, called *shift*, defined as: for all $w \in A^{\pm\omega}$

$$(Tw)(i) = w(i+1) \text{ for all } i \in \mathbb{Z}.$$

One has, trivially, that w and Tw describe the same unindexed symbolic sequence.

A word $u \in A^*$ is a finite *factor* of $w \in A^{\pm\omega}$ if $u = \epsilon$ or there exist integers $i, j \in \mathbb{Z}, i \leq j$, such that $u = w_i \cdots w_j$; the sequence $w[i, j] = w_i \cdots w_j$ is also called an *occurrence* of u in w.

An *one-sided* (from left to right) infinite word, or simply *right infinite* word, over A is any map

$$w : \mathbb{N}_+ \to A.$$

For each $n > 0$ the factor $w[1, n] = w_1 \cdots w_n$ of length n is called the *prefix* of w of length n and will be simply denoted by $w[n]$. The empty word ϵ is assumed to be the prefix of w of length 0.

In a symmetric way one can define a *left infinite* word as any map

$$w : \mathbb{N}_- \to A.$$

For each $n > 0$ the factor $w[-n, -1] = w_{-n} \cdots w_{-1}$ of length n is called the *suffix* of w of length n and will be simply denoted by $w[-n]$. The empty word is assumed as the suffix of w of length 0. The set of all right (left) infinite words over A will be denoted by A^ω (by $A^{-\omega}$).

If $u = u_1 \cdots u_n \in A^*$, $u_i \in A$, $(i = 1, \ldots, n)$, and $v \in A^\omega$, then we denote by uv the word $w \in A^\omega$, which we call *concatenation* or *product* of u and v, defined as

$$w_i = u_i \quad \text{for } i \leq n, \quad w_{i+n} = v_i, \quad \text{for } \; i > 0.$$

In a symmetric way if $v \in A^{-\omega}$ one defines the product $vu \in A^{-\omega}$.

For any finite or infinite word w, $F(w)$ will denote the set of all its finite factors and $\text{alph}(w)$ the set of all letters of the alphabet A occurring in w, $\text{alph}(w) = F(w) \cap A$. If $w \in A^* \cup A^\omega$, then $P(w)$ denotes the set of its finite prefixes. If $w \in A^* \cup A^{-\omega}$, then $S(w)$ denotes the set of its finite suffixes.

If $w \in A^{\pm\omega}$, then one can associate with it the one-sided infinite word $w_+ \in A^\omega$ defined for all $n > 0$ as $w_+(n) = w(n)$; trivially, one has that $F(w_+) \subseteq F(w)$.

If $\phi : A^* \to B^*$ is a nonerasing morphism of free monoids, then one can extend ϕ to a map

$$\phi : A^\omega \to B^\omega,$$

where for any $w \in A^\omega$, $\phi(w)$ is the infinite word

$$\phi(w) = \phi(w_1)\phi(w_2) \cdots \phi(w_n) \cdots.$$

In a similar way ϕ can be extended to left infinite words.

The following lemma is essentially the famous König lemma for infinite trees formulated in terms of words.

Lemma 1.2.1. *Let $L \subseteq A^*$ be an infinite language. Then there exists an infinite word $s \in A^\omega$ ($s \in A^{-\omega}$) such that $P(s) \subseteq P(L)$ ($S(s) \subseteq S(L)$).*

Proof. Let L be an infinite language on the alphabet A. The finiteness of A implies from the pigeon-hole principle that there exist infinitely many words of L starting with the same letter, say a_1. The set $L_1 = L \cap a_1 A^*$ is then infinite. Suppose by induction that there exist letters $a_1, \ldots, a_n$ such that

$$L_n = L \cap a_1 \cdots a_n A^*$$

is infinite. Since

$$L_n = \bigcup_{a \in A} (a_1 \cdots a_n a A^* \cap L)$$

and A is a finite set, it follows that there exists at least one letter such that one of the sets in the above union is infinite. Let us choose one letter a_{n+1} such that the set $L_{n+1} = L \cap a_1 \cdots a_n a_{n+1} A^*$ is infinite. It follows that there exists an infinite word $s \in A^\omega$

$$s = a_1 a_2 \cdots a_n \cdots$$

such that $L_n = L \cap a_1 \cdots a_n A^*$ is infinite for all $n > 0$. From this one has $P(s) \subseteq P(L)$. In a symmetric way one proves that there exists an infinite word $s \in A^{-\omega}$ such that $S(s) \subseteq S(L)$.

From the preceding lemma, one derives that if L is an infinite language over A, then there exists an infinite word $s \in A^\omega$ such that $F(s) \subseteq F(L)$. The following lemma, concerning bi-infinite words, is a slight generalization of König's lemma. The proof is very similar to the preceding one.

Lemma 1.2.2. *Let $L \subseteq A^*$ be an infinite language. Then there exists an infinite word $s \in A^{\pm\omega}$ such that $F(s) \subseteq F(L)$.*

Proof. For any word $w \in A^+$ such that $|w| \geq 2$ we set $r_w = \lfloor |w|/2 \rfloor$. Since L is an infinite language then from the pigeon-hole principle there will exist infinitely many words w of L having the same letter, say a_0 at the position r_w in w. Let us denote by L_1 this infinite subset. By using the same principle there will exist an infinite subset L_2 of L_1 whose words w are such that

$$w[r_w, r_w + 1] = a_0 a_1,$$

for a suitable letter $a_1 \in A$. By using again the same argument there will exist a letter a_{-1} and an infinite subset L_3 of L_2 such that all the words of L_3 satisfy the condition

$$w[r_w - 1, r_w + 1] = a_{-1} a_0 a_1.$$

Continuing in this way one can construct the two-sided infinite word

$$s = \cdots a_{-2} a_{-1} a_0 a_1 a_2 \cdots,$$

which is such that any of its factors is a factor of infinitely many words of L.

A bi-infinite word $w \in A^{\pm\omega}$ can be regarded as a partition of $\mathbb{Z}$ into a finite number of (possibly empty) classes C_i, $(i = 1, \ldots, k)$

$$\mathbb{Z} = C_1 \cup C_2 \cup \cdots \cup C_k,$$

where $k = \text{card}(A)$. Two integers $p, q \in \mathbb{Z}$ lie in the same class if and only if $w_p = w_q$. Interpreting the letters of the alphabet A as *colors* one says also that p and q have the same *color*. A set K of integers is said to be *monochromatic* if all the elements of K have the same color.

In a similar way an infinite word $w \in A^\omega$ corresponds to a finite partition of $\mathbb{N}_+$ and a finite word w of length n to a partition of the set $[n] = \{1, \ldots, n\}$.

1.3 Metric and topology

In this section we define a *metric* in the set of finite and infinite words over a finite alphabet (see [62], Chap. 13). This metric induces a topology that allows us to define the concept of *limit* of a sequence of words. This will be useful later, when we will introduce some infinite words as the limit of an infinite sequence of finite words.

Let us consider the set $A^\infty = A^* \cup A^\omega$. We introduce in A^∞ a metric d as follows:

Definition 1.3.1. *Let $u, v \in A^\infty$. Then $d(u, v)$ is 0 if $u = v$, otherwise*

$$d(u, v) = 2^{-(n+1)},$$

where n is the length of the maximal common prefix of u and v.

It is easy to prove that the function d, defined above, is a *distance*, i.e., for all $u, v, w \in A^\infty$,

i. $d(u, v) = 0$ if and only if $u = v$,
ii. $d(u, v) = d(v, u)$,
iii. d satisfies the triangle inequality:

$$d(u, w) \leq d(u, v) + d(v, w).$$

Then the pair (A^∞, d) is a *metric space* and thus a *topological space* with the topology induced by the metric d. We call d also the *prefix metric*.

Let us observe that, according to the above definition of distance, a finite word $u \in A^*$ can be regarded as an infinite word

$$uxxxxx \cdots x \cdots,$$

on the alphabet $A \cup \{x\}$, where x is an additional symbol not belonging to the alphabet A. It is well known that the metric d provides a topology to A^ω that coincides with the product topology obtained from the discrete topology of A. As a consequence of a well known result of topology, one derives that A^ω is a compact space, that is, each infinite sequence of elements of A^ω contains a convergent subsequence. One can easily prove that A^∞ is compact too.

According to the above definition an infinite word appears naturally as the limit sequence of its prefixes. Conversely, let us consider a sequence of finite words $(t_i)_{i\geq 1}$, such that for almost all i, t_i is a proper prefix of t_{i+1}, i.e., this latter condition is satisfied for all $i \geq i_0$, where i_0 is a suitable positive integer. We can construct an infinite word $t \in A^\omega$ as follows. For any integer $n \geq 1$, let t_i be any word of the given sequence such that $i \geq i_0$, $|t_i| \geq n$ and set $a_n = t_i(n)$; then

$$t = a_1 a_2 \cdots a_n \cdots .$$

One can easily verify that

$$t = \lim_{i\to\infty} t_i,$$

where the limit is taken with respect to the topology induced by the metric d.

In a symmetric way let $A^{-\infty} = A^* \cup A^{-\omega}$. One introduces in $A^{-\infty}$ a metric d', that we call *suffix metric*, as follows:

Definition 1.3.2. *Let $u, v \in A^{-\infty}$. Then $d'(u,v)$ is 0 if $u = v$, otherwise*

$$d'(u,v) = 2^{-(n+1)},$$

where n is the length of the maximal common suffix of u and v.

Let $(t_i)_{i\geq 1}$ be a sequence of words of A^* such that t_i is a proper suffix of t_{i+1}, for any $i \geq 1$. As above, one can construct a word $t \in A^{-\omega}$ such that

$$t = \lim_{i\to\infty} t_i$$

where the limit is taken with respect to the topology induced by the metric d'.

An infinite word can be often constructed by iterating a morphism as follows. Let $\phi : A^* \to A^*$ be a nonerasing morphism such that there exists a letter $a \in A$ such that

$$\phi(a) = au, \quad \text{with} \quad u \in A^+.$$

For all $n \geq 0$ one has

$$\phi^{n+1}(a) = \phi^n(au) = \phi^n(a)\phi^n(u),$$

so that $\phi^n(a)$ is a proper prefix of $\phi^{n+1}(a)$. Thus the sequence $(\phi^n(a))_{n\geq 0}$ converges to a limit denoted by $\phi^\omega(a)$

$$\phi^\omega(a) = \lim_{n\to\infty} \phi^n(a).$$

One says that $x = \phi^\omega(a)$ is the infinite word obtained by ***iterating the morphism*** ϕ ***on the letter*** a. Moreover, one has

$$x = \phi(x),$$

that is, x is a fixed point for ϕ. Indeed, if x' is a any prefix of x, then $\phi(x')$ is a prefix of $\phi(x)$. Hence, all $\phi^n(a)$ with $n > 0$ are prefixes of $\phi(x)$; moreover, x begins with the letter a. Thus $\phi(x) = \lim_{n\to\infty} \phi^n(a) = x$.

1.4 Periodicity and conjugacy

In this section we shall consider some basic results concerning the important combinatorial notions of ***periodicity*** and ***conjugacy*** of words.

Definition 1.4.1. *Let* $w = w_1 \cdots w_n$, $w_i \in A$, $1 \leq i \leq n$, *be a word. A positive integer p is called a* period *of w when* $p \geq |w|$ *or if* $p < |w|$, *then the following condition is satisfied:*

$$w_i = w_{i+p}$$

for all $i \in [1, n-p]$.

This definition is also equivalent to saying that w can be represented as $w = (xy)^r x$ with $x, y \in A^*$, $r \geq 0$, and $|xy| = p$. A word w can have several periods. A period p is called ***proper*** if $p < |w|$. The minimal period of a word is called the ***period*** of w. A word w is called ***primary*** if $|w|$ is equal to the minimal period of w.

The word *abaababa* of length 8 has the proper period 7 and the minimal period 5. The word *aabab* is primary.

If a word w has a period p, then p is also a period of any nonempty factor u of w.

The following important theorem, due to N. J. Fine and M. S. Wilf [64, 104], is a basic tool to deal with periods of words.

Theorem 1.4.1. *(Fine and Wilf) Let w be a word having periods p and q and denote by* $\gcd(p,q)$ *the greatest common divisor of p and q. If* $|w| \geq p + q - \gcd(p,q)$, *then w has also the period* $\gcd(p,q)$.

Proof. We first prove the theorem when p and q are coprimes, i.e., when $\gcd(p,q) = 1$. Let us suppose that $p \leq q - 1$. By hypothesis there exist $x, y \in A^*$ such that $|x| = p$, $|y| = q$ and

$$w = x^n x' = y^m y',$$

with $n, m > 0$, $x' \in P(x)$, $y' \in P(y)$. The statement will be achieved if we prove that x has 1 as a period. In order to do this we have to prove that for any $i \in \{1, \ldots, p-1\}$, one has

$$w_i = w_p.$$

Let $i \in \{1, \dots, p-1\}$ and let $\lambda = p-i$. Since $1 = \gcd(p,q)$, there exist $l_1, l_2 \in \mathbb{Z}$ such that $1 = l_1 p + l_2 q$; therefore, $p - i = \lambda l_1 p + \lambda l_2 q$ and so $p \equiv i + \lambda l_2 q \pmod p$. Let m be a positive integer such that $m \equiv \lambda l_2 \pmod p$, then

$$p \equiv i + mq \pmod p.$$

We prove by induction on m that $w_i = w_p$. If $m = 1$, then there exists a positive integer k such that $p = i + q - kp$. We first observe that $p + kp = i + q \leq p + q - 1$, since $i \leq p - 1$. Then we can write $w_p = w_{p+kp} = w_{i+q} = w_i$, thus, $w_i = w_p$. Let $m > 1$ and consider the integer $i' \in \{1, \dots, p\}$ such that $i' \equiv i + q \pmod p$. Then there exists a positive integer k such that $i' = i + q - kp$, i.e., $i' + kp = i + q$. Thus $w_{i'} = w_{i'+kp} = w_{i+q} = w_i$. If $i' = p$, then the result follows. Otherwise, since $p \equiv i' + (m-1)q \pmod p$, by the induction hypothesis one has $w_p = w_{i'}$; hence, $w_i = w_p$.

Suppose now that $\gcd(p,q) = d > 1$ and suppose, without loss of generality, that $|w| = p + q - d$. In such a case we represent the word w as a word w' over the alphabet $B = A^d$. One has that w' has periods $p' = p/d$ and $q' = q/d$. Moreover, $\gcd(p', q') = 1$ and $|w'| = p' + q' - 1$. Then w' has period 1, as a word over B, and so w has period d.

Remark 1.4.1. Let us observe that the bound $p + q - \gcd(p,q)$ given by the preceding theorem is optimal. For instance, the word $w = abaababaaba$ of length 11 on the alphabet $A = \{a, b\}$ has periods 5 and 8 but not the period $1 = \gcd(5,8)$. We mention that the words w having two periods p and q which are coprimes and such that $|w| = p + q - 2$ (so that one cannot apply the Fine and Wilf theorem) constitute an important class of words showing interesting and surprising combinatorial properties. This class can be considered [33] as the "kernel" of the theory of Sturmian words (see Sect. 1.6).

Definition 1.4.2. *A word $w \in A^*$ is called* primitive *if for any $u \in A^+$ and $h > 1$ one has $w \neq u^h$.*

The set of all primitive words of A^* is denoted by $\pi(A^*)$. Any word w can always be expressed as a power of a primitive word. The following proposition will show that this representation is unique.

Proposition 1.4.1. *Let $x, y \in A^+$ and n, m be positive integers such that $x^n = y^m$. Then there exists $t \in A^+$ such that $x, y \in t^*$. In particular, for any word $x \in A^+$, there exists a unique primitive word t such that $x \in t^*$.*

Proof. We may assume that $n, m \geq 2$, otherwise the statement is trivially true. Let $w = x^n = y^m$ and $d = \gcd(|x|, |y|)$. The word w has periods $|x|, |y|$ and $|w| \geq |x| + |y| > |x| + |y| - d$. By the theorem of Fine and Wilf, w, as well as x and y, has period d. Therefore, if t is the prefix of length d of x, which is also a prefix of y, one has $x, y \in t^*$.

We refer to any nonempty word w of the kind $u^k u'$, with $k \geq 0$, $u \in \pi(A^*)$, $u' \in P(u)$, and $0 \leq |u'| < |u|$, as a *sesquipower*. A sesquipower w has obviously a period $p = |u|$. The integer k is called *exponent* of w. A sesquipower is called a *strong sesquipower* if the exponent k is > 1.

Any word $w \in A^+$ can be, obviously, represented as a sesquipower. We call *order* of w the maximal value of the exponent k in all the representations of w as a sesquipower. For instance, the word $w = aabaabaa$ can be represented as $w = (aabaaba)a = (aab)^2 aa$, that is, by a sesquipower of exponent 1 and a sesquipower of exponent 2. The order of w is 2.

Lemma 1.4.1. *Let $w = u^k u'$ be a strong sesquipower. Then $|u|$ is the minimal period of w.*

Proof. Suppose that q is a period of w such that $q < |u|$. Since $|w| \geq 2|u| > |u| + q$, by the theorem of Fine and Wilf, the word w, as well as u, has the period $d = \gcd(|u|, q)$. Since $|u| = rd$ and $q = sd$, with $r, s \geq 1$, one can write $u = z^r$, with $z \in A^*$ and $|z| = d$. Since $q < |u|$, it follows that $r > s \geq 1$ so that u will not be primitive.

Corollary 1.4.1. *Any word can be uniquely represented as a strong sesquipower.*

Proof. Let $w = u^k u' = v^h v'$ with $u, v \in \pi(A^*)$, $u' \in P(u)$, $v' \in P(v)$, and $k, h > 1$. From the preceding lemma one has that $|u| = |v| = p$, where p is the minimal period of w. This implies $u = v$ and then $h = k$ and $u' = v'$.

Definition 1.4.3. *Let $x, y \in A^+$. We say that x, y are* conjugate *if there exist $u, v \in A^*$ such that $x = uv$ and $y = vu$.*

The conjugation relation (*conjugacy*) is trivially reflexive and symmetric. Moreover, it is also transitive. Indeed, suppose that x, y, and z are words such that y is a conjugate of x and z is a conjugate of y. One has

$$x = uv, \quad y = vu = u'v', \quad z = v'u'.$$

If $|v| \geq |u'|$, then $v = u'k$, with $k \in A^*$, so that $ku = v'$, $x = uu'k$, and $z = kuu'$. If $|v| < |u'|$, then $u' = vk$ so that $u = kv'$, $x = kv'v$, and $z = v'vk$. In both cases one has that x and z are conjugate. Thus conjugacy is an equivalence relation.

Proposition 1.4.2. *If w is primitive, then so will be any conjugate w' of w.*

Proof. Let us suppose that w' is non-primitive, i.e., $w' = z^h$, with $h > 1$. One has then

$$w' = z^h = uv, \; w = vu.$$

It follows that

$$u = z^{h_1} z_1, \; v = z_2 z^{h_2},$$

with $h = h_1 + h_2 + 1$ and $z = z_1 z_2$. Hence,

$$w = z_2 z^{h_1+h_2} z_1 = (z_2 z_1)^h,$$

which is a contradiction.

The following important theorem on conjugate words is due to R. C. Lyndon and M. P. Schützenberger [107].

Theorem 1.4.2. *Two words $x, y \in A^+$ are conjugate if and only if there exists a word $w \in A^*$ such that $xw = wy$. Moreover, in this case there exist words $u, v \in A^*$ and an integer $p \geq 0$ such that*

$$x = uv, \;\; y = vu, \;\; w = (uv)^p u.$$

Proof. Let $x, y \in A^+$. Suppose that x and y are conjugate and let $u, v \in A^*$ be such that $x = uv$ and $y = vu$. Then for $w = u$ one has $xw = wy$. Conversely, if $xw = wy$, then for any $n > 0$ one has

$$x^n w = w y^n.$$

Let n be the least integer such that $n|x| > |w| \geq (n-1)|x|$. One has then

$$x^n = wv, \;\; vw = y^n, \;\; v \in A^*.$$

As $|w| \geq (n-1)|x|$, from the preceding equation one derives

$$w = x^{n-1}u \;\text{ and }\; x = uv,$$

for a suitable $u \in A^*$. Thus

$$y^n = vx^{n-1}u = (vu)^n.$$

Since $|y| = |x|$, one has $y = vu$. Moreover, $w = (uv)^{n-1}u$.

Proposition 1.4.3. *Let $x, y \in A^+$ be such that $xy = yx$. Then there exists $t \in A^+$ such that $x, y \in t^*$.*

Proof. Let $x, y \in A^+$ and $xy = yx$. We give the proof by induction on $|xy|$. If $|xy| = 2$, then the only possibility is $x, y \in A$ and the statement is true for $t = x = y$. Let $|xy| > 2$ and suppose $|x| \leq |y|$. If $|x| = |y|$, then $x = y$ and also in this case one can take $t = x = y$. If $|x| < |y|$, then there exists $\mu \in A^+$ such that $x\mu = y = \mu x$. Since $|\mu x| < |xy|$, by the induction hypothesis there exists $t \in A^+$ such that $x, \mu \in t^*$. Since $y = \mu x$, one has also $y \in t^*$.

A different proof, based on periods, is the following. From Theorem 1.4.2 one derives that the word $w = xy = yx$ can be written as

$$w = x^p x' = y^q y',$$

where $x' \in P(x)$, $y' \in P(y)$, and $p, q > 0$. Hence, w has the periods $|x|$ and $|y|$. Since $|w| = |x| + |y|$, by Theorem 1.4.1, w has also the period $d = \gcd(|x|, |y|)$. If t is the common prefix of x and y of length d it follows that $x, y \in t^*$.

The following proposition is a straightforward consequence of Theorem 1.4.2 and Proposition 1.4.3.

Proposition 1.4.4. *Let $x, y \in A^+$ be such that $xw = wy$. Then there exist words $\lambda, \mu \in A^*$ and a unique positive integer q such that $\lambda\mu \in \pi(A^*)$ and*

$$x = (\lambda\mu)^q, \ y = (\mu\lambda)^q, \ w \in (\lambda\mu)^*\lambda.$$

Moreover, if $x \neq y$, then $\lambda, \mu \in A^+$ and the pair (λ, μ) is uniquely determined.

Proof. From Theorem 1.4.2 there exist $u, v \in A^*$ and an integer $p \geq 0$ such that

$$x = uv, \ y = vu, \ w = (uv)^p u.$$

By Proposition 1.4.1 the word x can be uniquely expressed as $x = z^q$ with $z \in \pi(A^*)$. As $x = uv$, one has

$$u = z^{q_1}\lambda, \ v = \mu z^{q_2}, \ z = \lambda\mu, \ q_1 + q_2 + 1 = q.$$

Hence

$$x = (\lambda\mu)^q, \ y = (\mu\lambda)^q, \ w = (\lambda\mu)^{qp+q_1}\lambda.$$

If $x \neq y$, then one has $\lambda, \mu \in A^+$. Moreover, the pair (λ, μ) is uniquely determined. Indeed, suppose that (λ', μ') is another pair, such that $\lambda'\mu' \in \pi(A^*)$, $x \in (\lambda'\mu')^+$, and $y \in (\mu'\lambda')^+$. From Propositions 1.4.1 and 1.4.2 one has

$$\lambda\mu = \lambda'\mu', \text{ and } \mu\lambda = \mu'\lambda'.$$

We may always suppose that $|\lambda| > |\lambda'|$. One would have $\lambda = \lambda'\zeta$ and $\zeta\mu = \mu'$. This implies

$$\mu\lambda'\zeta = \zeta\mu\lambda'.$$

By Proposition 1.4.3 it follows $\mu\lambda = t^k$, with $k > 1$, which is absurd since $\mu\lambda$ is primitive.

The following two periodicity lemmas are interesting consequences of the Fine and Wilf theorem. They will be used in Chap. 3.

Lemma 1.4.2. *Let w be a word having as suffix and as prefix two strong sesquipowers of orders r and s and minimal periods p and q, respectively. If $(r-1)p + (s-1)q \geq |w| - 1$, then $p = q$ and w is a strong sesquipower of order $\geq \max\{r, s\}$.*

Proof. By hypothesis we can write

$$w = f'u^r = v^s f'',$$

where $f', f'' \in A^*$, u, v are primitive words, $r, s > 1$, $|u| = p$, and $|v| = q$. Let us set $n = |w|$. One has

$$n = |f'| + rp = |f''| + sq.$$

Moreover, $|f'| < sq$. In fact, otherwise, one would have from the hypothesis made $n \geq rp + sq \geq n - 1 + p + q \geq n + 1$ which is a contradiction. Hence, one has

$$v^s = f'\Delta, \quad u^r = \Delta f'', \quad w = f'\Delta f'', \tag{1.1}$$

where $\Delta \in A^*$ and

$$|\Delta| = sq - |f'| = sq + rp - n.$$

By hypothesis $(s-1)q + (r-1)p \geq n - 1$, so that $|\Delta| \geq p + q - 1$. Since Δ has periods p and q, by Theorem 1.4.1 it follows that Δ has a period equal to $d = \gcd(p, q)$. If $p = hd$ and $q = kd$, then from equation (1.1) one derives $v = z_1^k$ and $u = z_2^h$ with $|z_1| = |z_2| = d$. Since u and v are primitive, it follows that $h = k = 1$ and $p = q = d$. From this and (1.1) one derives that $\Delta = u^i u_1 = v_2 v^i$, with $0 < i \leq \min\{r, s\}$, $u = u_1 u_2 = v_2 v_1$, and $v = v_1 v_2$. It follows that $u_1 = v_2$ and $u_2 = v_1$. If $i < \min\{r, s\}$, then one has $f'' = u_2 u^{r-i-1}$, $f' = v^{s-i-1} v_1$, and

$$w = v^s f'' = (v_1 v_2)^{s+r-i-1} v_1,$$

that is, w is a sesquipower of order $t = r + s - i - 1 \geq \max\{r, s\}$. Let us suppose $i = \min\{r, s\}$; if $r < s$ $(r \geq s)$, one derives that $f'' = \epsilon$ $(f' = \epsilon)$ and $w = v^s$ $(w = u^r)$. Thus w is a sesquipower of order equal to $\max\{r, s\}$.

For any $n > 0$, $\mathcal{S}_n$ denotes the group (*symmetric group*) of all the permutations on the set $\{1, \ldots, n\}$.

Lemma 1.4.3. *Let $k > 3$ and $x_1, x_2, \ldots, x_k$ be k words of A^*. For any permutation $\sigma \in \mathcal{S}_k$ we denote by w_σ the word $w_\sigma = x_{\sigma(1)} \cdots x_{\sigma(k)}$. Let $T = \{w_\sigma \mid \sigma \in \mathcal{S}_k\}$. If* card$(T) = k!$, *then there exists an element of T which is not a sesquipower of order* > 3.

Proof. We can always suppose that $|x_1| \geq \cdots \geq |x_k|$. We prove the lemma by contradiction. In fact let us consider the words $w = x_1 \cdots x_{k-2} x_{k-1} x_k$ and $w' = x_1 \cdots x_{k-2} x_k x_{k-1}$. Since card$(T) = k!$, one has that $w \neq w'$. Let us then suppose that w and w' are sesquipowers of order > 3:

$$w = (\lambda\mu)^h \lambda, \ w' = (\alpha\beta)^r \alpha, \tag{1.2}$$

where $\lambda\mu$, $\alpha\beta$ are primitive words and $h, r > 3$. Let us set $u = x_1 \cdots x_{k-2}$. Since $k > 3$ and $|x_1| \geq \cdots \geq |x_k|$, one has that

$$|u| \geq (1/2)|w| = (1/2)h|\lambda\mu| + (1/2)|\lambda| = (1/2)r|\alpha\beta| + (1/2)|\alpha|.$$

Thus $|u| \geq 2\max\{|\lambda\mu|, |\alpha\beta|\}$. Let us set $p = |\lambda\mu|$ and $q = |\alpha\beta|$. One has then that u has the periods p and q and, moreover, $|u| \geq p + q$. By Theorem 1.4.1, u has a period equal to $\gcd(p, q)$. It follows that $\lambda\mu = z^m$, $\alpha\beta = z^n$,

$m, n > 0$, and $|z| = \gcd(p,q)$. Since $\lambda\mu$, $\alpha\beta$ are primitive words, one derives $\lambda\mu = \alpha\beta = z$, so that $p = q$. From equation (1.2), one has moreover

$$|w| = |w'| = hp + |\lambda| = rp + |\alpha|. \tag{1.3}$$

From this one derives that $|(h-r)p| = ||\alpha| - |\lambda|| \leq p$. If $||\alpha| - |\lambda|| < p$, then one has $h = r$ and, by equation (1.3), $|\alpha| = |\lambda|$ which implies $\alpha = \lambda$, $\beta = \mu$, and, by (1.2), $w = w'$ which is a contradiction. Let us now suppose that $||\alpha| - |\lambda|| = p$. One has either $|\alpha| = p$ and $\lambda = \beta = \epsilon$ or $\alpha = \mu = \epsilon$ and $|\lambda| = p$. In the first case one has $h = r+1$, $\alpha = \mu$, and in the second case $r = h+1$, $\beta = \lambda$. In both cases from (1.2) one reaches again the contradiction $w = w'$.

Definition 1.4.4. *Let $s \in A^\omega$ be an infinite word. We say that s is* ultimately periodic *if $s = uv^\omega$, where $u \in A^*$, $v \in A^+$, and v^ω is the infinite word $v^\omega = vvvv\cdots v\cdots$. If $u = \epsilon$, then the word $w = v^\omega$ is called* periodic *of* period $p = |v|$. *The period p is* minimal *if v is primitive.*

In a symmetric way a word $s \in A^{-\omega}$ is called *ultimately periodic* if $s = v^{-\omega}u$, where $u \in A^*, v \in A^+$, and $v^{-\omega}$ denotes the word of $A^{-\omega}$ given by

$$v^{-\omega} = \cdots v \cdots vvvv.$$

A bi-infinite word $w \in A^{\pm\omega}$ is *periodic* of *period p* if for all $i, j \in \mathbb{Z}$, $|i-j| = p$ implies that $w_i = w_j$. From the theorem of Fine and Wilf one easily derives the following:

Proposition 1.4.5. *Let $w, w' \in A^\omega$ be two periodic words of periods p and q respectively. If w and w' have the same prefix of length $p + q - \gcd(p,q)$, then $w = w'$.*

Proof. Let $w = u^\omega$ and $w' = v^\omega$ with $|u| = p$ and $|v| = q$. By hypothesis $w = \zeta t$, $w' = \zeta t'$ with $t, t' \in A^\omega$ and $\zeta \in A^*$ is such that $|\zeta| \geq p + q - d$, where $d = \gcd(p,q)$. By the theorem of Fine and Wilf ζ has the period d. Since $|\zeta| \geq p, q$ then also the words u and v will have the period d. We can write $p = rd$, $q = sd$, and $u = z^r$, $v = z^s$ with $z \in A^*$ and $|z| = d$. It follows $w = w' = z^\omega$.

Definition 1.4.5. *A factor u of an infinite word $w \in A^\omega$ ($w \in A^{\pm\omega}$) is* recurrent *if the set of all $i \in \mathbb{N}_+$ ($i \in \mathbb{Z}$) such that $u = w[i, i+|u|-1]$ has not an upper (upper and lower) bound. The word w is* recurrent *if any of its factors is recurrent.*

Thus in a recurrent word w any factor of w occurs infinitely often in w (in both sides if the word w is bi-infinite). Any periodic word is obviously recurrent. However, this is not the case for non-periodic words of A^ω which are ultimately periodic.

Lemma 1.4.4. *Let $w \in A^\omega$ be an ultimately periodic word. If w is recurrent, then w is periodic.*

Proof. Let $w = uv^\omega$ be an ultimately periodic word. If $u = \epsilon$, then w is periodic and the result is achieved. Let us then assume that $u \neq \epsilon$. We can always suppose that v is primitive. Since w is recurrent then its prefix uv has to occur in w infinitely often. Hence, there exists an integer $p > 1$ such that

$$v^p = \lambda uv\mu,$$

with $\lambda, \mu \in A^*$. This implies

$$uv = v_2 v^q v_1, \tag{1.4}$$

with $v_2 \in S(v)$, $v_1 \in P(v)$, $q \geq 0$. Hence, $v = v_1 v_2' = v_1' v_2$. Moreover, from equation (1.4) one derives

$$v = v_1 v_2' = v_2' v_1.$$

If $0 < |v_1| < |v|$, then v is not primitive which is a contradiction. If $v = v_1$, then $u = v_2 v^q = v_2(v_1' v_2)^q$ and

$$w = uv^\omega = (v_2 v_1')^\omega.$$

Finally, if $|v_1| = 0$, then $uv = v_2 v^q$ and $v = v_1' v_2$ so that $uv^\omega = v_2(v_1' v_2)^\omega = (v_2 v_1')^\omega$, which concludes the proof.

1.5 Lyndon words

This section is devoted to present some combinatorial properties of Lyndon words over an alphabet A. These are primitive words of A^* minimal with respect to the lexicographic order $\leq$, in their conjugation classes. Thus a Lyndon word represents a class of conjugation of a primitive word. A basic theorem (see Theorem 1.5.1) shows that any word $w \in A^*$ can be uniquely factorized in a non-increasing product of Lyndon words,

$$w = l_1 l_2 \cdots l_n, \quad \text{with } l_1 \geq l_2 \geq \cdots \geq l_n,$$

where the l_i's are Lyndon words. Thus Lyndon words form a factorization of the free monoid [104] called *Lyndon factorization.* Some results on Lyndon factorizations which are of interest in relation to the "division properties" of the words are proved.

Let A be a totally ordered alphabet. One can extend this order to A^* by the *lexicographic order* $<_{lex}$, or simply $<$, defined as: for all $u, v \in A^*$

$$u < v \Longleftrightarrow v \in uA^+ \text{ or } u = hx\xi,\ v = hy\eta,\ h, \xi, \eta \in A^*,\ x, y \in A, \text{ and } x < y.$$

One easily verifies that $<$ is a total order of A^*. The lexicographic order satisfies the two following basic properties whose proof is straightforward:

(L1) $\forall w \in A^*,\ \ u < v \Longleftrightarrow wu < wv.$
(L2) If $v \notin uA^*$, then $u < v \Longrightarrow \forall w, z \in A^*,\ \ uw < vz.$

In the following we shall denote by $\leq$ the reflexive closure of $<$ and by $>$ and $\geq$ the inverse relations of $<$ and $\leq$, respectively.

Definition 1.5.1. *A word $w \in A^+$ is called a* Lyndon word *if it is primitive and minimal with respect to the lexicographic order in the class of its conjugates.*

The set of all Lyndon words on the alphabet A will be denoted by $\mathcal{L}_A$, or simply $\mathcal{L}$ when there is no confusion. For $A = \{a, b\}$ and $a < b$ a list of Lyndon words of increasing length is:

$$a, b, ab, aab, abb, aaab, aabb, abbb, aaaab, aaabb, aabab, \ldots.$$

The following proposition gives a different equivalent definition of a Lyndon word.

Proposition 1.5.1. *A word $w \in A^+$ is a Lyndon word if and only if it is strictly less than any of its proper (nonempty) right factors.*

Proof. Let us suppose that $w \in \mathcal{L}$, $w = uv$, $u, v \in A^+$, and $v < w$. If $w \notin vA^*$ then by property $(L2)$ of the lexicographic order one has $vu < w$ which is a contradiction. Let us then suppose that $w \in vA^*$. One has then $w = uv = vu'$. By the theorem of Lyndon and Schützenberger (see Theorem 1.4.2) there exist words $\alpha, \beta \in A^*$ and a natural integer n such that

$$u = \alpha\beta,\ u' = \beta\alpha,\ v = (\alpha\beta)^n\alpha,\ w = (\alpha\beta)^{n+1}\alpha.$$

Since $v \neq \epsilon$ and w is primitive one has $\alpha \neq \epsilon$. Moreover, since w is Lyndon it follows that

$$w = (\alpha\beta)^{n+1}\alpha < \alpha(\alpha\beta)^{n+1}.$$

By property $(L1)$ of the lexicographic order one derives

$$(\beta\alpha)^{n+1} < (\alpha\beta)^{n+1}.$$

By property $(L2)$ we can multiply on the right both sides of the preceding equation by α, obtaining

$$(\beta\alpha)^{n+1}\alpha < (\alpha\beta)^{n+1}\alpha = w,$$

which is a contradiction.

Conversely, let us suppose that w is a word such that for any of its factorizations $w = uv$, $u \neq \epsilon$ one has

$$w < v < vu.$$

Hence, w is strictly less than any proper conjugate. Thus it is primitive and minimal in its conjugation class and, therefore, is a Lyndon word.

From the preceding proposition it follows that a Lyndon word is primary since it cannot have a period less than its length.

Proposition 1.5.2. *If x, y are Lyndon words and $x < y$, then xy is a Lyndon word and $x < xy < y$.*

Proof. Suppose that $x, y \in \mathcal{L}$ and $x < y$. One derives that $w = xy < y$. Indeed, this is trivially true by ($L2$) if x is not a prefix of y. Let us then suppose that x is a prefix of y, i.e., $y = xy'$. Since $y \in \mathcal{L}$, by Proposition 1.5.1 one has $y < y'$ and by ($L1$)

$$w = xy < xy' = y.$$

We now prove that w is strictly less than every proper right factor $u \in A^+$. In fact, if u is a nonempty proper right factor of y, then since $y \in \mathcal{L}$, one has $y < u$ so that $w = xy < y < u$. If $u = y$, then the result has been already shown. Let us now suppose that $u = u'y$, where u' is a proper suffix of x. Since $x \in \mathcal{L}$, one has $x < u'$, so that by ($L2$) $w = xy < u'y = u$. By Proposition 1.5.1 it follows $w \in \mathcal{L}$. The remaining part of the proof is trivial.

Proposition 1.5.3. *If x, y are Lyndon words and $x < y$, then, for any $i, j \geq 1$, $x^i y^j$ is a Lyndon word and $x < x^i y^j < y$.*

Proof. By Proposition 1.5.2, xy is a Lyndon word and $x < xy < y$. Again by Proposition 1.5.2 one has $x < xy < xy^2 < y$ and xy^2 is a Lyndon word. Thus, by iteration, one has $x < xy < xy^j < y$ and xy^j is a Lyndon word. Starting from the inequality $x < xy^j$ and applying Proposition 1.5.2, as above, one derives $x < x^i y^j < y$ and $x^i y^j$ is a Lyndon word.

The following basic theorem on Lyndon words is due to K. T. Chen, R. H. Fox, and R. Lyndon (compare [104]).

Theorem 1.5.1. *Any word $w \in A^+$, can be written uniquely as $w = l_1 l_2 \cdots l_n$, where l_i is a Lyndon word, $1 \leq i \leq n$, and $l_1 \geq l_2 \geq \cdots \geq l_n$.*

Proof. Let $w \in A^+$. Since the letters are Lyndon words, then w can always be factorized, and possibly in more ways, as

$$w = l_1 l_2 \cdots l_n,$$

with $l_i \in \mathcal{L}$, $i = 1, \ldots, n$. Let us call n the rank of the factorization of w. We can always consider a factorization of w of minimal rank n. We prove that for such a factorization $l_i \geq l_{i+1}$ ($i = 1, \ldots, n-1$). Indeed, if there exists an integer $i \in [1, n-1]$ for which $l_i < l_{i+1}$ one would have from Proposition 1.5.2, $l_i l_{i+1} \in \mathcal{L}$ which contradicts the minimality of the rank. This shows the existence of at least one non-increasing factorization of w in Lyndon words. In order to prove the uniqueness suppose that there exist two factorizations of w by Lyndon words such that

$$w = l_1 l_2 \cdots l_n = l'_1 l'_2 \cdots l'_m$$

where $l_i \in \mathcal{L}$ $(i = 1, \ldots, n)$, and $l'_j \in \mathcal{L}$ $(j = 1, \ldots, m)$

$$l_1 \geq l_2 \geq \cdots \geq l_n, \; l'_1 \geq l'_2 \geq \cdots \geq l'_m.$$

We prove now that $n = m$ and that for all $i \in [1, n]$, $l_i = l'_i$. Indeed, suppose by contradiction that $|l_1| > |l'_1|$. One would have

$$l_1 = l'_1 \cdots l'_p \xi,$$

where $1 \leq p < m$ and ξ is a prefix of l'_{p+1}. If $\xi \neq \epsilon$, then from Proposition 1.5.1 one has

$$l_1 < \xi \leq l'_{p+1} \leq l'_1 < l_1,$$

which is a contradiction. If $\xi = \epsilon$, then one has

$$l_1 < l'_p \leq l'_1 < l_1,$$

that is again a contradiction. In a similar way one reaches a contradiction if one supposes that $|l_1| < |l'_1|$. Hence, $l_1 = l'_1$. By cancellation one has $l_2 \cdots l_n = l'_2 \cdots l'_m$. By iterating the same argument as before, one derives $n = m$ and $l_i = l'_i$ $(i = 1, \ldots, n)$, which concludes the proof.

One can easily prove that in the unique factorization of a word $w \in A^+$ in Lyndon words:

$$w = l_1 \cdots l_n,$$

with $l_1 \geq l_2 \geq \cdots \geq l_n$, the word l_i $(i = 1, \ldots, n)$ is the minimal right factor, with respect to the lexicographic order, of the word $w_i = l_1 \cdots l_i$. This fact gives a simple algorithm to construct the unique non-increasing factorization of a word in Lyndon words. For instance, in the case of the word $w = aabaababbaaaba$ one has the factorization $w = (aabaababb)(aaab)(a)$.

Let A be a finite totally ordered alphabet, $a = \min(A)$, and $<_A$ the lexicographic order in A^*. Let us consider the set

$$Y = a^+(A \setminus \{a\})^+.$$

Y is a code on the alphabet A, that is, the base of a free submonoid of A^* (see [8]). Any subset of Y is by itself a code over A. Let X be a finite subset of Y and B be an alphabet such that $\text{card}(B) = \text{card}(X)$. If δ is a bijection of B and X, then this can be extended to a monomorphism $\delta : B^* \to Y^*$. We can then totally order B by setting for $x, y \in B$

$$x <_B y \Longleftrightarrow \delta(x) <_A \delta(y). \tag{1.5}$$

The total order of B can be extended to the lexicographic order $<_B$ of B^*.

Lemma 1.5.1. *For $u, v \in B^*$,*

$$u <_B v \Longrightarrow \delta(u) <_A \delta(v).$$

Proof. Suppose first that u is a left factor of v, i.e., $v = u\xi$ with $\xi \in B^*$. One has then $\delta(v) = \delta(u)\delta(\xi)$, i.e., $\delta(u)$ is a prefix of $\delta(v)$, so that $\delta(u) <_A \delta(v)$. Let us then suppose that

$$u = hx\xi, \ v = hy\eta,$$

with $h, \xi, \eta \in B^*$, $x, y \in B$, and $x <_B y$. From equation (1.5), $x <_B y \Rightarrow \delta(x) <_A \delta(y)$. We have to consider two cases:

Case 1. $\delta(x) = rbs$, $\delta(y) = rct$ with $r, s, t \in A^*$, $b, c \in A$, and $b < c$. One has then $\delta(u) = \delta(h)rbs\delta(\xi)$, $\delta(v) = \delta(h)rct\delta(\eta)$, and $\delta(u) <_A \delta(v)$.

Case 2. $\delta(x)$ is a proper prefix of $\delta(y)$, i.e., $\delta(y) = \delta(x)\zeta$ with $\zeta \in A^+$. Since $\delta(x), \delta(y) \in X$, one has $\delta(x) = a^h f$, $\delta(y) = a^k g$ with $h, k > 0$ and $f, g \in (A \setminus \{a\})^+$. It follows that ζ begins with a letter $b > a$. Hence, we can write $\zeta = b\zeta'$. One has then $\delta(u) = \delta(h)\delta(x)\delta(\xi)$ and $\delta(v) = \delta(h)\delta(x)b\zeta'\delta(\eta)$. Hence, if $\delta(\xi) = \epsilon$ then $\delta(u)$ is a prefix of $\delta(v)$ so that $\delta(u) <_A \delta(v)$. If, on the contrary, $\delta(\xi) \neq \epsilon$ then $\delta(\xi) \in aA^*$ and again $\delta(u) <_A \delta(v)$.

Proposition 1.5.4. *If $w \in \mathcal{L}_B$, then $\delta(w) \in \mathcal{L}_A$.*

Proof. Let $w = x_1 \cdots x_n \in \mathcal{L}_B$ with $x_i \in B$ $(i = 1, \ldots, n)$. From Proposition 1.5.1 one has

$$x_1 \cdots x_n <_B x_i \cdots x_n,$$

for all $i \geq 2$. This implies by Lemma 1.5.1

$$\delta(x_1) \cdots \delta(x_n) <_A \delta(x_i) \cdots \delta(x_n). \tag{1.6}$$

To prove that $\delta(w) \in \mathcal{L}_B$ we have to show that for any factorization $\delta(w) = uv$, with $u \neq \epsilon$ and $v \in A^+$, one has $\delta(w) <_A v$. This is the case, by equation (1.6), when $v = \delta(x_i) \cdots \delta(x_n)$ for a suitable $i \geq 2$. Let us then suppose that there exist an integer i and words $\delta' \in A^+$, $\delta'' \in A^*$ such that $0 < i \leq n$ and

$$\delta(x_i) = \delta'\delta'', \ v = \delta''\delta(x_{i+1}) \cdots \delta(x_n),$$

(if $i = n$, then $v = \delta''$). By definition $\delta(x_i) = \delta'\delta'' = a^k f$ with $k > 0$ and $f \in (A \setminus \{a\})^+$. If $|\delta'| \geq k$, then δ'' begins with a letter $b > a$, i.e., $\delta'' = b\zeta$; thus

$$v = b\zeta\delta(x_{i+1}) \cdots \delta(x_n).$$

Since $\delta(w)$ begins with the letter a the result in this case trivially follows. Let us then suppose that $|\delta'| < k$ so that

$$v = a^{k-p} f\delta(x_{i+1}) \cdots \delta(x_n),$$

with $p > 0$. Now $a^k f <_A a^{k-p} f$ so that by $(L2)$ one derives

$$\delta(x_i) \cdots \delta(x_n) = a^k f \delta(x_{i+1}) \cdots \delta(x_n) <_A a^{k-p} f \delta(x_{i+1}) \cdots \delta(x_n) = v.$$

Therefore,

$$\delta(x_1) \cdots \delta(x_n) <_A \delta(x_i) \cdots \delta(x_n) <_A v,$$

that concludes the proof.

Definition 1.5.2. *Let $<$ be a total order in A^*. A sequence $(u_1, u_2, \ldots, u_n)$ of n words of A^+ is called an* n-division, *with respect to $<$, if for any nontrivial permutation σ of S_n one has*

$$u_{\sigma(1)} u_{\sigma(2)} \cdots u_{\sigma(n)} < u_1 u_2 \cdots u_n.$$

A word u is called n-divided, *with respect to $<$, if there exists an n-division $(u_1, u_2, \ldots, u_n)$ of u, i.e., such that $u = u_1 u_2 \cdots u_n$.*

An n-division $(u_1, u_2, \ldots, u_n)$ with respect to the relation $>$, which is the inverse of $<$, is also called ***inverse n-division*** with respect to the order $<$. A word which is n-divided with respect to $>$ is also called ***inversely n-divided*** with respect to $<$.

In the following we shall consider n-divisions (inverse n-divisions) and n-divided (inversely n-divided) words only with respect to the lexicographic order, so that the above terms will be used without specifying the total order.

Example 1.5.1. Let $A = \{a, b\}$ be a two-letter alphabet ordered by setting $a < b$. The word $w = ababbaba$ is 3-divided by the sequence $(ababb, ab, a)$ and inversely 3-divided by $(a, ba, bbaba)$. Moreover, one can easily verify that $(ababb, ab, a)$ is the only 3-division of w and that w does not admit 4-divisions.

Lemma 1.5.2. *Let $l_1, l_2, \ldots, l_n$ be n Lyndon words with $l_1 > l_2 > \cdots > l_n$. Let $w_i = l_i^{k_i}$, with $k_i \geq 1$, $1 \leq i \leq n$. The word $w = w_1 w_2 \cdots w_n$ is n-divided and $(w_1, w_2, \ldots, w_n)$ is an n-division.*

Proof. We observe that, by Proposition 1.5.3, $w_j w_i$ is a Lyndon word so that $w_j w_i < w_i w_j$, for $1 \leq i < j \leq n$. The statement then follows observing that for any nontrivial permutation $\sigma \in S_n$ the word $w_{\sigma(1)} w_{\sigma(2)} \cdots w_{\sigma(n)}$ can be obtained from $w_1 w_2 \cdots w_n$ by a sequence of transpositions of the kind $w_i w_j \to w_j w_i$, where $1 \leq i < j \leq n$.

1.6 Factorial languages and subword complexity

A language L is called *factorial*, or *closed by factors*, if any factor of any word of L belongs to L, that is, $L = F(L)$.

Lemma 1.6.1. *A language $L \subset A^*$ is factorial if and only if $L = A^* \setminus J$, where J is a two-sided ideal of A^*.*

Proof. Suppose first that $L = F(L)$ and set $J = A^* \setminus L$. Since L is factorial, then for any $j \in J$ and $w \in A^*$ the products jw and wj do not belong to L. This implies that J is a two-sided ideal of A^*. Conversely, suppose that J is a two-sided ideal of A^* and set $L = A^* \setminus J$. If $w \in L$ and u is any factor of w, then we can write $w = huk$ with $h, k \in A^*$. One has that $u \notin J$; in fact, otherwise $w \in J$. It follows that $u \in L$, hence, L is a factorial language.

Let $L \subseteq A^*$ be a language. The *structure or enumeration function* s_L of L is the map $s_L : \mathbb{N} \to \mathbb{N}$ defined as: for all $n \geq 0$

$$s_L(n) = \text{card}(L \cap A^n).$$

Thus the map s_L counts, for any $n \geq 0$, the number of distinct words of L of length n. The *growth function* g_L of L is defined as

$$g_L(n) = \text{card}(L \cap A^{[n]}) = \sum_{i=0}^{n} s_L(i).$$

For any $n \geq 0$, $g_L(n)$ gives the number of words of L of length $\leq n$.

For any $L \subseteq A^*$ the structure function $s_{F(L)}$ of $F(L)$ is called the *subword complexity* of L and will be denoted by f_L. Thus for any $n \geq 0$, $f_L(n)$ gives the number of distinct subwords of words of L, of length n. If L is factorial, then $f_L = s_L$.

When w is an infinite word the subword complexity and the growth function of the language $F(w)$ will be simply denoted by f_w and g_w, respectively.

Definition 1.6.1. *A language L over A is called* prolongable (right prolongable, left prolongable) *if for any $w \in L$ there exist letters $a, b \in A$ such that*

$$awb \in L \ (\ wb \in L, \ aw \in L).$$

Example 1.6.1. 1. Let L be a language over A and L^* be the submonoid of A^* generated by L. Then $F(L^*)$ is factorial and prolongable. Let $w \in A^{\pm\omega}$ be a bi-infinite word on the alphabet A. The set $F(w)$ is factorial and prolongable.

2. Let $w \in A^\omega$ ($w \in A^{-\omega}$) be a right (left) infinite word. The set $F(w)$ is factorial and right (left) prolongable.

3. Let L be the language $L = a^* \cup a^*b \cup ca^* \cup ca^*b$. The language L is factorial but neither right nor left prolongable. In fact the words ca^n, $n \geq 0$, are not prolongable on the left and the words a^mb, $m \geq 0$ are not prolongable on the right.

The following proposition gives a characterization of ultimately periodic words in terms of the subword complexity (see [25]).

Proposition 1.6.1. *Let $w \in A^\omega$ have the subword complexity f_w and let $k = \text{card}(\text{alph}(w))$. The following conditions are equivalent:*

i. The word w is ultimately periodic.
ii. There exists a constant c such that

$$f_w(n) < c, \text{ for all } n \geq 0.$$

iii. There exists an integer $n > 0$ such that

$$f_w(n) \leq n + k - 2.$$

iv. There exists an integer n such that

$$f_w(n) = f_w(n+1).$$

Proof. *i* $\Rightarrow$ *ii*: let $w \in A^\omega$ be ultimately periodic, then, there exist words $u, v \in A^*$, $v \neq \epsilon$ such that $w = uv^\omega$. We observe first that the subword complexity f_w is a non-decreasing function of n. Indeed, the language $F(w)$ is right prolongable so that for any factor x of w of length n there exists at least one letter $a \in A$ such that $xa \in F(w)$. Hence, $f_w(n+1) \geq f_w(n)$. Let $n_0 = |u| + |v|$ and $c = f_w(n_0) + 1$. Any factor x of w of length $n \geq n_0$ can be factorized as $x = \lambda x'$, where x' is the suffix of x of length $|v|$. Since x' is a factor of the periodic word v^ω having period $|v|$, there exists a unique letter $a \in A$ such that $x'a \in F(v^\omega)$. This implies that x can be prolonged on the right only by the letter a. Hence, for all $n \geq n_0$

$$f_w(n+1) = f_w(n) = f_w(n_0) < c.$$

ii $\Rightarrow$ *iii*: it is trivial. It is sufficient to take $n \geq c + 1 - k$.

iii $\Rightarrow$ *iv*: suppose that $f_w(n+1) > f_w(n)$ for all $n > 0$. One can write $f_w(n+1) = f_w(n) + d(n)$, where $d(n) \geq 1$ for all $n > 0$. One has then, by iteration,

$$f_w(n) = f_w(1) + \sum_{i=1}^{n-1} d(i) \geq n + k - 1,$$

for all $n > 0$, which is a contradiction.

iv $\Rightarrow$ *i*: let n be the least integer for which $f_w(n) = f_w(n+1)$. This condition implies that there does not exist any factor of w of length n which is prolongable on the right in $F(w)$ with two distinct letters. Thus for $j \geq n+1$ the letter w_j of the word w is uniquely determined by the n letters immediately preceding it in the word w. This implies that w is ultimately periodic. In fact, since w is infinite, there exists a factor v of length n having two occurrences in w:

$$v = w[r, r+n-1] = w[s, s+n-1].$$

Let $r < s$. One has then $w_{r+n} = w_{s+n}$. Since $w[r+1, r+n] = w[s+1, s+n]$, one has also $w_{r+n+1} = w_{s+n+1}$. By repeating the same argument one derives that $w_{r+m} = w_{s+m}$ for all $m \geq 0$. Hence, setting $\alpha = w[1, r-1]$ and $\beta = w[r, s-1]$ one has $w = \alpha\beta^\omega$ and w is ultimately periodic with a period $|\beta| = s - r$.

In terms of the growth function of a word, the property of being ultimately periodic can be expressed as follows:

Corollary 1.6.1. *Let $w \in A^\omega$ have a growth function g_w. Then w is ultimately periodic if and only if there exists an integer n such that*

$$g_w(n) \leq \frac{n(n+3)}{2}.$$

Proof. If w is not ultimately periodic, then $k = \text{card}(\text{alph}(w)) > 1$, so that by condition *iii* of the preceding proposition one has $f_w(n) \geq n + 1$ for all $n > 0$. Hence,

$$g_w(n) = f_w(0) + \sum_{i=1}^{n} f_w(i) \geq 1 + \sum_{i=1}^{n} (i+1) = 1 + \frac{n(n+3)}{2}.$$

Thus, if there exists an integer n for which $g_w(n) \leq n(n+3)/2$, then w is ultimately periodic. Conversely, if w is ultimately periodic, then by condition *ii* of Proposition 1.6.1 there exists a constant $c > 0$ such that $f_w(n) < c$ for all $n > 0$. Thus, for $n > 0$ $g_w(n) \leq cn$, so that certainly there will exist a value of n such that $g_w(n) \leq n(n+3)/2$.

From Proposition 1.6.1 one has that if an infinite word w is not ultimately periodic, then for all n, $f_w(n) \geq n + 1$. ***Sturmian words*** are infinite words w whose subword complexity f_w is such that

$$f_w(n) = n + 1$$

for all $n \geq 0$, so that they have the minimal possible value for subword complexity without being ultimately periodic (see, e.g., [33]). Moreover, since $f_w(1) = 2$, one has that these words are in a two-letter alphabet. It is worth noting that between ultimately periodic and Sturmian words there are no other words. The term "Sturmian" was first used by M. Morse and G. Hedlund [119]. A recent survey of J. Berstel on Sturmian words is in [105].

The most famous Sturmian word is the ***Fibonacci word***

$$f = abaababaabaababaababaabaababaabaab\cdots,$$

which is the limit of the sequence of words $(f_n)_{n\geq 0}$, inductively defined as

$$f_0 = b, \; f_1 = a, \; f_{n+1} = f_n f_{n-1},$$

for all $n > 0$. The words f_n of this sequence are called the *finite Fibonacci words*. The name Fibonacci is due to the fact that for each n, $|f_n|$ is equal to the $(n+1)$ th term of the Fibonacci numerical sequence:

$$1, 1, 2, 3, 5, 8, 13, \ldots.$$

Standard (or characteristic) Sturmian words can be defined in the following way which is a natural generalization of the definition of the Fibonacci word. Let $c_0, c_1, \ldots, c_n, \ldots$ be any sequence of natural integers such that $c_0 \geq 0$ and $c_i > 0$ $(i = 1, \ldots, n)$. We define, inductively, the sequence of words $(s_n)_{n \geq 0}$, where

$$s_0 = b,\ s_1 = a, \text{ and } s_{n+1} = s_n^{c_{n-1}} s_{n-1}, \text{ for } n \geq 1.$$

The sequence $(s_n)_{n \geq 0}$ converges to a limit s which is an infinite standard Sturmian word. Any standard Sturmian word is obtained in this way. The sequence $(s_n)_{n \geq 0}$ is called the *approximating sequence* of s and $(c_0, c_1, c_2, \ldots)$ the *directive sequence* of s. The Fibonacci word f is the standard Sturmian word whose directive sequence is $(1, 1, \ldots, 1, \ldots)$.

The words s_n are also called *finite standard words* or *generalized (finite) Fibonacci words.*

Let us denote by St the factorial language whose words are the finite factors of all infinite Sturmian words. One can prove (see [115]) that *for any Sturmian word x there exists a standard Sturmian word s having the same set of factors*, that is, $F(s) = F(x)$. Hence, St coincides with the set of factors of all standard Sturmian words. It was proven [116, 33] that the subword complexity f_{St} of St is given for all $n \geq 1$ by the formula

$$f_{St}(n) = 1 + \sum_{i=1}^{n} \phi(i)(n - i + 1), \tag{1.7}$$

where ϕ is the totient Euler's function.

There exist several infinite words w having the maximal possible value for the subword complexity, i.e., $f_w(n) = q^n$, with $q = \text{card}(A)$, for all $n \geq 0$. An example on the alphabet $A = \{0, 1\}$ is the *Champernowne sequence* c (see [122]) which can be simply constructed by writing consecutively each integer in base 2:

$$c = 0110111001011101111000\cdots.$$

One can easily prove that for all $n > 0$, $f_c(n) = 2^n$. Moreover, c is *normal*, that is, every subword of c of length n occurs with a frequency 2^{-n}.

We say that a function $f : \mathbb{N} \to \mathbb{N}$ is *polynomially upper-bounded* if there exists a polynomial p with non-negative coefficients such that for all $n \geq 0$,

$$f(n) \leq p(n).$$

This is also equivalent to the statement that a constant $c > 0$ and a positive integer k exist such that for all $n > 0$

$$f(n) \leq cn^k.$$

When $k = 1$, we say that f is linearly upper-bounded and, when $k = 2$, quadratically upper-bounded.

Proposition 1.6.2. *Let L be a factorial language. If the subword complexity f_L of L is linearly upper-bounded, then the growth function g_L is quadratically upper-bounded. If L is right or left prolongable, then also the converse is true.*

Proof. Let us suppose that $f_L(n) \leq dn$, for all $n > 0$, with $d > 0$. One has then for all $n > 0$

$$g_L(n) = 1 + \sum_{i=1}^{n} f_L(i) \leq 1 + d\sum_{i=1}^{n} i = 1 + (d/2)n(n+1).$$

Conversely, let us suppose that L is right or left prolongable and that for all $n > 0$, $g_L(n) \leq cn^2$, with $c > 0$. We want to prove that the subword complexity f_L is linearly upper-bounded. Suppose, by contradiction, that for any $d > 0$ there exists an integer $n_0 > 0$ such that $f_L(n_0) > dn_0$. Since L is right or left prolongable one has also $f_L(n) \geq f_L(n_0)$ for all $n > n_0$. Hence, one has

$$4cn_0^2 \geq g_L(2n_0) = 1 + \sum_{i=1}^{2n_0} f_L(i) > dn_0^2,$$

that implies $4c > d$. If we take $d \geq 4c$, then we reach a contradiction.

Let us observe that it has been proved in [5] that there exist factorial languages having a subword complexity which is not linearly upper-bounded and a growth function which is quadratically upper-bounded. Thus in the preceding proposition the hypothesis that L is right (or left) prolongable cannot be dropped. However, in the case of factorial languages the following partial converse holds.

Proposition 1.6.3. *Let L be a factorial language. If $g_L(n) \leq cn^2$ for all $n > 0$, then there exists a positive integer d such that for any pair n, m of integers, with $0 \leq m \leq n$, there exists an integer $n' \in [(3/4)n, n]$ such that $f_L(n' + m) \leq d(n' + m)$. The constant d can be taken equal to $22c$.*

Proof. Suppose that for any $d > 0$ there exist integers m and n such that $f_L(n' + m) > d(n' + m)$ for all integers n' in the interval $[(3/4)n, n]$. One would have

$$4cn^2 \geq g_L(2n) \geq g_L(n+m) \geq \sum_{n' \in [(3/4)n,n]} f_L(n' + m)$$

$$> d \sum_{n' \in [(3/4)n,n]} (n' + m) \geq (3/16)dn^2.$$

If $d > (64/3)c$, then one reaches a contradiction. One can then assume $d = 22c$.

Let L be a factorial language and u, v be two given words. We introduce for any $m \geq 0$ the set

$$L_{u,v,m} = uA^m v \cap L$$

of all the words of L of length $|u|+|v|+m$ which begin with the prefix u and terminate with the suffix v.

The following lemma shows that if $L_{u,v,m}$ with $|v| \geq |u| = n$, contains a sufficiently large number of elements with respect to the value $f_L(n+m)$ of the subword complexity of L, then u has a suffix which is a sesquipower of a suitably high order and small period.

Lemma 1.6.2. *Let L be a factorial language and u, v be words such that $|v| \geq |u| = n$. If $\text{card}(L_{u,v,m}) > k$ ($\text{card}(L_{v,u,m}) > k$) and $f_L(n+m)/k \leq n$, then u has a suffix (prefix) which is a sesquipower of order*

$$h \geq \lfloor nk/f_L(n+m) \rfloor,$$

period $p < f_L(n+m)/k$, and such that $ph > n - f_L(n+m)/k$.

Proof. We shall consider only the case $\text{card}(L_{u,v,m}) > k$; the case $\text{card}(L_{v,u,m}) > k$ is dealt with similarly. By hypothesis $\text{card}(L_{u,v,m}) = q > k$. The elements of $L_{u,v,m}$ are of the kind $uw_a v$ $(a = 1, \ldots, q)$ with $|w_a| = m$, $|u| = n$, and $|v| \geq n$. For any j in the interval $[1, n]$ denote by u_j the suffix of u of length $n-j+1$ and by v_j the prefix of v of length $j-1$. For any fixed j the words $u_j w_a v_j$ $(a = 1, \ldots, q)$ are elements of L (since L is closed by factors) of length $n+m$ and pairwise distinct so that their number is q. Let us now consider all the words that one can generate, with possible repetitions, for j ranging in the interval $[1, \lceil f_L(n+m)/k \rceil]$. One yields a number of words of length $n+m$ given by

$$q\lceil f_L(n+m)/k \rceil \geq (q/k) f_L(n+m) > f_L(n+m).$$

Since the number of words of L of length $n+m$ is equal to $f_L(n+m)$, there must exist integers i, j, $i \neq j$, in the interval

$$[1, \lceil f_L(n+m)/k \rceil]$$

and integers a, b in the interval $[1, q]$ such that

$$u_i w_a v_i = u_j w_b v_j. \tag{1.8}$$

We may always suppose that $i < j$. By the definition u_j is a proper suffix of u_i, so that we can write $u_i = \alpha u_j$, with $\alpha \in A^+$; moreover from equation (1.8) one has $u_i = u_j \beta$, with $\beta \in A^+$. Hence, $u_i = \alpha u_j = u_j \beta$.

By Proposition 1.4.4, the solution of this equation can be written as

$$\alpha = (\lambda\mu)^s, \quad \beta = (\mu\lambda)^s, \quad u_j = (\lambda\mu)^r \lambda, \quad u_i = (\lambda\mu)^{r+s}\lambda, \tag{1.9}$$

where $\lambda\mu$ and $\mu\lambda$ are primitive words, $s \geq 1$, $r \geq 0$. Let us write $u = p_i u_i = p_j u_j$, where p_i and p_j denote the prefixes of u of length $i-1$ and $j-1$, respectively. One derives

$$u = \rho(\mu\lambda)^h,$$

where $h = r + s \geq 1$ and $\rho = p_i\lambda$. Thus, setting $p = |\lambda\mu| = |\mu\lambda|$, it follows that

$$|u| = n = |\rho| + hp. \tag{1.10}$$

Moreover, from equation (1.9) it follows that

$$p_i\lambda(\mu\lambda)^{s-1}\mu = \rho(\mu\lambda)^{s-1}\mu = p_j,$$

so that $|\rho| \leq |p_j| = j - 1 \leq \lceil f_L(n+m)/k \rceil - 1 < f_L(n+m)/k$. Moreover, from (1.9) one has that

$$|u_i| - |u_j| = sp = j - i < f_L(n+m)/k$$

and then

$$p < f_L(n+m)/sk \leq f_L(n+m)/k.$$

Hence, from equation (1.10) one has

$$n < (1 + h/s) f_L(n+m)/k$$

and then $h \geq h/s > (nk/f_L(n+m)) - 1$, so that, $h \geq \lfloor nk/f_L(n+m) \rfloor$. Finally, from (1.10) one has

$$ph = n - |\rho| > n - f_L(n+m)/k.$$

By Lemmas 1.6.2 and 1.6.3 we derive the following proposition [38].

Proposition 1.6.4. *Let L be a factorial language whose growth function is quadratically upper-bounded, i.e., $g_L(n) \leq cn^2$, for all $n > 0$. Let $d = 22c$ and $u, v \in A^*$ be words such that $|v| \geq |u| = n$. Let, moreover, m and k be integers such that $0 \leq m \leq (1/2)n$ and $k/2d > 1$. If $\mathrm{card}(L_{u,v,m}) > k$, ($\mathrm{card}(L_{v,u,m}) > k$), then u has a suffix (prefix) which is a sesquipower of order $h \geq \lfloor k/2d \rfloor$, period $p < 3nd/2k$, and such that $ph > (3/4)n(1 - 2d/k)$.*

Proof. If $\mathrm{card}(L_{u,v,m}) > k$, then $\mathrm{card}(L_{u',v,m}) > k$ for any suffix u' of u. By Proposition 1.6.3 there exists an integer $n' \in [(3/4)n, n]$ such that

$$f_L(n' + m) \leq d(n' + m).$$

Let us then consider the suffix u' of length $n' \geq (3/4)n$ such that $f_L(n'+m) \leq d(n' + m)$. Under the hypotheses made one has

$$f_L(n' + m)/k \leq n' 2d/k < n'.$$

By Lemma 1.6.2 one has that u' has a suffix which is a sesquipower of order

$$h \geq \lfloor n'k/f_L(n'+m) \rfloor \geq \lfloor k/d(1+m/n') \rfloor \geq \lfloor k/2d \rfloor,$$

period

$$p < f_L(n'+m)/k \leq d(n'+m)/k \leq d(n+m)/k \leq 3nd/2k,$$

and such that

$$ph > n' - f_L(n'+m)/k \geq n' - d(n'+m)/k \geq n'(1-2d/k) \geq (3/4)n(1-2d/k).$$

The case $\mathrm{card}(L_{v,u,m}) > k$ is dealt with similarly.

An application of the preceding proposition to finitely generated semigroups whose growth function is quadratically upper-bounded, will be given in Chap. 3. We give now a further consequence of Lemma 1.6.2 which is interesting for the "completion properties" of subwords of an infinite word.

A finite or infinite word is called *p-power-free*, $p > 1$, if for any $u \in A^+$, $u^p \notin F(v)$ (see Sect. 2.6). Let w be an infinite word and r be a positive integer. Let us define for any fixed word $u \in A^*$

$$L_{u,r} = \{uv \in F(w) \mid |v| = r|u|\}.$$

One has

$$L_{u,r} = \bigcup_{v \in A^{|u|} \cap F(w)} L_{u,v,(r-1)|u|}. \tag{1.11}$$

Proposition 1.6.5. *The subword complexity of an infinite word w is linearly upper-bounded if and only if the following property (*completion property*) is satisfied. For any fixed $p > 1$ and p-power-free word $u \in A^*$ there exists an integer D, which does not depend on the length of u, such that for all $r > 0$*

$$\mathrm{card}(L_{u,r}) \leq Dr.$$

Proof. Let us first prove the "if" part. For any letter $a \in \mathrm{alph}(w)$ one has for all $n > 1$ that $\mathrm{card}(L_{a,n-1}) \leq D(n-1)$. Hence, setting $q = \mathrm{card}(A)$, one has

$$f_w(n) \leq f_w(1)D(n-1) \leq qDn.$$

Let us now prove the "only if" part. If u is not a factor of w, then for all $r > 0$, $F_{u,r} = \emptyset$ and the result is, trivially, true. Let us then suppose that $u \in F(w)$. By hypothesis the subword complexity f_w is linearly upper-bounded, i.e., there exists a constant c such that for all $n > 0$

$$f_w(n) \leq cn.$$

Suppose, by contradiction, that an integer $p > 1$ and a p-power-free word u, of length n, exist such that for any $D > 0$ there exists a positive integer r for which

$$\text{card}(L_{u,r}) > Dr.$$

Let us take D such that $D/(cf_w(n)) > p$. By equation (1.11) there exists a word $v \in A^n \cap F(w)$ such that

$$f_w(n)\,\text{card}(L_{u,v,(r-1)|u|}) \geq \text{card}(L_{u,r}).$$

Thus

$$\text{card}(L_{u,v,(r-1)|u|}) > (Dr)/f_w(n) > rcp.$$

Moreover, $f_w(rn) \leq cnr$. Hence, $f_w(nr)/(rcp) \leq n/p < n$. By Lemma 1.6.2 the word u has a suffix which is a sesquipower of order $h \geq p$ which is a contradiction.

A different and more direct proof of the preceding proposition, where there is also an evaluation of the constant D, is in [42]. A straightforward consequence of the preceding proposition is the following:

Corollary 1.6.2. *Let w be an infinite p-power-free word ($p > 1$). The subword complexity of w is linearly upper-bounded if and only if for any $u \in A^*$ and $r > 0$,*

$$\text{card}(L_{u,r}) \leq Dr,$$

where D is a constant which does not depend on the length of u.

2. Unavoidable Regularities

This chapter deals with some properties, known as unavoidable regularities, which are always satisfied by sufficiently long words over a finite alphabet. The study of these regularities, as we shall see in the following sections and chapters, is of great interest in combinatorics on words both for the importance of the subject itself and for the applications in many areas of algebra and theoretical computer science.

Informally a property P defined in a free monoid A^* is *unavoidable* if it is not possible to construct arbitrarily long words not satisfying P.

Definition 2.0.2. *Let P be a property defined in the free monoid A^*. P is called an* unavoidable regularity *if the set $L_P = \{x \in A^* \mid x$ does not satisfy $P\}$ is finite. P is called* avoidable *if it is not unavoidable, i.e., there exist infinitely many words which do not satisfy P.*

A property P is called an *ideal property* if for all $x, u, v \in A^*$

$$P(x) \quad \text{implies} \quad P(uxv).$$

The attribute *ideal* is due to the fact that the set

$$J_P = \{x \in A^* \mid P(x)\},$$

if nonempty, is a two-sided ideal of A^*. By Lemma 1.6.1 one has that the set $L_P = A^* \setminus J_P$ is closed by factors. For this reason the negation $\neg P$ of P is also called a property *stable for factors.*

Example 2.0.2. We consider the property P in A^* defined as follows: x satisfies P if and only if x contains a square as a factor, that is, $x = \lambda uu\mu$, with $\lambda, \mu \in A^*, u \in A^+$. Trivially P is an ideal property. If A has only two letters, then it is easy to see that any word whose length is greater than three contains a square, so in this case P is unavoidable. If A has three letters, then the property P is avoidable. Indeed, as proved by A. Thue [153], there exist infinitely many square-free words in a three-letter alphabet (see Sect. 2.6).

Proposition 2.0.6. *Let P be an ideal property. The following statements are equivalent:*

i. There exists $w \in A^{\pm\omega}$ such that no factor of w satisfies P.

ii. P is avoidable.

Proof. Clearly i implies ii. Conversely, if P is avoidable, then $L = \{x \in A^* \mid \neg P(x)\}$ is infinite. By Lemma 1.2.2 there exists an infinite word $w \in A^{\pm\omega}$ such that $F(w) \subseteq F(L)$. Since P is an ideal property, then L is closed by factors. Thus one has $L = F(L)$ and $F(w) \subseteq L$.

2.1 Ramsey's theorem

One of the most important results in combinatorics is a famous theorem that was formulated and proved in 1930 by F. P. Ramsey in some investigations on decision procedures for logical systems [134]. The theorem of Ramsey can be considered as a deep generalization of the simple pigeon-hole principle. There are several formulations and proofs of this theorem and many questions and open problems are related to it (see [70]). Nowadays there exists a discipline, part of combinatorial analysis, called *Ramsey theory* which collects several combinatorial results (such as van der Waerden's theorem) based on the same kind of ideas as Ramsey's theorem. In our context we are only interested in different applications of Ramsey's theorem in combinatorics on words.

Let E be a finite set and $P_m(E)$ the set of all subsets of E having cardinality m. We shall use the following version of Ramsey's theorem (see [70]):

Theorem 2.1.1. *Let m, n, k be positive integers, with $m \leq n$. There exists an integer $R(m,n,k)$ such that for any finite set E with* $\text{card}(E) \geq R(m,n,k)$ *and for any partition θ of $P_m(E)$ in k classes there exists $F \subseteq E$ such that* $\text{card}(F) = n$ *and $P_m(F)$ is contained in a single class modulo θ.*

In the following we call *m-subset* of E any subset of E of cardinality m. Hence, $P_m(E)$ is the set of all m-subsets of E. A partition θ of $P_m(E)$ in k classes is called a *k-coloring* and a subset of $P_m(E)$ included in a single equivalence class is called *monochromatic*. Then Theorem 2.1.1 can be equivalently restated as:

Theorem 2.1.2. *Let m, n, k be positive integers, with $m \leq n$. There exists an integer $R(m,n,k)$ such that for any integer $N \geq R(m,n,k)$ and for any k-coloring of $P_m([1,N])$, there exists an n-subset F of $[1,N]$ such that all the m-subsets of F have the same color, that is, $P_m(F)$ is monochromatic.*

Proof. We prove the statement by induction on m. If $m = 1$, then the statement trivially follows from the pigeon-hole principle, taking $R(1,n,k) = (n-1)k+1$. Now we assume that the theorem holds for some $m \geq 1$, for all n, k, with $n \geq m$. We define a function $f_m(j,q,k)$, $j \leq q$, satisfying the following condition. For any $N \geq f_m(j,q,k)$ and for any k-coloring of the $(m+1)$-subsets of $[1,N]$, there exists a subset H of $[1,N]$ such that $H = X \cup Y$, $\text{card}(X) = j$, $\text{card}(H) = q$, and $x < y$, for $x \in X$ and $y \in Y$.

Moreover, for any $(m+1)$-subset K of H, if $K \cap X \neq \emptyset$, then the color of K is uniquely determined by $\min\{x \mid x \in K \cap X\}$.

We define $f_m(j,q,k)$ by induction on j. Setting $f_m(0,q,k) = q$, the function f_m trivially satisfies the required conditions for $j = 0$. Suppose now that we have defined $f_m(j,q,k)$ satisfying the statement, for a given $j \geq 0$ and for any $q \geq j$. Then we set

$$f_m(j+1,q,k) = f_m(j, R(m,q-j-1,k)+j+1,k).$$

We observe that in the above definition we use the inductive hypothesis for f and R. Suppose that $N \geq f_m(j+1,q,k)$ and a k coloring of the $(m+1)$-subsets of $[1,N]$ is given. By the inductive hypothesis there exists a subset H' of $[1,N]$, $H' = X' \cup Y'$, $\mathrm{card}(X') = j$, $\mathrm{card}(H') = R(m,q-j-1,k)+j+1$, and $x < y$ for $x \in X'$, $y \in Y'$. Moreover, H' satisfies the required conditions.

Let $x = \min\{y \mid y \in Y'\}$. We set $X = X' \cup \{x\}$. We consider the k-coloring of the m-subsets of $Y' \setminus \{x\}$ induced by the given k-coloring of the $(m+1)$-subsets of $[1,N]$. In fact we define the color of Z, $Z \in P_m(Y' \setminus \{x\})$, as the color of $Z \cup \{x\}$. We observe that $\mathrm{card}(Y' \setminus \{x\}) = R(m,q-j-1,k)$. Thus, there exists a $(q-j-1)$-subset Y of $Y' \setminus \{x\}$ such that all the m-subsets of Y have the same color. This implies that each $(m+1)$-subset of $X \cup Y$, having nonempty intersection with X, has a color uniquely determined by its minimal element. We set then $H = X \cup Y$. In fact we have $\mathrm{card}(X) = j+1$, $\mathrm{card}(Y) = q-j-1$, and $\mathrm{card}(H) = q$ so that H satisfies the statement for $j+1$ and the induction step is complete.

Now we set

$$R(m+1,n,k) = f_m((n-1)k+1,(n-1)k+1,k).$$

Let $N \geq R(m+1,n,k)$ and consider a k-coloring of the $(m+1)$-subsets of $[1,N]$. By the properties of f_m, there exists an $((n-1)k+1)$-subset X of $[1,N]$ such that the color of any $(m+1)$ subset of X is uniquely determined by its minimal element. Since X contains $(n-1)k+1$ elements, by the pigeonhole principle, there exists a set Y of n elements of X which determine the same color. Thus, Y is an n-subset whose $(m+1)$-subsets have the same color. Thus the proof of the induction step is terminated and the statement is completely proved.

Let S be a set and $\phi : A^* \to S$ be a map from A^* in S. A word $w = w_1 \cdots w_k$ with $w_i \in A^+$, $i = 1, \ldots, k$, $k \geq 1$, is called a *k-power modulo ϕ* if

$$\phi(w_1) = \phi(w_2) = \cdots = \phi(w_k).$$

A k-power modulo ϕ is said to be *uniform* if in addition

$$|w_1| = |w_2| = \cdots = |w_k|.$$

Definition 2.1.1. *Let A^* be a free monoid, S a set, and k an integer > 1. A map $\phi : A^* \to S$ is called* k-repetitive *if there exists a positive integer L, depending on ϕ and k, such that any word w, with $|w| \geq L$, has a factor which is a k-power modulo ϕ. One says that ϕ is* repetitive *if ϕ is k-repetitive for all $k > 1$.*

Two notions stronger than that of a repetitive map are those of *ramseyan* and *uniformly* repetitive maps.

Definition 2.1.2. *Let S be a set and k be an integer > 1. A map $\phi : A^* \to S$ is called* k-ramseyan *if there exists a positive integer L, depending on ϕ and k, such that any word w, with $|w| \geq L$, has a factor*

$$w_1 w_2 \cdots w_k, \text{ with } w_i \in A^+, \ i = 1, \ldots, k$$

such that for all $i, j, i', j' \in \{1, \ldots, k\}$ with $i \leq j$ and $i' \leq j'$ one has

$$\phi(w_i \cdots w_j) = \phi(w_{i'} \cdots w_{j'}).$$

The map ϕ is called uniformly k-repetitive *if any sufficiently large word has a factor which is a uniform k-power modulo ϕ. One says that ϕ is* ramseyan (uniformly repetitive) *if ϕ is k-ramseyan (uniformly k-repetitive) for all $k > 1$.*

Trivially any ramseyan map is repetitive, but the converse is not true in general. Moreover, it is easy to show that any map from A^* in a finite set is repetitive; however, the following stronger result holds:

Theorem 2.1.3. *Let A^* be a free monoid and S be a finite set. Then any map $\phi : A^* \to S$ is ramseyan.*

Proof. Let k be a positive integer. Let $n = \text{card}(S)$ and $S = \{1, 2, \ldots, n\}$. We set $L = R(2, k+1, n)$, where R is the function of the Ramsey theorem. Let $w \in A^*$, $|w| = L$, and $E = \{1, \ldots, L+1\}$. The set $P_2(E)$ can be partitioned in the n classes $\theta_1, \theta_2, \ldots, \theta_n$, where for any $l \in \{1, \ldots, n\}$, θ_l is defined by

$$\{i, j\} \in \theta_l \iff \phi(w[i, j-1]) = l,$$

for $1 \leq i < j \leq L+1$. By Ramsey's theorem, there exists a subset $Y = \{i_1, i_2, \ldots, i_{k+1}\}$ of E with $i_1 < i_2 < \cdots < i_{k+1}$, and an integer $s \in \{1, 2, \ldots, n\}$ such that $P_2(Y)$ is contained in θ_s. Then for any $i, j, i', j' \in \{i_1, i_2, \ldots, i_{k+1}\}$, with $i < j$ and $i' < j'$, one has $\phi(w[i, j-1]) = \phi(w[i', j'-1]) = s$. Thus if we set

$$w_j = w[i_j, i_{j+1} - 1],$$

for $1 \leq j \leq k$, the word $w_1 w_2 \cdots w_k$ is a factor of w, such that $\phi(w_i \cdots w_j) = \phi(w_{i'} \cdots w_{j'})$, for $1 \leq i \leq j \leq k$, $1 \leq i' \leq j' \leq k$.

The notion of repetitivity will be reconsidered in Sect. 3.9, where some finiteness conditions for semigroups, based on this and related concepts, will be given.

Definition 2.1.3. *Let $(w_1, \ldots, w_m)$ be a sequence of words. We say that $(u_1, \ldots, u_n)$ is a* derived sequence *of $(w_1, \ldots, w_m)$ if there exist $n+1$ integers $j_1, j_2, \ldots, j_{n+1}$ such that $1 \leq j_1 < j_2 < \cdots < j_{n+1} \leq m+1$, and*

$$u_1 = w_{j_1} \cdots w_{j_2 - 1}, \ldots, u_n = w_{j_n} \cdots w_{j_{n+1}-1}.$$

From the preceding definition one has that the word $w = w_1 \cdots w_m$ can be rewritten as

$$w = xu_1 \cdots u_n y \text{ with } x = w_1 \cdots w_{j_1 - 1} \text{ and } y = w_{j_{n+1}} \cdots w_m.$$

Note that if $j_1 = 1$ ($j_{n+1} = m+1$), then $x = \epsilon$ ($y = \epsilon$).

The following lemma, whose proof we omit since it is a straightforward consequence of Def. 2.1.3, holds.

Lemma 2.1.1. *Let $\sigma = (w_1, \ldots, w_m)$ be a sequence of words and $(u_1, \ldots, u_n)$ be the derived sequence of σ, corresponding to the sequence $1 \leq j_1 < j_2 < \cdots < j_{n+1} \leq n+1$. Let $i_1 < i_2 < \cdots < i_{k+1}$ be an increasing subsequence of $j_1, j_2, \ldots, j_{n+1}$. The derived sequence $(v_1, \ldots, v_k)$ of σ corresponding to the sequence $i_1 < i_2 < \cdots < i_{k+1}$ is the derived sequence of $(u_1, \ldots, u_n)$ corresponding to the sequence of integers $1 \leq h_1 < h_2 < \cdots < h_{k+1} \leq k+1$, where $i_r = j_{h_r}$ for any $r = 1, \ldots, k+1$.*

The following lemma is another application of the Ramsey theorem.

Lemma 2.1.2. *Let $<$ be a total ordering on the free monoid A^*. For any integer $n > 1$ there exists a positive integer $r(n)$ with the property that, if $(w_1, \ldots, w_h)$ is a sequence of h words with $h \geq r(n)$, then there exists a derived sequence $(u_1, \ldots, u_n)$ of $(w_1, \ldots, w_h)$ such that either*

i. $\forall j \in [1, n-1], \quad u_j u_{j+1} < u_{j+1} u_j$, *or*
ii. $\forall j \in [1, n-1], \quad u_j u_{j+1} > u_{j+1} u_j$, *or*
iii. $\forall j \in [1, n-1], \quad u_j u_{j+1} = u_{j+1} u_j$.

Proof. We set $r(n) = R(3, n+1, 3) - 1$ where R is the function of the Ramsey theorem. Let $(w_1, \ldots, w_h)$ be a sequence of h words with $h \geq r(n)$ and let $E = \{1, \ldots, h+1\}$. Then $P_3(E)$ can be partitioned in three classes $\theta_1, \theta_2, \theta_3$ as follows. Let i, j and p be three integers such that $1 \leq i < j < p \leq h+1$, then

(a) $\{i, j, p\} \in \theta_1 \Longleftrightarrow w_i \cdots w_{j-1} w_j \cdots w_{p-1} < w_j \cdots w_{p-1} w_i \cdots w_{j-1}$,
(b) $\{i, j, p\} \in \theta_2 \Longleftrightarrow w_i \cdots w_{j-1} w_j \cdots w_{p-1} > w_j \cdots w_{p-1} w_i \cdots w_{j-1}$,
(c) $\{i, j, p\} \in \theta_3 \Longleftrightarrow w_i \cdots w_{j-1} w_j \cdots w_{p-1} = w_j \cdots w_{p-1} w_i \cdots w_{j-1}$.

By Ramsey's theorem there exists a subset $F = \{j_1, \ldots, j_{n+1}\}$ of E, with $j_1 < j_2 < \cdots < j_{n+1}$ such that $P_3(F)$ is contained in one of the θ_i $(i = 1, 2, 3)$. The sequence $j_1, \ldots, j_{n+1}$ allows us to construct the derived sequence

$$u_1 = w_{j_1} \cdots w_{j_2 - 1}, \ldots, u_n = w_{j_n} \cdots w_{j_{n+1}-1}.$$

Now if $P_3(F) \subseteq \theta_1$ ($P_3(F) \subseteq \theta_2$), then for any $i \in [1, n-1]$, $u_i u_{i+1} < u_{i+1} u_i$ ($u_i u_{i+1} > u_{i+1} u_i$). Moreover, if $P_3(F) \subseteq \theta_3$ then for any $i \in [1, n-1]$, $u_i u_{i+1} = u_{i+1} u_i$.

2.2 Van der Waerden's theorem

The following theorem was proved by B. L. van der Waerden in 1927 [155].

Theorem 2.2.1. *For all positive integers k, r there exists a positive integer $n(k, r)$ such that if the set $\{1, \ldots, n(k, r)\}$ is partitioned in k classes, then at least one class contains an arithmetic progression of length r.*

There exist several and conceptually different proofs of van der Waerden's theorem (see [70, 104]). An historical account in which van der Waerden describes the circumstances of the theorem's discovery is in [156].

The theorem of van der Waerden can be equivalently reformulated in terms of words as follows. Let $w = w_1 w_2 \cdots w_n$, $w_i \in A$, $i = 1, \ldots, n$, be a word of length n over the alphabet A with $\mathrm{card}(A) = k$. A *cadence* of w is a sequence $T = (t_1, t_2, \ldots, t_r)$ of integers such that

$$0 < t_1 < t_2 < \cdots < t_r \leq n \quad \text{and} \quad w_{t_1} = w_{t_2} = \cdots = w_{t_r}.$$

The number r is called the *order* of the cadence. The notion of cadence can also be given in the same way in the case of infinite words.

The cadence T is called *arithmetic* if there exists a positive integer d, called *rate*, such that, for any $i \in \{1, \ldots, r-1\}$, one has $t_{i+1} = t_i + d$.

Since any word $w \in A^n$ can be regarded as a partition of the set $[n] = \{1, 2, \ldots, n\}$ in k-classes, then Theorem 2.2.1 shows the existence of the following unavoidable regularity: all sufficiently long sequences of elements belonging to a finite set contain arbitrarily long subsequences of identical elements in arithmetic progressions.

Theorem 2.2.2. *Let A be an alphabet with k letters. For any positive integer r, there exists a positive integer $n(k, r)$, such that any word $w \in A^*$ of length $|w| \geq n(k, r)$ contains at least one arithmetic cadence of order r.*

Proof. The proof we report is along the lines of the proof given in [97]. The theorem is trivially true for $r = 2$ whatever the value of k is. Indeed, from the pigeon-hole principle one can take $n(k, 2) = k + 1$. The proof is by induction. One assumes that for any k the statement is true up to r and one proves that the statement is true for $r+1$; in other words one supposes that the integer $n(k,r)$ exists for all $k>0$ and some $r \geq 2$ and prove that there exists $n(k,r+1)$.

We introduce two sequences $q_s, n_s, s \geq 0$ of natural numbers recursively defined as: $q_0 = 1, n_0 = n(k,r)$, and for $s > 0$

$$q_s = 2n_{s-1}q_{s-1}, \quad n_s = n(k^{q_s}, r).$$

We shall prove that for each k one can take as $n(k,r+1)$ the integer q_k. In the following we denote $r+1$ simply by r'. We articulate the proof in three steps.

Step 1. Let A be an alphabet of cardinality k and w be a word

$$w = w_1 \cdots w_{q_k}, \quad w_i \in A, \quad 1 \leq i \leq q_k.$$

Since $q_k = 2n_{k-1}q_{k-1}$ we can factorize w as $w = uv$ with $|u| = |v| = n_{k-1}q_{k-1}$. Hence, we can decompose u in n_{k-1} blocks:

$$u = W_1 \cdots W_{n_{k-1}}$$

such that $|W_i| = q_{k-1}$, $1 \leq i \leq n_{k-1}$. Each block can be regarded as a letter in the alphabet $A_{k-1} = A^{q_{k-1}}$ whose cardinality is $k^{q_{k-1}}$. The length of u as a word on the alphabet A_{k-1} is $n_{k-1} = n(k^{q_{k-1}}, r)$; by induction it follows that there exists a word $U \in A^{q_{k-1}}$ such that u can be factorized as

$$u = \lambda U \xi_1 U \cdots U \xi_{r-1} U \mu,$$

where $\lambda, \mu, \xi_1, \ldots, \xi_{r-1} \in A_{k-1}^*$ and $|\xi_1| = |\xi_2| = \cdots = |\xi_{r-1}| = c \in \mathbb{N}$. We shall denote the r different occurrences of U in u by

$$U_1, U_2, \ldots, U_r.$$

The distance of the first letters of two consecutive such occurrences is a constant d_1. Let us observe that $|\mu v| \geq n_{k-1}q_{k-1}$ and that $c < (n_{k-1}-1)q_{k-1}$ (indeed, any ξ_i $(i = 1, \ldots, r-1)$ contains at most $n_{k-1} - 2$ letters of the alphabet A_{k-1}). Hence, the word μv can be factorized as

$$\mu v = \xi_r V \mu',$$

with $|\xi_r| = c$ and $|V| = |U|$. We shall denote by U_{r+1} this occurrence of V in $w = uv$.

Step 2. As $|U| = |V| = 2n_{k-2}q_{k-2}$ we repeat the same construction as above in the case of the words U and V. Since we are interested in the occurrences of U and V in w, we fix an occurrence U_i, $(i = 1, \ldots, r)$, of U. By the previous inductive argument U_i contains r occurrences

$$U_{i1}, \ldots, U_{ir}$$

of the same word U' of length q_{k-2}. Let d_2 be the distance of the occurrences of the first letters of two consecutive such blocks. We add, moreover, to the above sequence a further term $U_{ir'}$ such that $|U_{ir'}| = |U'|$ and the distance of its first letter from the first letter of U_{ir} is still d_2.

We carry on the same construction for all U_i, $(i = 1, \ldots, r)$, and also for $U_{r'} = V$. In this way we have produced a set of occurrences of factors of w depending on two indices $U_{i_1 i_2}$, $1 \le i_1, i_2 \le r'$. We shall denote $U_{i_1 i_2}$ simply by $[i_1 i_2]$. For $1 \le i_1, i_2, j_1, j_2 \le r'$, we say that $[i_1 i_2]$ is *equivalent* to $[j_1 j_2]$ and we write $[i_1 i_2] \equiv [j_1 j_2]$ if and only if they are occurrences of the same factor of w. Hence, from the construction above one has

$$[i_1 i_2] \equiv [j_1 j_2], \quad 1 \le i_1, i_2, j_1, j_2 \le r.$$

Let us repeat the same construction k-times. After the k th step the results of our construction are occurrences of factors of length $q_0 = 1$, i.e., *letters* of w, depending on k indices that we denote by

$$[i_1 i_2 \ldots i_k], \quad 1 \le i_1, i_2, \ldots, i_k \le r'.$$

Any $[i_1 i_2 \ldots i_k]$ denotes then a position of a letter in the word w. We shall then identify $[i_1 i_2 \ldots i_k]$ with the integer representing this position.

From the above construction one has

(i). $[i_1 i_2 \ldots i_k] \equiv [j_1 j_2 \ldots j_k]$ for $1 \le i_1, i_2, \ldots, i_k, j_1, j_2, \ldots, j_k \le r$

(ii). For any $1 \le s \le r$ one has

$$[i_1 \ldots i_s i_{s+1} \ldots i_k] \equiv [j_1 \ldots j_s i_{s+1} \ldots i_k]$$

with $1 \le i_1, \ldots, i_s, j_1, \ldots, j_s \le r$ and $1 \le i_{s+1}, \ldots, i_k \le r'$.

In fact $[i_1 \ldots i_s]$ and $[j_1 \ldots j_s]$ are two occurrences of the same factor f of w, so that $[i_1 \ldots i_s i_{s+1} \ldots i_k]$ and $[j_1 \ldots j_s i_{s+1} \ldots i_k]$ determine the same letter in f and then in w.

(iii). For $s \le k$

$$[i_1 \ldots i_{s-1}(i_s + 1) i_{s+1} \ldots i_k] - [i_1 \ldots i_{s-1} i_s i_{s+1} \ldots i_k] = d_s.$$

In fact $[i_1 \ldots i_{s-1}(i_s+1)]$ and $[i_1 \ldots i_{s-1} i_s]$ are two consecutive occurrences of neighboring blocks in the s-step construction so that their first letters are at the distance d_s; thus the numbers

$$[i_1 \ldots i_{s-1}(i_s + 1) i_{s+1} \ldots i_k] \quad \text{and} \quad [i_1 \ldots i_{s-1} i_s i_{s+1} \ldots i_k]$$

appear in the same position in such neighboring blocks.

Step 3. For $1 \le i \le r'$ we denote by $(i)^n$ the sequence

$$(i)^n = \underbrace{ii \ldots i}_{n-times}.$$

Let us then consider the $k+1$ integers

$$b_0 = [(r')^k] = [\underbrace{r'r'\dots r'}_{k-times}],\ b_1 = [1(r')^{k-1}] = [1\ \underbrace{r'r'\dots r'}_{(k-1)-times}]$$

$$\dots\dots\dots$$

$$b_{k-1} = [(1)^{k-1}r'],\ b_k = [(1)^k].$$

Since card$(A) = k$, two of these integers have to represent two occurrences of the same letter in w, therefore, there exist integers s, t such that $1 \le t < s \le k$ and

$$[(1)^t(r')^{k-t}] \equiv [(1)^s(r')^{k-s}]. \tag{2.1}$$

Let us then consider the $r+1$ integers

$$c_i = [(1)^t(i)^{s-t}(r')^{k-s}],\ 1 \le i \le r'.$$

These $r' = r+1$ numbers represent occurrences in w of the same letter. Indeed, from (ii) one has $c_1 \equiv c_2 \equiv \dots \equiv c_r$. Moreover, from equation (2.1) it follows $c_{r+1} \equiv c_1$. We shall prove now that these numbers are in an arithmetic progression, i.e., the difference $c_{i+1} - c_i$, $1 \le i \le r$, does not depend on i. Let us set

$$c_{i,m} = [(1)^t(i+1)^m(i)^{s-t-m}(r')^{k-s}],\ 0 \le m \le s-t,$$

so that $c_{i,0} = c_i$ and $c_{i,s-t} = c_{i+1}$. We can write

$$c_{i+1} - c_i = \sum_{m=1}^{s-t}(c_{i,m} - c_{i,m-1}).$$

From (iii) one has $c_{i,m} - c_{i,m-1} = d_{t+m}$, so that

$$c_{i+1} - c_i = \sum_{m=1}^{s-t} d_{t+m},$$

and this quantity does not depend on i. This concludes the proof.

For $k, r > 0$ we denote by $W(k,r)$ the minimal value of $n(k,r)$ for which Theorem 2.2.2 holds. The map W is also called the *van der Waerden function*. The evaluation of W, as well as the function of the Ramsey theorem, has proved to be extremely difficult. The only nontrivial exact values of $W(k,r)$ are: $W(2,3) = 9$, $W(2,4) = 35$, $W(3,3) = 27$, $W(4,3) = 76$, $W(2,5) = 178$. Let us fix $k = 2$ and consider the function $W(r) = W(2,r)$. There are upper bounds to $W(r)$ which are not "reasonable," in the sense that they are expressed by functions which are not even primitive recursive [70].

We give now some equivalent formulations of the van der Waerden theorem in the case of infinite words.

Proposition 2.2.1. *Let $k, r > 0$ and A be an alphabet with* $\mathrm{card}(A) = k$. *The following statements hold.*

i. There exists a positive integer $n(k,r)$ such that for any $w \in A^\omega$, if $u \in F(w)$ and $|u| \geq n(k,r)$, then u has at least one arithmetic cadence of order r.
ii. Any $w \in A^\omega$ has at least one arithmetic cadence of order r.
iii. For any $w \in A^\omega$ there exists a letter $a \in A$ such that w has an arithmetic cadence of the letter a of an arbitrary order.

Proof. The statement *i* is true from the van der Waerden theorem. Moreover, trivially $i \Rightarrow ii$. Let us now prove that $ii \Rightarrow iii$. Let $w \in A^\omega$. By *ii* for any $r > 0$ there exists a letter $x \in A$ such that w admits an arithmetic cadence of order r of the letter x. Let us denote by λ a map $\lambda : \mathbb{N}_+ \to A$, where for any $r \in \mathbb{N}_+$, $\lambda(r) \in A$ is a letter such that w has an arithmetic cadence of order r of $\lambda(r)$. Since A is a finite set from the pigeon-hole principle there exist infinitely many integers $n_1 < n_2 < \cdots < n_i < n_{i+1} < \cdots$ and $a \in A$ such that

$$\lambda(n_1) = \lambda(n_2) = \cdots = \lambda(n_i) = \lambda(n_{i+1}) = \cdots = a \in A.$$

Let now p be an arbitrary positive integer. There will exist an integer n_i of the above sequence such that $n_i > p$. Since $\lambda(n_i) = a$ one has that w has an arithmetic cadence of the letter a of order n_i and so of order p.

An infinite word $w \in A^\omega$, with $\mathrm{card}(A) = k$, represents a partition of $\mathbb{N}_+$ in k classes (see Sect. 1.2); hence, statements *i*, *ii*, and *iii* can be restated also in terms of partitions of $\mathbb{N}_+$. In particular, *iii* becomes: *if $\mathbb{N}_+$ is partitioned in k classes, then at least one must contain arbitrarily long arithmetic progressions.*

The following proposition shows that there exist infinite words such that any of their arithmetic cadences is not arbitrarily large.

Proposition 2.2.2. *A word $c \in \{a,b\}^\omega$ and a map $h : \mathbb{N} \to \mathbb{N}$ exist such that, for any $d > 0$ any arithmetic cadence in c of rate $\leq d$ has order $\leq h(d)$.*

Proof. For any real z we denote by $\{z\}$ the fractional part of z, $\{z\} = z - \lfloor z \rfloor$ where $\lfloor z \rfloor$ is the integer part of z. Let α be a fixed irrational number. We define an infinite word $c \in \{a,b\}^\omega$ as follows: for any $n > 0$,

$$c(n) = \begin{cases} a & \text{if } \{n\alpha\} \in [0, 1/2) \\ b & \text{if } \{n\alpha\} \in [1/2, 1). \end{cases}$$

Let $(x, x+p, \ldots)$ be an arithmetical cadence of c of rate $p \leq d$ and T be the set $T = \{x, x+p, \ldots\}$. From a classical result of number theory one has that the set $\{\{mp\alpha\} \mid m \in \mathbb{N}\}$ is dense and uniformly distributed in the interval $[0,1]$ (see [122]). The same will occur to any translate of this set by $\{x\alpha\}$ mod 1. Hence, an integer $f(p)$ exists, independent of x, such that $\text{card}(T) \leq f(p)$. The remaining part of the proof is achieved by defining $h(d) = \max\{f(p) \mid 1 < p \leq d\}$.

Some important applications of van der Waerden's theorem to repetitive morphisms and semigroups will be given in Sect. 3.9.

2.3 Uniformly recurrent words

In this section we introduce uniformly recurrent words. These words are very important, since, as we shall see later, they present many interesting regularities.

Let t be an infinite word. We recall (see Def. 1.4.5) that a factor u of t is *recurrent* if it occurs infinitely often in t (in both sides of t when t is bi-infinite). The word t is called recurrent when any of its factors is recurrent. However, when a word is recurrent one cannot, in general, limit the distances, or gaps, between two consecutive occurrences of the same factor. A stronger notion of recurrence is given by the following definitions:

Definition 2.3.1. *A factor u of an infinite word t occurs* syndetically *in t if there exists an integer k such that in any factor of t of length k there is at least one occurrence of u.*

For an infinite word t, we define for all $u \in F(t)$ the quantity

$$k(t,u) = \sup\{|w| \mid w \in F(t) \;\text{ and }\; u \notin F(w)\}.$$

It is clear that u occurs syndetically in t if and only if $k(t,u) < \infty$.

Definition 2.3.2. *An infinite word t is called* uniformly recurrent, *or with bounded gaps, when all its factors occur syndetically in t.*

A uniformly recurrent word is obviously recurrent, whereas the converse is not generally true. When t is uniformly recurrent then $k(t,u)$ takes a finite value for all $u \in F(t)$. We can then introduce the map $k_t : F(t) \to \mathbb{N}$ defined for all $u \in F(t)$ as

$$k_t(u) = k(t,u) + 1.$$

It follows that in any factor of t of length $k_t(u)$ there will be at least one occurrence of u.

It is clear that if $t \in A^{\pm\omega}$ is uniformly recurrent, then $t_+ \in A^\omega$ will be so. For any uniformly recurrent word t we denote by K_t, or simply by K, the map $K_t : \mathbb{N} \to \mathbb{N}$, defined for all $n \in \mathbb{N}$ as

$$K_t(n) = \max\{k_t(u) \mid u \in F(t) \cap A^n\}.$$

If $v \in F(t)$ and $|v| \geq K_t(n)$, then

$$F(t) \cap A^n \subseteq F(v),$$

that is, all factors of t of length n will occur in v. The functions k_t and K_t will be called the *word recurrency function* and the *recurrency function* of t, respectively.

The following proposition shows the existence of a basic inequality satisfied by the recurrency function and the subword complexity of any uniformly recurrent word.

Proposition 2.3.1. *Let $w \in A^\omega$ be a uniformly recurrent word having subword complexity f_w and recurrency function K_w. For all $n \geq 0$ one has*

$$K_w(n) \geq f_w(n) + n - 1.$$

If w is periodic, of minimal period p, then $K_w(n) \leq n + p - 1$. Moreover, if $n \geq p$, then $K_w(n) = n + p - 1$. If w is not periodic, then $K_w(n) \geq 2n$.

Proof. Let u be any factor of w of length $K_w(n)$. The number of all occurrences of factors of length n in u is trivially given by $K_w(n) - n + 1$. Since all the factors of w of length n have to occur in u, one has

$$K_w(n) - n + 1 \geq f_w(n). \tag{2.2}$$

If w is not periodic, then by Proposition 1.6.1, $f_w(n) \geq n + 1$ so that $K_w(n) \geq 2n$.

Let us then suppose that w is periodic and of minimal period p. One has $w = v^\omega$ with v primitive word such that $|v| = p$. In view of the p-periodicity any factor u of w of length $n + p - 1$ will certainly contain all the factors of w of length n whose number $f_w(n)$ is $\leq p$. Hence, $K_w(n) \leq n + p - 1$. Let us now suppose $n \geq p$. In this case there are in w exactly p factors of length n. Indeed, any factor f of w of length n can be written as $f = (v')^q v_1$, where $q > 0$, v' is a conjugate of v and v_1 is a proper, possibly empty, prefix of v'. Since v is primitive, all its conjugates are distinct and p in number, so that $f_w(n) = p$. From this and equation (2.2) it follows that $K_w(n) = n + p - 1$.

The next theorem shows that for any infinite language L there exists a uniformly recurrent word whose factors are factors of the words of L. This result can be derived by arguments of symbolic-dynamics [65]. We report here a proof based on a simple argument even though it is non-constructive [91].

Theorem 2.3.1. *Let $L \subseteq A^*$ be an infinite language. There exists an infinite word $x \in A^{\pm\omega}$ such that*

i. x is uniformly recurrent,
ii. $F(x) \subseteq F(L)$.

Proof. In view of Lemma 1.2.2 it is sufficient to prove that if $s \in A^\omega$, then there exists a uniformly recurrent word $x \in A^{\pm\omega}$ such that $F(x) \subseteq F(s)$. Let $s \in A^\omega$ be an infinite word and let $w_1, w_2, \ldots, w_n, \ldots$ be an arbitrary enumeration of the factors of s. We define an infinite sequence $(t_n)_{n \geq 0}$ of infinite words as follows:

$$t_0 = s;$$

for every $i > 0$ we consider the set

$$E_i = \{v \in F(t_{i-1}) \mid w_i \notin F(v)\}.$$

By definition E_i is closed by factors. Moreover, E_i is a finite set if and only if $k(t_{i-1}, w_i) < \infty$. We then set $t_i = t_{i-1}$ if E_i is a finite set. If E_i is infinite, then, by Lemma 1.2.2, there exists an infinite word, that we take as t_i, such that $F(t_i) \subseteq E_i$. Let us observe that in the latter case $w_i \notin F(t_i)$. Moreover, in any case $F(t_i) \subseteq F(t_{i-1})$. Hence, if $w_r \in F(t_j)$ and $j \geq r$, then one derives that $w_r \in F(t_r)$, $t_r = t_{r-1}$, and $k(t_{r-1}, w_r) < \infty$. Further, since $F(t_j) \subseteq F(t_r)$ it follows that $k(t_j, w_r) \leq k(t_{r-1}, w_r) < \infty$.

Let us now choose in each t_i, $i \geq 0$, a factor u_i of length $|u_i| = i$. Let us denote by U the infinite set $U = \{u_i \mid i \geq 0\}$. By using again Lemma 1.2.2 one has that there exists an infinite word $x \in A^{\pm\omega}$ such that any factor of x is a factor of infinitely many words of U. This implies, of course, $F(x) \subseteq F(s)$. Suppose now that $w \in F(x)$. One has $w = w_r$ for a suitable $r > 0$. Since there are infinitely many $j \geq r$ such that $w = w_r \in F(t_j) \subseteq F(t_r)$, it follows that $k(t_j, w) \leq k(t_{r-1}, w) < \infty$. Thus $k(x, w) < \infty$ since by construction $F(x) \subseteq F(t_j)$. Hence, x is uniformly recurrent.

Definition 2.3.3. *An infinite word $x \in A^{\pm\omega}$ ($x \in A^\omega$) is* minimal *if there is no $y \in A^{\pm\omega}$ ($y \in A^\omega$) such that $F(y) \subset F(x)$.*

Corollary 2.3.1. *An infinite word $x \in A^{\pm\omega}$ ($x \in A^\omega$) is uniformly recurrent if and only if it is minimal.*

Proof. Let $x \in A^{\pm\omega}$ be a uniformly recurrent word having the word recurrency function k_x and suppose, by contradiction, that there exists $y \in A^{\pm\omega}$ such that $F(y) \subset F(x)$. Let $w \in F(x) \setminus F(y)$. Thus any factor of x of length $k_x(w)$ will contain w as a factor. Since $F(y) \subset F(x)$ it follows that any factor of y of length $k_x(w)$ will have w as a factor, which is a contradiction. Hence, x has to be minimal.

Conversely, suppose that $x \in A^{\pm\omega}$ is minimal. From Theorem 2.3.1 there exists a uniformly recurrent word $y \in A^{\pm\omega}$ such that $F(y) \subseteq F(x)$. From the minimality one derives $F(y) = F(x)$, so that x is uniformly recurrent having as recurrency function that of y. The proof in the case of a word $x \in A^\omega$ is similar.

Some classical results on unavoidable regularities in words are an easy consequence of Theorem 2.3.1.

Corollary 2.3.2. *Let J be a two-sided ideal of A^*. If for any uniformly recurrent word $w \in A^\omega$, $F(w) \cap J \neq \emptyset$, then there exists an $n > 0$ such that $A^n A^* \subseteq J$.*

Proof. Suppose that there exist infinitely many words which belong to the set $C = A^* \setminus J$. Since C is closed by factors, then by Theorem 2.3.1 there exists a uniformly recurrent word $w \in A^{\pm\omega}$ such that $F(w) \subseteq C$. Hence, $F(w_+) \subseteq C$ which is a contradiction since $w_+ \in A^\omega$ is uniformly recurrent.

The following result due to T. C. Brown [16] is reminiscent of van der Waerden's theorem.

Corollary 2.3.3. *Let $t \in A^\omega$ be an infinite word. Then there exist a letter $a \in A$ and an integer k such that for any $n > 0$ there are positive integers $i_1 < i_2 < \cdots < i_n$ (which depend on n) with*

$$a = a_{i_1} = a_{i_2} = \cdots = a_{i_n}$$

and $i_{j+1} - i_j \leq k$, $1 \leq j \leq n-1$.

Proof. From Theorem 2.3.1 there exists a uniformly recurrent word $x \in A^\omega$ such that $F(x) \subseteq F(t)$. From the uniform recurrence of x one has that for any letter $a \in \mathrm{alph}(x)$ there exists a positive integer k_a with the property that any factor of x of length k_a contains at least one occurrence of the letter a. Let n be an arbitrary positive integer. Any factor w of x of length $\geq nk_a$ will have at least n occurrences of the letter a and the distance between two consecutive such occurrences is $\leq k_a$. Since w is a factor of t, the result follows.

A further interesting application of Theorem 2.3.1 to unavoidable regularities will be given in Sect. 2.7 (see Theorem 2.7.1).

Proposition 2.3.2. *Sturmian words are uniformly recurrent.*

Proof. Let us first prove that any Sturmian word x is recurrent. Indeed, suppose, by contradiction, that there exists a factor u of x which occurs in x only finitely many times. We can then write

$$x = x'y, \text{ with } x' \in A^+, \ y \in A^\omega,$$

and $u \notin F(y)$. Thus, $F(y) \subset F(x)$ and if $|u| = n$, then one has

$$f_y(n) < f_x(n) = n + 1.$$

Hence, $f_y(n) \leq n$. By Proposition 1.6.1, y is ultimately periodic, $y = y'v^\omega$ with $y' \in A^*$ and $v \in A^+$. Thus, $x = x'y'v^\omega$ and x is ultimately periodic, which is a contradiction.

Let us now prove that x is uniformly recurrent. Indeed, suppose, by contradiction, that there exists a factor u of x which does not occur syndetically in x, i.e., there exist arbitrarily large factors of x in which u does not occur. Let $|u| = n$. Since x is recurrent, there exists a factor of a suitable large length m containing $n+1$ occurrences of u and having u as prefix. Let

$$x[i, i+m-1]$$

be an occurrence of this factor. For any factor $x[j, j+m-1]$, with $j \geq i$, of length m, we denote by $\delta(j)$ the number of occurrences of u in $x[j, j+m-1]$. Since u does not occur syndetically in x, there will exist a positive integer j such that the factor $x[i+j, i+j+m-1]$ has 0 occurrences of u. We shall take the minimal value of j for which this condition is satisfied. Let us then consider all the factors $x[i+h, i+h+m-1]$ with $h = 0, 1, \ldots, j$, and the sequence

$$\delta(i), \delta(i+1), \ldots, \delta(i+j).$$

One has $\delta(i) = n+1$ and $\delta(i+j) = 0$. Let us observe that for all $h = 0, \ldots, j-1$, $|\delta(i+h+1) - \delta(i+h)| \leq 1$. Moreover, if $\delta(i+h+1) - \delta(i+h) = -1$, then $x[i+h, i+h+m-1]$ begins with u. Since for h running in the interval $[0, j]$, δ has to reach all the integer values from $n+1$ to 0, there exist integers $h_1, h_2, \ldots, h_n$ such that $h_1 < h_2 < \cdots < h_n$ and

$$\delta(i+h_j) = n-j+1,\ \delta(i+h_j+1) < \delta(i+h_j), \quad j = 1, \ldots, n,$$

that is for $j = 1, \ldots, n$, $x[i+h_j, i+h_j+m-1]$ contains exactly $n-j+1$ occurrences of u and has u as prefix. Thus there exist $n+2$ factors of x of length m having u as prefix and each different from the others.

Let us now consider a factor f of x of length m having a unique occurrence of u as suffix (this factor exists since u is not syndetical in x). We can write for $\lambda \in A^*$ and for a suitable $i > 0$

$$f = \lambda u = x[i, i+m-1].$$

Let us now consider the $m-n$ factors $x[i+j, i+j+m-1]$ with $0 \leq j < m-n$. All these factors are distinct. Indeed, suppose that there exist r, s such that $0 \leq r < s < m-n$ and $x[i+r, i+r+m-1] = x[i+s, i+s+m-1] = w$. We can write

$$w = \xi_r u \eta_r = \xi_s u \eta_s \text{ and } f = \lambda_r \xi_r u = \lambda_s \xi_s u$$

with $\lambda_r, \lambda_s, \xi_r, \xi_s, \eta_r, \eta_s \in A^*$ and $|\xi_r| > |\xi_s|$. Hence, $\xi_r u = \xi_s u \zeta$, $\zeta \in A^+$, and

$$f = \lambda_r \xi_s u \zeta = \lambda_s \xi_s u,$$

so that f has two occurrences of u which is a contradiction. Moreover, all these $m-n$ factors, by construction, do not begin with u. Thus $f_x(m) \geq n+2+m-n = m+2$. This implies that x is not Sturmian which is a contradiction.

The following proposition is useful to show that some infinite words, constructed by iterating a morphism, are uniformly recurrent.

Proposition 2.3.3. *Let $\phi : A^* \to A^*$ be a nonerasing morphism such that $\phi(a) \in aA^+$, where $a \in A$ and $x = \phi^\omega(a)$ be the infinite word obtained by iterating ϕ on the letter a. If the letter a occurs syndetically in x, then x is uniformly recurrent.*

Proof. Let u be any factor of x. There exists an integer n, depending on u, such that u is a factor of $\phi^n(a)$. We recall that the word x is a fixed point of ϕ, so that $x = \phi(x) = \phi^n(x)$. Since the number of symbols between two consecutive occurrences of the letter a is upper-bounded by a constant c, one derives that the distance between two consecutive occurrences of u in x is upper-bounded by an integer depending on the factor u and on the value of c. Thus any factor of x occurs syndetically in x and, therefore, x is uniformly recurrent.

2.4 Shirshov's theorem

Recently many authors have rediscovered a combinatorial theorem of A. I. Shirshov [145, 146], which has, as we shall see in the next chapter, interesting applications for the Burnside problem and related questions. Originally this theorem was used for proving some properties of the algebras with polynomial identities (compare [104, 142]). In the following A will be a totally ordered finite alphabet.

Lemma 2.4.1. *Let $m \in A^\omega$. There exists an infinite sequence of positive integers $(k_n)_{n\geq 1}$ such that for any $n \geq 1$, m contains as a factor a word of the kind $l_n l_{n-1} \cdots l_1$, where l_i is a Lyndon word, $|l_i| \leq k_i$, $1 \leq i \leq n$, and $l_n \geq l_{n-1} \geq \cdots \geq l_1$.*

Proof. We can suppose that m is uniformly recurrent, otherwise we can apply Theorem 2.3.1. Let a be the minimal letter that appears in m. We can factorize m as

$$m = w_1 w_2 \cdots w_n \cdots, \tag{2.3}$$

where the words w_i, $i \geq 1$, are defined inductively as follows. The word w_1 is the maximal prefix of m such that w_1 and all its prefixes are Lyndon words. Let $m = w_1 w_2 \cdots w_j t$, with $t \in A^\omega$ then w_{j+1} is the maximal prefix of t such that w_{j+1} and all its prefixes are Lyndon words. Since m is uniformly recurrent, there exists a positive integer n_2 such that any factor w of m, with $|w| > n_2$, contains the letter a. We prove that for any $i \geq 1$, $|w_i| \leq n_2$. In fact, otherwise, we can write $w_i = uav$, where u and ua are Lyndon words and $u \geq a$. Then the word ua has two different factorizations as a product of non-increasing Lyndon words and, by Theorem 1.5.1, this is a contradiction.

Call x_i the first letter of w_i. For any $i \geq 1$, one has $w_i \geq x_{i+1}$; in fact if $w_i < x_{i+1}$ then, by Proposition 1.5.2, $w_i x_{i+1}$ is a Lyndon word and this is a contradiction.

We set $k_1 = 1$ and $k_2 = n_2$. Let $l_2 = w_i$ and $l_1 = x_{i+1}$ for some i. One has then $l_2 \geq l_1$, $|l_1| \leq k_1$, and $|l_2| \leq k_2$. Moreover, $l_2 l_1$ is a factor of m.

Let $Y = \{w_i \mid i \geq 1\}$. The set Y is finite, since the words of Y have length bounded by n_2. We consider the word m as an infinite word on the alphabet Y. As before, we can suppose that m is uniformly recurrent, as a word over Y, otherwise we can apply Theorem 2.3.1. Let w be the minimal word in Y with respect to the lexicographic order.

We write

$$m = v_1 v_2 \cdots v_n \cdots,$$

where the words v_i, $i \geq 1$, are defined inductively as follows:

$$v_1 = w_1 w_2 \cdots w_{j_1},$$

where $w_1 \cdots w_i$ is a Lyndon word (over the alphabet A), for $1 \leq i \leq j_1$, and $w_1 \cdots w_{j_1} w_{j_1+1}$ is not a Lyndon word. Let $m = v_1 \cdots v_i w_{j_i+1} w_{j_i+2} \cdots$, then we set $v_{i+1} = w_{j_i+1} \cdots w_{j_{i+1}}$, where $w_{j_i+1} \cdots w_s$ is a Lyndon word, for $j_i + 1 \leq s \leq j_{i+1}$, and $w_{j_i+1} \cdots w_{j_{i+1}} w_{j_{i+1}+1}$ is not a Lyndon word.

Considering m as a word over Y, for any factor v of m, we denote by $|v|_Y$ its length as a word over the alphabet Y. Since m is uniformly recurrent, as a word over Y, there exists a positive integer n_3 such that any factor v of m, with $|v|_Y \geq n_3$, contains as a factor the word w. As before we can prove that $|v_i|_Y \leq n_3$, for any $i \geq 1$. This implies that $|v_i| \leq n_3 n_2$, for any $i \geq 1$, since the words of Y have length bounded by n_2.

As before, from Proposition 1.5.2, one derives that for any $i \geq 1$

$$v_i \geq w_{j_i+1} \geq x_{j_i+2}.$$

Let us set $k_3 = n_3 n_2$. The word m has factors of the kind $v_i w_{j_i+1} x_{j_i+2}$, with $v_i \geq w_{j_i+1} \geq x_{j_i+2}$, for any $i \geq 1$. Thus, setting $l_3 = v_i$, $l_2 = w_{j_i+1}$, $l_1 = x_{j_i+2}$, for some $i \geq 1$, one has $|l_3| \leq k_3$, $|l_2| \leq k_2$, $|l_1| \leq k_1$, and $l_3 \geq l_2 \geq l_1$. Moreover, l_1, l_2, and l_3 are Lyndon words and $l_3 l_2 l_1$ is a factor of m.

Iterating the same procedure, as above, one constructs a sequence of positive integers $(k_n)_{n \geq 1}$ satisfying the statement of the lemma.

Theorem 2.4.1. *For any $m \in A^\omega$ there exists a left infinite word s such that*

i. $F(s) \subseteq F(m)$,
ii. $s = \cdots l_n \cdots l_2 l_1$, *where for any $i \geq 1$, l_i is a Lyndon word and $l_{i+1} \geq l_i$.*

Proof. Let $(k_n)_{n \geq 1}$ be the sequence of integers for which Lemma 2.4.1 holds. Let us consider the set M of all sequences $(l_1, \ldots, l_n)$, $n \geq 1$, such that m

contains as a factor the word $l_n l_{n-1} \cdots l_1$, l_i is a Lyndon word, $|l_i| \le k_i$, $1 \le i \le n$, and $l_n \ge l_{n-1} \ge \cdots \ge l_1$. By Lemma 2.4.1, the set M is infinite. Moreover, for any fixed $i \ge 1$, the words l_i's are finitely many, since $|l_i| \le k_i$. We can then choose a Lyndon word l_1 such that infinitely many sequences of M have l_1 as a first word. Hence, there exists a Lyndon word l_2, such that infinitely many sequences of M have l_1, l_2 as the first two words, and so on. The left infinite word $s = \cdots l_n \cdots l_1$ satisfies the statement.

Definition 2.4.1. *A word $s \in A^\omega$ is* ω-divided *if s can be factorized as $s = w_1 \cdots w_n \cdots$, with $w_i \in A^+$, $i > 0$, and for any $k > 0$, $(w_1, \ldots, w_k)$ is a k-division. A word $t \in A^{-\omega}$ is* ω-divided *if t can be factorized as $t = \cdots w_n \cdots w_1$, with $w_i \in A^+$, $i > 0$, and for any $k > 0$, $(w_k, \ldots, w_1)$ is a k-division.*

Remark 2.4.1. In a similar way one can introduce the notion of an infinite word $s \in A^\omega$ ($t \in A^{-\omega}$) which is inversely ω-divided. In this case one requires that for any $k > 0$, the sequence $(w_1, \ldots, w_k)$ (sequence $(w_k, \ldots, w_1)$) is an inverse k-division. An example of ω-divided word is given by the Fibonacci word. Indeed, f can be factorized as

$$f = \tilde{f_3}\tilde{f_5} \cdots \tilde{f}_{2n+1} \cdots,$$

where $\sim$ denotes the reversal operation. If we set $b < a$ ($a < b$), then one can prove [32] that this factorization is an ω-division (inverse ω-division) of f.

Corollary 2.4.1. *For any $t \in A^\omega$ there exists $s \in A^{-\omega}$ such that*

i. $F(s) \subseteq F(t)$,
ii. s is ultimately periodic or s is ω-divided.

Proof. By Theorem 2.4.1 there exists $s \in A^{-\omega}$ such that $F(s) \subseteq F(t)$ and $s = \cdots l_n \cdots l_1$ where l_i is a Lyndon word and $l_{i+1} \ge l_i$ for any $i \ge 1$. Now either

(a) there exists $k > 0$ such that $l_j = l_k$ for any $j \ge k$, in such a case s is ultimately periodic,

or

(b) one can write $s = \cdots l_n'^{p_n} \cdots l_2'^{p_2} l_1'^{p_1}$, where $p_i > 0$, l_i' is a Lyndon word and $l_{i+1}' > l_i'$ for any $i \ge 1$. By Lemma 1.5.2, for any $n > 0$ the word $l_n'^{p_n} \cdots l_2'^{p_2} l_1'^{p_1}$ is n-divided and $(l_n'^{p_n}, \ldots, l_2'^{p_2}, l_1'^{p_1})$ is an n-division so that, by definition, s is ω-divided.

Theorem 2.4.1 gives a left infinite word s, having an *infinite Lyndon factorization*, whose factors are contained in the set of factors of t. Using *Viennot factorizations* (compare [104]) which are a generalization of the Lyndon factorization, one can prove the following theorem, that generalizes a result of C. Reutenauer [141]. For the proof see [158].

Theorem 2.4.2. *For any $t \in A^\omega$ there exists $s \in A^\omega$ such that*

i. $F(s) \subseteq F(t)$,
ii. $s = l_1 \cdots l_n \cdots$, *where for any* $i \geq 1$, l_i *is a Lyndon word and* $l_i \geq l_{i+1}$.

Then, similarly to Corollary 2.4.1, we derive the following:

Corollary 2.4.2. *For any* $t \in A^\omega$ *there exists* $s \in A^\omega$ *such that*

i. $F(s) \subseteq F(t)$,
ii. s *is ultimately periodic or* s *is* ω*-divided.*

Theorem 2.4.3. *Let* A *be a totally ordered alphabet with* k *letters and* p, n *be integers* ≥ 1. *There exists a positive integer* $N(k,p,n)$ *such that any word* $w \in A^*$ *of length at least* $N(k,p,n)$ *contains as a factor an* n*-divided word or a* p*-power of a nonempty word.*

Proof. We suppose by contradiction that for some integers $k, n, p \geq 1$, there are arbitrarily large words in A^*, with $\text{card}(A) = k$, containing, as a factor, neither a p-power of a nonempty word nor an n-divided word. We denote by L the infinite factorial language of these words. By Lemma 1.2.2 there exists an infinite word $m \in A^\omega$ such that $F(m) \subseteq L$. By Corollary 2.4.1 there exists $s \in A^{-\omega}$ such that $F(s) \subseteq F(m) \subseteq L$ and s is ultimately periodic or ω-divided. If s is ω-divided, then $F(s)$ contains an n-divided word. If s is ultimately periodic, then $F(s)$ contains a p-power of a nonempty word. In any case we reach a contradiction, since no word of L is of this kind.

The above proof is different from the original one (see [104]); another proof will be given later as a consequence of a much stronger result (namely Theorem 2.7.2).

Corollary 2.4.3. *Let* A *be a totally ordered alphabet with* k *letters and* p, n *be integers* ≥ 1, *with* $p \geq 2n$. *There exists a positive integer* $N(k,p,n)$ *such that any word in* A^* *of length at least* $N(k,p,n)$ *contains as a factor an* n*-divided word or a* p*-power of a nonempty word* v *with* $|v| < n$.

Proof. We take the integer $N = N(k,p,n)$ which satisfies Theorem 2.4.3. Let w be a word of length at least N and suppose that w does not contain as a factor any n-divided word. By Theorem 2.4.3, w contains a factor of the kind $u = v^p$, with $v \in A^+$. We may assume that v is primitive. The statement is proved if we show that $|v| < n$. Suppose by contradiction that $|v| \geq n$. Since v is primitive, there are n distinct conjugates $v_1 > v_2 > \cdots > v_n$ of v. It is evident that for any $i \in \{1, 2, \ldots, n\}$, v_i is a factor of v^2. Since $p \geq 2n$, u has the factor $(vv)^n$ and so u contains as a factor a word of the kind

$$f = (v_1 z_1)(v_2 z_2) \cdots (v_n z_n),$$

with $z_i \in A^*$, $(i = 1, \ldots, n)$. Let $\sigma \in S_n$ be a nontrivial permutation and $i > 0$ be the first integer such that $i \neq \sigma(i)$. Since $\sigma(i) > i$, one has $v_i > v_{\sigma(i)}$

with $|v_i| = |v_{\sigma(i)}|$, so that from property ($L2$) of the lexicographic order one has

$$(v_{\sigma(i)}z_{\sigma(i)})\cdots(v_{\sigma(n)}z_{\sigma(n)}) < (v_iz_i)\cdots(v_nz_n).$$

From property ($L1$) of the lexicographic order, multiplying both the sides of the above equation by

$$(v_1z_1)\cdots(v_{i-1}z_{i-1}) = (v_{\sigma(1)}z_{\sigma(1)})\cdots(v_{\sigma(i-1)}z_{\sigma(i-1)}),$$

one obtains

$$(v_{\sigma(1)}z_{\sigma(1)})\cdots(v_{\sigma(n)}z_{\sigma(n)}) < f.$$

Thus f is n-divided and this contradicts the fact that $F(w)$ does not contain any n-divided word.

The following proposition [102] will be used in Chap. 6 (see Theorem 6.6.7).

Proposition 2.4.1. *Let A be a totally ordered alphabet with k letters and p be a positive integer ≥ 1. There exists an integer $r(k,p)$ such that any word $w \in A^*$ of length at least $r(k,p)$ contains as a factor a 3-divided word or a p-power of a letter of A.*

Proof. Suppose, without loss of generality, that $p \geq 3$. Let us set $r(k,p) = N(k,p,3)$, where N is the function in Theorem 2.4.3. Let $w \in A^*$, with $|w| \geq r(k,p)$. Suppose that w does not contain as a factor any 3-divided word. By Theorem 2.4.3, w contains as a factor a p-power of a nonempty word v. We can then write

$$w = xv^py,$$

with $x, y \in A^*$ and $v \in A^+$. We will now prove that v is a word on a one-letter alphabet. First we prove that in v cannot occur three distinct letters. In fact, suppose that the letters a, b, c, with $a < b < c$, occur in v. In such a case, since $p \geq 3$, the word $w = xv^py$ contains a factor of the kind $(cv_1)(bv_2)(a)$ which is 3-divided and this is a contradiction. Suppose now that only two distinct letters a and b, with $a < b$, occur in v. In such a case, as one easily verifies, the word $w = xv^py$ contains a factor of the kind $(b^t)(a^sb^r)(a^q)$, $t, s, r, q \geq 1$. Such a factor is 3-divided and this is a contradiction. In conclusion, if the word w does not contain any 3-divided factor, then it has a factor which is a p-power of a letter, and this proves the statement.

2.5 Bounded languages

We give now a remarkable application of Shirshov's theorem to language theory, that we shall use in Chaps. 3 and 6. Let us recall the following [68]:

Definition 2.5.1. *A language L over the alphabet A is called* bounded *if there exists a finite number of words $u_1, \ldots, u_q$ such that*

$$L \subseteq u_1^* \cdots u_q^*.$$

As easily follows from the definition, the family of bounded languages is closed under union and product. Moreover, any subset of a bounded language is bounded. The following lemma gives a useful characterization of bounded languages.

Lemma 2.5.1. *A language L over A is bounded if and only if there exist two positive integers n_L and q_L such that for any $w \in L$, there exist q words, $u_1, \ldots, u_q \in A^{[n_L]}$ such that $q \leq q_L$ and $w \in u_1^* \cdots u_q^*$.*

Proof. The "only if" part is trivial. Let us prove the "if" part. Let $q = q_L$ and $n = n_L$. Set $m = \mathrm{card}(A^{[n]})$ and let $v_1, \ldots, v_m$ be an arbitrary enumeration of the words of $A^{[n]}$. We have then

$$L \subseteq (v_1^* \cup \cdots \cup v_m^*)^q.$$

From the closure properties of bounded languages it follows that L is bounded.

The following theorem, due to Restivo and Reutenauer [136], gives a characterization of bounded languages in terms of n-divided words:

Theorem 2.5.1. *A language L is bounded if and only if there exists a positive integer n such that $F(L)$ does not contain n-divided words.*

The proof is based on Lemma 2.5.2 below. We shall set in the following $N = N(k, 2n, n)$ and $k = \mathrm{card}(A)$.

Lemma 2.5.2. *Let $w \in A^*$ be a word of length $|w| \geq N$ which has no n-divided factors. Then w can be factorized as*

$$w = ux^p v,$$

with $|u| < N + n$, $0 < |x| < n$, $p \geq 2n$, and either v is empty or $Iv \neq Ix$ (where Iv denotes the first letter of v).

Proof. Let us factorize w as $w = w'w''$ with $|w'| = N$. Since w has no n-divided factors, by Corollary 2.4.3 there exist words $\alpha, \beta, y \in A^*$ such that

$$w' = \alpha y^{2n} \beta,$$

with $|\alpha| < N$ and $0 < |y| < n$. We can then write

$$w = \alpha y^p \gamma,$$

where $\gamma \in A^*$, $p \geq 2n$, and y is not a prefix of γ. Let y' be the largest prefix common to y and to γ. One has $y = y'y''$, $\gamma = y'v$, $y'' \neq \epsilon$, where either $v = \epsilon$ or $Iv \neq Iy''$. Hence, one derives

$$w = \alpha y^p \gamma = \alpha(y'y'')^p y'v = \alpha y'(y''y')^p v.$$

Setting $u = \alpha y'$ and $x = y''y'$, the result follows.

Corollary 2.5.1. *Let w be a word which has no n-divided factors. Then there exists an integer $q \geq 0$ such that w admits a factorization*

$$w = u_0 x_1^{p_1} u_1 x_2^{p_2} \cdots u_{q-1} x_q^{p_q} u_q, \tag{2.4}$$

with $|u_0|, |u_i| < N + n$, $0 < |x_i| < n$, $p_i \geq 2n$, for $1 \leq i \leq q$, $Ix_i \neq Iu_i x_{i+1}$ if $1 \leq i \leq q-1$ and either $u_q = \epsilon$ or $Ix_q \neq Iu_q$.

Proof. If $|w| < N$, then the result is, trivially, true (in this case $q = 0$). Let us then suppose $|w| \geq N$. By Lemma 2.5.2, one has $w = u_0 x^p v$, where $|u_0| < N + n$, $0 < |x| < n$, $p \geq 2n$, and $v = \epsilon$ or $Iv \neq Ix$. By induction, the result holds for v and then for w.

Proof of Theorem 2.5.1. We first prove that if L is a language such that $F(L)$ does not contain n-divided words for a suitable positive integer n, then L is bounded. Let $w \in F(L)$; from Corollary 2.5.1 we can factorize w as in equation (2.4). We shall prove that $q \leq Q$, where

$$Q = 1 + knm \quad \text{with} \quad m = \text{card}\{x \in A^* \mid 0 < |x| < n\}.$$

Indeed, suppose by contradiction that $q > Q$. By a simple combinatorial argument based on the pigeon-hole principle one has that there exist integers $i_1, \ldots, i_n$ such that $1 \leq i_1 < \cdots < i_n \leq q$, $x_{i_1} = \cdots = x_{i_n} (= x)$, and $Iu_{i_1} x_{i_1+1} = \cdots = Iu_{i_n} x_{i_n+1} (= b)$. Let us set $a = Ix$. One has $a \neq b$. Since $p_i \geq 2n$, $1 \leq i \leq q$, we can rewrite w as

$$w = v_0 x^n b v_1 x^n b \cdots v_{n-1} x^n b v_n.$$

Let us first suppose $a < b$. Then w has the factor $u = xbv_1 x^n b \cdots x^n b v_n$ which has the n-division

$$u = (xbv_1 x^{n-2})(x^2 b v_2 x^{n-3}) \cdots (x^{n-1} b v_n).$$

If we suppose $a > b$, then w has the factor $u' = x^n b v_1 x^n b \cdots v_{n-1} x^n b v_n$ which has the n-division

$$u' = (x^n b v_1 x)(x^{n-1} b v_2 x^2) \cdots (xbv_n).$$

Hence, in both cases, w has a factor which is n-divided and this is a contradiction. This shows that any word w which has no n-divided factors admits a factorization as in equation (2.4) with $q \le Q$, so that, by Lemma 2.5.1, L is a bounded language.

In order to prove the "only if" part of the theorem it is sufficient to show that if L is the language $L = u_1^* \cdots u_q^*$, $q \ge 1$, then an integer n exists such that $F(L)$ does not contain n-divided words. Let us set

$$n = q \ \max\{2|u_i| + 1 \mid 1 \le i \le q\} \tag{2.5}$$

and suppose that $w \in L$ contains an n-divided factor. One can write

$$w = u_1^{n_1} \cdots u_q^{n_q} = u x_1 \cdots x_n v, \tag{2.6}$$

where $u, v \in A^*$, $n_i \ge 0$ $(i = 1, \dots, q)$, and $(x_1, \dots, x_n)$ is an n-division of the word $x_1 \cdots x_n$. From the previous equations one derives that for some i, $1 \le i \le q$, and some j, k, $1 \le j < k \le n$, one has

$$u_i^{n_i} = u' x_{j+1} \cdots x_k v',$$

with $u', v' \in A^*$ and $k - j \ge 2|u_i| + 1$. Indeed, otherwise, one would contradict the condition in equation (2.5). By the pigeon-hole principle one derives that three "parsing lines" of the factorization $x_{j+1} \cdots x_k$ are such as to cut the word u_i in the same positions, i.e., there exist words u_i' and u_i'' and positive integers k_1, k_2, and k_3 such that $k_1 < k_2 < k_3$, $u_i = u_i' u_i''$, and

$$x_{k_1+1} \cdots x_{k_2}, \ x_{k_2+1} \cdots x_{k_3} \in (u_i'' u_i')^*.$$

Hence, the words $x_{k_1+1} \cdots x_{k_2}$ and $x_{k_2+1} \cdots x_{k_3}$ commute which is a contradiction with the hypothesis that $(x_1, \dots, x_n)$ is an n-division.

As a consequence of the preceding theorem, one derives a simple proof of the following result on infinite words [130]:

Corollary 2.5.2. *Let $x \in A^\omega$. If, for some positive integer n, the word x does not contain n-divided factors, then x is ultimately periodic.*

Proof. By Theorem 2.5.1, the set $P(x)$ of the prefixes of x is a bounded language. Then there exist words $u_1, \dots, u_q \in A^+$ such that

$$P(x) \subseteq u_1^* \cdots u_q^*,$$

that is, for any prefix u of x, there exist some non-negative integers $k_1, \dots, k_q$ such that

$$u = u_1^{k_1} \cdots u_q^{k_q}.$$

Since the prefixes of x are infinitely many, one can take the least integer $j \in \{1, \dots, q\}$ such that u_j appears, with arbitrarily large exponents, as a factor of x. This means that one can construct an infinite sequence of prefixes of x, $(\lambda_n u_j^n)_{n>0}$, with $\lambda_n \in u_1^* \cdots u_{j-1}^*$. Since j is the least integer in $\{1, \dots, q\}$ satisfying the above condition, there exists a positive integer k such that $|\lambda_n| \le k$, for $n > 0$. Therefore, one easily derives that $x = uv^\omega$, where u is the prefix of x of length k, and v is some conjugate of u_j.

Proposition 2.5.1. *Let $L \subseteq A^*$ be a bounded language. Then the growth function g_L of L is polynomially upper-bounded.*

Proof. Since L is bounded, there exist q nonempty words $u_1, \ldots, u_q$ such that

$$L \subseteq u_1^* \cdots u_q^*.$$

Any word $w \in L \cap A^{[n]}$ is of the kind

$$w = u_1^{t_1} u_2^{t_2} \cdots u_q^{t_q},$$

where $t_i \geq 0$ $(i = 1, \ldots, q)$ and $t_1 + t_2 + \cdots + t_q \leq n$. Let

$$\phi(q, n) = \mathrm{card}(\{(t_1, t_2, \ldots, t_q) \mid t_1 + t_2 + \cdots + t_q \leq n\}).$$

Clearly $g_L(n) \leq \phi(q, n)$. The result follows by observing that, for any constant q, $\phi(q, n)$ is polynomial in n.

2.6 Power-free words

In this section we shall consider power-free words, that is, finite or infinite words which do not have consecutive occurrences of a same nonempty factor.

As we mentioned in the introduction to this chapter, in an alphabet having only two letters one cannot construct words of length greater than three without producing squares, i.e., two consecutive repetitions of a same nonempty factor. On the contrary, in the case of an alphabet having at least three letters, as shown by Thue, one can construct infinitely many finite words or, what is the same an infinite word, without squares. This result is of great interest since it has several applications in various domains (for instance, Burnside's problem, physics, and game theory).

In the analysis of repetitions of a factor in a word one can fix the number of consecutive occurrences of the factor. A word w is called p-power-free, $p > 1$, if in w there are not p consecutive occurrences of a same nonempty factor. In the case of an infinite word a weaker notion is that of ω-power-free word. A word w satisfies this property if for any nonempty factor u of w there exists a suitable power of u which is not a factor of w. Any p-power-free word is trivially ω-power-free, whereas the contrary is not, in general, true.

The notion of ω-power-free word will be of great interest in the following. As we shall see (compare Lemmas 2.6.2 and 2.6.3) any uniformly recurrent word which is not periodic is ω-power-free.

Definition 2.6.1. *Let p be an integer > 1. A finite or infinite word w is called* p-power-free *if for any $u \in A^+$, $u^p \notin F(w)$.*

It is clear from the definition that if w is p-power-free it is also $(p+1)$-power-free, so that it is convenient to consider the minimal value of p such that w is p-power-free.

A factor f of a word w is called an *overlapping factor* of w if $f = avava$, with $a \in A$ and $v \in A^*$.

Definition 2.6.2. *Let $w \in A^*$. We say that w is* overlap-free, *if w does not contain overlapping factors.*

For instance, the word $abbabaab$ is overlap-free, whereas the word $aabaababb$ has the overlapping factor $aabaaba$. We observe that if a word is overlap-free, then it is also cube-free, but the converse is not generally true.

Let us recall the following important examples of p-power-free infinite words. Let $A = \{a, b\}$ be a two-symbol alphabet and f the infinite Fibonacci word introduced in Sect. 1.6:

$$f = abaababaabaababaababa \cdots .$$

The word f is also the limit sequence obtained by iterating, on the letter a, the morphism $\phi : A^* \to A^*$, defined as

$$\phi(a) = ab, \ \phi(b) = a.$$

One has, in fact, $\phi(a) = ab$, $\phi^2(a) = aba$, $\phi^3(a) = abaab$, etc., and $\phi^{n+1}(a) = \phi^n(a)\phi^{n-1}(a)$, for all $n > 1$. The word f has squares and cubes. However, it is possible to show [95] that it is 4-power-free.

Another very famous infinite word is the *Thue–Morse word t on two symbols*

$$t = abbabaabbaababba \cdots ;$$

t can be introduced by iterating, on the letter a, the morphism $\tau : A^* \to A^*$, defined as

$$\tau(a) = ab, \ \tau(b) = ba.$$

Let $B = A \cup \{c\}$. The *Thue–Morse word m on three symbols*

$$m = abcacbabcbacabcacb \cdots$$

can be introduced by iterating, on the letter a, the morphism $\mu : B^* \to B^*$, defined as

$$\mu(a) = abc, \ \mu(b) = ac, \ \mu(c) = b.$$

The words t and m were introduced by A. Thue in two papers [153, 154] of 1906 and 1912 and, subsequently, rediscovered by Morse [118] and several other authors. We mention that the word t is the infinite word associated with a suitable partition of the set $\mathbb{N}$ in two sets considered by M. E. Prouhet [133] in 1851.

Now we prove that the Thue–Morse words t and m are overlap-free and square-free, respectively. We need the following:

Lemma 2.6.1. *Let $X = \{ab, ba\}$. If $v \in X^*$, then $ava \notin X^*$ and $bvb \notin X^*$.*

Proof. By induction on $|v|$. If $|v| = 0$, then the statement is trivial, since $aa \notin X^*$ and $bb \notin X^*$. Let $|v| = r$, with r positive (even) integer. Suppose by contradiction that $ava \in X^*$ (the case $bvb \in X^*$ is similar). One can then write $ava = abuba$, with $|u| = r - 2$, $u \in X^*$, and $v = bub$. By the induction hypothesis $bub \notin X^*$ and this is in contradiction with $v \in X^*$.

Proposition 2.6.1. *The Thue–Morse word on two letters t is overlap-free.*

Proof. We prove that for any $u \in \{a, b\}^*$, if u is overlap-free, then $\tau(u)$ is overlap-free too. The result will be a consequence of the fact that a is overlap-free and $t = \lim_{n\to\infty} \tau^n(a)$.

Suppose by contradiction that for some overlap-free word u the word $w = \tau(u)$ contains an overlap. One can write

$$w = xavavay,$$

with $x, y, v \in \{a, b\}^*$ (the case $w = xbvbvby$ is similar). We observe that $|v|$ is odd, otherwise, as one can easily check, $v, ava \in \{ab, ba\}^*$ which is absurd by Lemma 2.6.1. We consider two subcases.

If $|x|$ is even, then one has $w = x(ab)v'(ab)v'(ab)z$, where $v = bv'$ and $y = bz$. Since $w = \tau(u)$, one has $u = x'av''av''az'$, where $\tau(x') = x$, $\tau(v'') = v'$, and $\tau(z') = z$. This is in contradiction with the fact that u is overlap-free.

If $|x|$ is odd, then $w = z(ba)v'(ba)v'(ba)y$, with $x = zb$ and $v = v'b$. In such a case $u = z'bv''bv''by'$, where $\tau(z') = z$, $\tau(v'') = v'$, and $\tau(y') = y$. We have again a contradiction.

Proposition 2.6.2. *The Thue–Morse word on three letters m is square-free.*

Proof. As well known (see [104], Chap. 2) the word t is the image of the Thue–Morse word m by the morphism $\delta : \{a, b, c\}^* \to \{a, b\}^*$ defined by $\delta(a) = a$, $\delta(b) = ab$, and $\delta(c) = abb$. Suppose, by contradiction, that m contains a square uu. We can write $m = xuum'$, with $x \in \{a, b, c\}^*$, $u \in \{a, b, c\}^+$, and $m' \in \{a, b, c\}^\omega$. Then $t = \delta(x)\delta(u)\delta(u)\delta(m')$ and, moreover, $\delta(m') \in a\{a, b\}^\omega$. Therefore, t contains a factor of the kind $avava$, with $av = \delta(u)$, which is a contradiction, since t is overlap-free.

The words t, m, and f are uniformly recurrent. Indeed, from Lemma 2.3.3 it is sufficient to show that the letter a occurs syndetically in the preceding words. Let us denote by c_t, c_m, and c_f the maximal number of symbols between two consecutive occurrences of the letter a in the words t, m, and f, respectively. Since t is cube-free one has $b^3 \notin F(t)$, so that $c_t \leq 2$. Since m is square-free it follows that $c_m \leq 3$. Finally, in the case of the Fibonacci word f, one easily derives from the definition of the morphism ϕ, that $c_f \leq 1$. The fact that f is uniformly recurrent may be also derived from Proposition 2.3.2.

Definition 2.6.3. *An infinite, or bi-infinite, word w is called* ω-power-free *if for any $u \in F(w)$, $u \neq \epsilon$, there exists an integer p such that $u^p \notin F(w)$. We shall denote by $p_w(u)$, or simply by $p(u)$, the minimal integer for which the above condition is satisfied.*

If w is ω-power-free, then we set for any $n > 0$

$$P_w(n) = \max\{p(u) \mid u \in F(w) \text{ and } 0 < |u| \leq n\}.$$

The map $P_w : \mathbb{N}_+ \to \mathbb{N}$, that we simply denote by P, will be called the *power-free function* of w. If $\sup\{P_w(n) \mid n > 0\} = p < \infty$, then w is p-power-free.

The following lemma shows that a uniformly recurrent word is ω-power-free if and only if it is not periodic.

Lemma 2.6.2. *A uniformly recurrent word $w \in A^\omega$ is either periodic or ω-power-free.*

Proof. Suppose by contradiction that $w \in A^\omega$ is a uniformly recurrent word which is neither periodic nor ω-power-free. Hence, there exists a word $u \in A^+$, that one can always take to be primitive, such that for all $n > 0$, $u^n \in F(w)$. Let us now consider an occurrence of u in w; this is determined by a word $\lambda \in A^*$ such that λu is a prefix of w. Since w is not ultimately periodic (see Lemma 1.4.4), there exist $n > 0$ and $v \in A^+$ such that $|v| = |u|$, $v \neq u$, and $\lambda u^n v$ is still a prefix of w. Let $m > 0$ be such that $|u^m| > K((n+1)|u|)$ where K is the recurrency function of w. Hence, u^m has as a factor the word $u^n v$. Since u is primitive and $|u| = |v|$, one easily derives $u = v$ which is a contradiction.

A different and simpler proof uses the minimality property of the uniformly recurrent words. In fact, if $w \in A^\omega$ is not ω-power-free, then there exists a word $u \in A^+$ such that for all $n > 0$, $u^n \in F(w)$. This implies $F(u^\omega) \subseteq F(w)$. If w is uniformly recurrent, then from the minimality (see Corollary 2.3.1) one has $F(u^\omega) = F(w)$. Hence, there exists a constant c such that for all $n > 0$, $\mathrm{card}(F(w) \cap A^n) < c$. From Proposition 1.6.1 one derives that w is ultimately periodic and then periodic by Lemma 1.4.4.

Lemma 2.6.3. *A uniformly recurrent word $w \in A^{\pm\omega}$ is either periodic or ω-power-free.*

Proof. Let w be a bi-infinite uniformly recurrent word. Suppose that w is not ω-power free and prove that w is periodic. Indeed, for any $i > 0$ consider the infinite word $u_i = w_{-i}w_{-i+1} \cdots w_{-1}w_0w_1 \cdots w_n \cdots$. For any $i > 0$ the word u_i is uniformly recurrent and, as one easily verifies, is not ω-power free. Therefore, by Lemma 2.6.2, the word u_i is periodic. Moreover, for any $i, j > 0$, from the theorem of Fine and Wilf (see Theorem 1.4.1), the words u_i and u_j are periodic and have the same minimal period p. This implies that the word w is periodic of minimal period p.

If a word is p-power free, then it is also ω-power free. However, the converse is not generally true even though the word is supposed to be uniformly recurrent; this is shown by the following:

Examples. Let $A = \{a, b, c, d\}$ and m be the Thue–Morse word on the alphabet $\{a, b, c\}$. Let us denote, for any $i > 0$, by p_i the prefix $m[i]$ of m of length i and construct the word

$$w = dp_1(dp_2)^2(dp_3)^3 \cdots (dp_n)^n \cdots .$$

For any $p > 1$, w has a factor which is a p-power. However, w is ω-power free. Indeed, let $f \in F(w)$. If $f \in F(m)$, then $f^2 \notin F(w)$. This is due to the fact that m is square-free and to the consideration that, by construction, if $f^2 \in F(w)$, then $f^2 \in F(m)$. Let us now suppose that $f \in F(w) \cap A^*dA^*$. We observe that, by construction, in w the sequence of distances between two consecutive occurrences of the letter d are

$$1, 2, 2, 3, 3, 3, 4 \ldots ,$$

so that a factor dud of w with $u \in (A \setminus \{d\})^*$ occurs exactly $|u|$ times in w. Let us first suppose that in f there are at least two occurrences of the letter d. We can write $f = \lambda dvd\mu$, where $\lambda, \mu \in (A \setminus \{d\})^*$; in other words we have insulated the first and the last occurrence of d in f. For any $p > 1$ one has

$$f^p = \lambda dv(d\mu\lambda dv)^{p-1}d\mu.$$

Since $\mu\lambda \in (A \setminus \{d\})^*$, it follows that if $f^p \in F(w)$, then $d\mu\lambda d$ occurs $p - 1$ times in w. Hence, if $p > |\mu\lambda| + 1$, then $f^p \notin F(w)$. One reaches a similar result if one supposes that in f there is only one occurrence of the letter d. This proves that the word w is ω-power free.

Let us observe that the word w is not recurrent. However, there exist uniformly recurrent ω-power free words which are not p-power free for any fixed $p > 1$. Indeed, let us consider any standard Sturmian word s having the directive sequence $(c_0, c_1, \ldots, c_n, \ldots)$ (see Sect. 1.6). We recall that s is the limit sequence of the infinite sequence $(s_n)_{n \geq 0}$ of words, where $s_0 = b$, $s_1 = a$, and for all $n \geq 1$

$$s_{n+1} = s_n^{c_{n-1}} s_{n-1}.$$

From Proposition 2.3.2 any Sturmian word is uniformly recurrent. If one supposes that $\sup\{c_i \mid i \geq 0\} = \infty$, then s has arbitrarily large powers as factors, so that s is not p-power free for any $p > 1$. However, since a Sturmian word is not periodic, by Lemma 2.6.2 one has that s is ω-power free. On the contrary, if $\sup\{c_i \mid i \geq 0\} < \infty$, then an integer $p > 1$ exists such that s is p-power free [115].

2.7 Bi-ideal sequences

In this section we shall concern ourselves with bi-ideal sequences which are special sequences of words showing many interesting combinatorial properties which have applications in various fields.

Bi-ideal sequences have been considered, with different names, by several authors in algebra and combinatorics (see [6, 24, 84, 150, 160]). The term bi-ideal sequence is due to M. Coudrain and M. P. Schützenberger [24].

A basic theorem due to Coudrain and Schützenberger shows that bi-ideal sequences are unavoidable in the sense that the terms of bi-ideal sequences of any order will appear as factors of all sufficiently long words over a finite alphabet. In the next subsections we analyze some interesting combinatorial properties of bi-ideal sequences and the links existing between bi-ideal sequences, recurrence, and n-divisions. From these results we will obtain an important improvement of the theorem of Shirshov.

Definition 2.7.1. *A sequence of words of* A^*

$$f_1, \ldots, f_n, \ldots$$

is called a bi-ideal sequence *if for all* $i > 0$

$$f_{i+1} \in f_i A^* f_i.$$

If the sequence is of finite length n*, then* $(f_1, \ldots, f_n)$ *is called bi-ideal sequence of order* n.

The term "bi-ideal sequence" is due to the fact that for any $i > 0$, $f_i A^* f_i$ is a bi-ideal of A^* (see Sect. 3.6.3).

If $f_1, \ldots, f_n, \ldots$ is a bi-ideal sequence, then there exists a unique sequence of words

$$g_1, g_2, \ldots, g_n, \ldots$$

such that for all $i > 0$

$$f_{i+1} = f_i g_i f_i.$$

Thus a bi-ideal sequence is any sequence of words $f_1, \ldots, f_n, \ldots$ satisfying the following requirements: for all $i > 0$

a) f_i is both a prefix and a suffix of f_{i+1}.
b) $2|f_i| \leq |f_{i+1}|$.

A sequence of words satisfying only the first requirement, that is,

$$f_{i+1} \in f_i A^* \cap A^* f_i$$

is called *quasi-ideal* sequence. A bi-ideal sequence is a quasi-ideal sequence, whereas the converse is not generally true.

A bi-ideal sequence $f_1, \ldots, f_n, \ldots$ will be called *proper* if $f_1 \neq \epsilon$ and *strict* if for all $i > 0$, $2|f_i| < |f_{i+1}|$, i.e., $g_i \neq \epsilon$. In the following the term bi-ideal sequence will be referred only to proper bi-ideal sequences.

If $(f_1, \ldots, f_n)$ is a bi-ideal sequence of order n, then $f = f_n$ will be called the *last term* of the sequence. Obviously, f is also the last term of a bi-ideal sequence of order k for all $k = 1, \ldots, n-1$. Thus to any $f \in A^+$ one can associate a positive integer, called *degree* of f, defined as the maximal order of any bi-ideal sequence having f as the last term. However, in general, for a given word $f \in A^+$ there can exist different bi-ideal sequences of order equal to the degree of f and having f as the last term. For instance, the word $f = ababababa$ of degree 3 is the last term of the two bi-ideal sequences (a, aba, f) and $(a, ababa, f)$.

Examples of infinite bi-ideal sequences are the sequences $(f_{2k})_{k>1}$ and $(f_{2k+1})_{k\geq 1}$ of the terms of even and odd index, respectively, of the Fibonacci sequence $(f_n)_{n\geq 0}$ (see Sect. 1.6). Indeed, one has

$$f_2 = ab \text{ and } f_{2k} = f_{2k-2}f_{2k-3}f_{2k-2}, \; k > 1,$$

and

$$f_1 = a \text{ and } f_{2k+1} = f_{2k-1}f_{2k-2}f_{2k-1}, \; k \geq 1.$$

Theorem 2.7.1. *(Coudrain and Schützenberger) Let A be a k-letter alphabet. For any $n > 0$ there exists a positive integer $M(k,n)$ such that any word of A^* of length at least $M(k,n)$ contains as a factor the n th term of a bi-ideal sequence.*

Proof. We define $M(k,n)$ by induction on n as follows. For all $k > 0$, $M(k,1) = 1$, and for $n \geq 1$

$$M(k, n+1) = M(k,n)(1 + k^{M(k,n)}).$$

For any fixed k let us simply denote $M(k,n)$ by m_n. The proof is by induction on n. For $n = 1$ the result is trivial. Let us then suppose that any word of length m_n contains a factor which is the n th term of a bi-ideal sequence. Let w be a word of length at least $m_{n+1} = m_n(1 + k^{m_n})$. Then w has a factor of the kind

$$w' = v_1 v_2 \cdots v_p,$$

where $|v_i| = m_n$, $i = 1, \ldots, p$, and $p = 1 + k^{m_n}$. Since there are only k^{m_n} distinct words of length m_n on A, there exist i, j such that $1 \leq i < j \leq p$ and $v_i = v_j = v$. Hence, w has a factor vuv with $|v| = m_n$. By the induction hypothesis $v = xv'y$, where $x, y \in A^*$ and v' is the n th term of a bi-ideal sequence. Hence, w has the factor $v'yuxv'$ which is the last term of a bi-ideal sequence of order $n + 1$.

A different simple, but non-constructive, proof based on recurrence is the following. Suppose, by contradiction, that for some integers k, n the integer $M(k,n)$ does not exist. Then by Proposition 2.0.6 one derives that there exist an alphabet A with k letters and an infinite word $m \in A^\omega$ such that no factor of m is the last term of a bi-ideal sequence of order n. By Theorem 2.3.1 there exists a uniformly recurrent word $t \in A^\omega$ such that $F(t) \subseteq F(m)$. Now we prove that for all $k \geq 1$, t contains, as a factor, the k th term of a bi-ideal sequence. For $k = 1$ the statement is trivial. Let $k > 1$ and suppose that $f_{k-1} \in F(t)$ is the $(k-1)$ th term of a bi-ideal sequence. Since t is recurrent, there exists $g \in A^*$ such that $f_k = f_{k-1} g f_{k-1} \in F(t) \subseteq F(m)$. Thus m contains the factor f_k which is the k th term of a bi-ideal sequence and this gives rise to a contradiction.

2.7.1 Canonical factorizations

In this section we investigate some interesting combinatorial properties of bi-ideal sequences which will be useful later in order to prove some extensions of the Shirshov Theorem.

We introduce some canonical factorizations of the last term of a bi-ideal sequence of any order n. These factorizations are expressible by special sequences of n words called *n-sequences* (and *inverse n-sequences*). These latter will be often considered in the following. An important property of them (see Proposition 2.7.5) is that for any $n > 0$ any sufficiently long m-sequence will admit a derived n-sequence which is either an n-division or an inverse n-division or all its terms are periodic (that is, powers of the same word).

Definition 2.7.2. *Let n be a positive integer and $(w_1, \ldots, w_n)$ be a sequence of n words of A^+. The sequence $(w_1, \ldots, w_n)$ is called* n-sequence *if for any $i = 1, \ldots, n-1$*

$$w_i \in w_{i+1} \cdots w_n A^*.$$

The sequence $(w_1, \ldots, w_n)$ is called inverse n-sequence *if*

$$w_{i+1} \in A^* w_1 \cdots w_i,$$

for any $i = 1, \ldots, n-1$.

We analyze now an important relationship existing between bi-ideal sequences of order n and n-sequences (inverse n-sequences). From this relation one derives two canonical factorizations of the last term of any bi-ideal sequence of order n.

Let $(f_i)_{i=1,\ldots,n}$ be a bi-ideal sequence of order n with $f_1 \in A^+$ and $f_{i+1} = f_i g_i f_i$, $g_i \in A^*$, $i = 1, \ldots, n-1$. We set $w_n = f_1$ and

$$w_{n-i} = f_i g_i, \ i \in [1, n-1].$$

From Def. 2.7.1, for all $i \in [1, n-1]$, $f_{i+1} = f_i g_i f_i = w_{n-i} f_i$, so that, by iteration, one has for all $i \in [0, n-1]$

$$f_{i+1} = w_{n-i} \cdots w_n. \tag{2.7}$$

Moreover, since $w_i = f_{n-i}g_{n-i}$, $i \in [1, n-1]$, from equation (2.7) one has

$$w_i = w_{i+1} \cdots w_n g_{n-i} \in w_{i+1} \cdots w_n A^*. \tag{2.8}$$

By iteration one obtains for $1 \leq i \leq j < n$,

$$w_i = w_{j+1} \cdots w_n g_{n-j} u, \tag{2.9}$$

with $u \in A^*$. From (2.7), $f_n = w_1 w_2 \cdots w_n$. The n-tuple $(w_1, w_2, \ldots, w_n)$ is called the *canonical factorization* of f_n.

One can also introduce another canonical factorization of f_n by setting $w'_1 = f_1$ and

$$w'_{i+1} = g_i f_i, \ i \in [1, n-1].$$

One easily derives that for all $i \in [1, n]$

$$w'_1 \cdots w'_i = f_i, \tag{2.10}$$

where for any $i \in [1, n-1]$

$$w'_{i+1} = g_i w'_1 \cdots w'_i \in A^* w'_1 \cdots w'_i. \tag{2.11}$$

From equation (2.10), $f_n = w'_1 \cdots w'_n$. The n-tuple $(w'_1, \ldots, w'_n)$ is called the *inverse canonical factorization* of f_n.

One has then, from equations (2.8) and (2.11), that the canonical (inverse canonical) factorization of the last term of a bi-ideal sequence of order n, is an n-sequence (inverse n-sequence).

Conversely, one easily verifies that if $(w_1, w_2, \ldots, w_n)$ is an n-sequence (inverse n-sequence), then the sequence of words $f_i = w_{n-i+1} \cdots w_n$ ($f_i = w_1 \cdots w_i$), $1 \leq i \leq n$, is a bi-ideal sequence of order n whose last term has a canonical (inverse canonical) factorization given by $(w_1, w_2, \ldots, w_n)$.

Example 2.7.1. Consider the bi-ideal sequence of order 3, $f_1 = a$, $f_2 = aba$, and $f_3 = ababaaba$. In this case $g_1 = b$ and $g_2 = ba$. The canonical factorization of f_3 is the 3-sequence $(ababa, ab, a)$. The inverse canonical factorization is the inverse 3-sequence $(a, ba, baaba)$.

Let us explicitly remark that since a word f can be the last term of more bi-ideal sequences having the same order, the canonical factorizations of f *are uniquely determined by the bi-ideal sequence having f as the last term.* As we have seen in the previous section, the word $f = abababababa$ is the last term of the two bi-ideal sequences of order 3

$$(a, aba, f) \text{ and } (a, ababa, f).$$

Thus f has the two canonical factorizations

$$(abababab, ab, a) \text{ and } (ababab, abab, a),$$

which uniquely correspond to the preceding bi-ideal sequences.

The following proposition, called *reciprocity law*, summarizes the links existing between the two canonical factorizations of the last term of a bi-ideal sequence of order n, expressed by equations (2.7) and (2.10).

Proposition 2.7.1. *(Reciprocity law) Let $(w_1, \ldots, w_n)$ and $(w'_1, \ldots, w'_n)$ be the canonical factorizations of the n th term f_n of a bi-ideal sequence. For any $i \in [0, n-1]$ one has*

$$w'_1 \cdots w'_{i+1} = f_{i+1} = w_{n-i} \cdots w_n.$$

From now on we shall suppose that the alphabet A, even though not explicitly stated, is totally ordered. By $<$ we denote the lexicographic ordering induced in A^*.

Proposition 2.7.2. *An n-sequence $(w_1, \ldots, w_n)$ is an n-division (inverse n-division) if and only if for all $i \in [1, n-1]$, $w_{i+1}w_i < w_iw_{i+1}$ ($w_{i+1}w_i > w_iw_{i+1}$).*

Proof. ($\Rightarrow$). Let us suppose that the n-sequence $(w_1, \ldots, w_n)$ is an n-division and prove that for any $i \in [1, n-1]$, $w_{i+1}w_i < w_iw_{i+1}$. Assume, by contradiction, that an integer $i \in [1, n-1]$ exists for which $w_{i+1}w_i \geq w_iw_{i+1}$. This implies, by ($L1$), that $w_1 \cdots w_{i-1}w_iw_{i+1} \leq w_1 \cdots w_{i-1}w_{i+1}w_i$. If $w_iw_{i+1} = w_{i+1}w_i$, then

$$w_1 \cdots w_{i-1}w_iw_{i+1}w_{i+2} \cdots w_n = w_1 \cdots w_{i-1}w_{i+1}w_iw_{i+2} \cdots w_n,$$

which is a contradiction. Let us then suppose $w_iw_{i+1} < w_{i+1}w_i$. Since $|w_iw_{i+1}| = |w_{i+1}w_i|$, from property ($L2$) of the lexicographic order it follows that

$$w_1 \cdots w_{i-1}w_iw_{i+1}w_{i+2} \cdots w_n < w_1 \cdots w_{i-1}w_{i+1}w_iw_{i+2} \cdots w_n$$

which is again a contradiction as $(w_1, \ldots, w_n)$ is an n-division.

($\Leftarrow$). We begin by proving that $w_jw_i < w_iw_j$ for any i, j with $1 \leq i < j \leq n$. If $i = j-1$, then the result follows from the hypotheses. Then let us suppose that $i < j-1$. We can write

$$w_i \in w_{j-1}w_j \cdots w_nA^*$$

and

$$w_iw_j = w_{j-1}w_j\mu, \ \mu \in A^*.$$

By hypothesis one has

$$w_jw_{j-1} < w_{j-1}w_j;$$

since $i \leq j-2$, we can write

$$w_i = w_{j-1}\lambda, \ \lambda \in A^*.$$

As $|w_j w_{j-1}| = |w_{j-1} w_j|$ and $w_j w_{j-1} < w_{j-1} w_j$, in view of property $(L2)$ of the lexicographic ordering, one has

$$w_j w_i = w_j w_{j-1} \lambda < w_{j-1} w_j \mu = w_i w_j.$$

Now we have to prove that for any nontrivial permutation $\sigma \in S_n$ one has $w_{\sigma(1)} w_{\sigma(2)} \cdots w_{\sigma(n)} < w_1 w_2 \cdots w_n$. Let us observe that it is possible to reorder the n-word sequence $w_{\sigma(1)} w_{\sigma(2)} \cdots w_{\sigma(n)}$ up to $w_1 w_2 \cdots w_n$ by a "sorting algorithm" which consists in a finite number of transpositions of consecutive words w_j, w_i with $i < j$. In such a way one can easily reach the result by observing that if $w_j w_i < w_i w_j$, then for any $u, v \in A^*$ one has $u w_j w_i v < u w_i w_j v$. The proof in the case of the inverse n-division is perfectly symmetric.

By an argument similar to that of the preceding proposition, one can prove the following:

Proposition 2.7.3. *An inverse n-sequence $(w_1, \ldots, w_n)$ is an n-division (inverse n-division) if and only if for all $i \in [1, n-1]$, $w_{i+1} w_i < w_i w_{i+1}$ $(w_{i+1} w_i > w_i w_{i+1})$.*

Proposition 2.7.4. *Let $(w_1, \ldots, w_n)$ and $(w'_1, \ldots, w'_n)$ be the canonical factorizations of the n th term of a bi-ideal sequence. For each $i \in [1, n-1]$ one has: $w'_i w'_{i+1} < w'_{i+1} w'_i$ if and only if $w_{n-i+1} w_{n-i} < w_{n-i} w_{n-i+1}$.*

Proof. Let $i \in [1, n-1]$ and suppose that

$$w'_i w'_{i+1} < w'_{i+1} w'_i. \tag{2.12}$$

By the definition of the canonical factorizations one has (see equations (2.8) and (2.11)) that there exists $g_i \in A^*$ such that

$$w_{n-i} = w_{n-i+1} \cdots w_n g_i, \; w'_{i+1} = g_i w'_1 \cdots w'_i, \tag{2.13}$$

so that by equation (2.12)

$$w'_i g_i w'_1 \cdots w'_i < g_i w'_1 \cdots w'_i w'_i.$$

Since the two words in the previous inequality have the same length, we may delete w'_i on the right; thus one has

$$w'_i g_i w'_1 \cdots w'_{i-1} < g_i w'_1 \cdots w'_i. \tag{2.14}$$

Now by the reciprocity law

$$w'_1 \cdots w'_{i-1} = w_{n-i+2} \cdots w_n \quad \text{and} \quad w'_1 \cdots w'_i = w_{n-i+1} \cdots w_n;$$

then substituting in equation (2.14) we obtain

$$w'_i g_i w_{n-i+2} \cdots w_n < g_i w_{n-i+1} w_{n-i+2} \cdots w_n.$$

As before we can delete $w_{n-i+2} \cdots w_n$ on the right, so that

$$w'_i g_i < g_i w_{n-i+1}.$$

Multiplying on the left both sides of the above inequality by $w_{n-i+1} w'_1 \cdots w'_{i-1}$ we have

$$w_{n-i+1} w'_1 \cdots w'_{i-1} w'_i g_i < w_{n-i+1} w'_1 \cdots w'_{i-1} g_i w_{n-i+1}; \qquad (2.15)$$

by the reciprocity law and equation (2.13) one derives

$$w_{n-i+1} w'_1 \cdots w'_{i-1} w'_i g_i = w_{n-i+1} w_{n-i+1} \cdots w_n g_i = w_{n-i+1} w_{n-i}$$

and

$$w_{n-i+1} w'_1 \cdots w'_{i-1} g_i w_{n-i+1} = w_{n-i+1} w_{n-i+2} \cdots w_n g_i w_{n-i+1} = w_{n-i} w_{n-i+1},$$

so that (2.15) becomes

$$w_{n-i+1} w_{n-i} < w_{n-i} w_{n-i+1}.$$

Conversely, if one supposes that $w_{n-i+1} w_{n-i} < w_{n-i} w_{n-i+1}$, then one derives in a similar way from (2.13) that $w'_i g_i < g_i w_{n-i+1}$. Hence, multiplying on the right both sides of this inequality by $w_{n-i+2} \cdots w_n w'_i$ and using the reciprocity law, one reaches the result.

Corollary 2.7.1. *Let $(w_1, \ldots, w_n)$ and $(w'_1, \ldots, w'_n)$ be the canonical factorizations of the n th term of a bi-ideal sequence. Then $(w_1, \ldots, w_n)$ is an n-division (inverse n-division) if and only if $(w'_1, \ldots, w'_n)$ is an inverse n-division (is an n-division).*

Proof. From Proposition 2.7.2, $(w_1, \ldots, w_n)$ is an n-division (inverse n-division) if and only if for all $i \in [1, n-1]$, $w_{i+1} w_i < w_i w_{i+1}$ ($w_{i+1} w_i > w_i w_{i+1}$). From Proposition 2.7.4 this occurs if and only if for all $i \in [1, n-1]$, $w'_i w'_{i+1} < w'_{i+1} w'_i$ ($w'_i w'_{i+1} > w'_{i+1} w'_i$). From Proposition 2.7.3 one has that for all $i \in [1, n-1]$, $w'_i w'_{i+1} < w'_{i+1} w'_i$ ($w'_i w'_{i+1} > w'_{i+1} w'_i$) if and only if $(w'_1, \ldots, w'_n)$ is an inverse n-division (n-division).

Let $(w_1, \ldots, w_m)$ be an m-sequence and $(u_1, \ldots, u_n)$ be a derived sequence of $(w_1, \ldots, w_m)$ (see Def. 2.1.3). As one easily verifies, for any $i \in [1, n-1]$, $u_i \in u_{i+1} \cdots u_n A^*$ so that the derived sequence $(u_1, \ldots, u_n)$ is an n-sequence. In the same way one can verify that a derived sequence $(u_1, \ldots, u_n)$ of an inverse m-sequence $(w'_1, \ldots, w'_m)$ is an inverse n-sequence.

Proposition 2.7.5. *For any integer $n > 1$ there exists an integer $r(n)$ with the property that if $(w_1, \ldots, w_m)$ is an m-sequence with $m \geq r(n)$, then there exists a derived sequence $(u_1, \ldots, u_n)$ of $(w_1, \ldots, w_m)$ such that either*

i. $(u_1, \ldots, u_n)$ *is an* n*-division or*
ii. $(u_1, \ldots, u_n)$ *is an inverse* n*-division or*
iii. $\forall j \in [1, n-1], \quad u_j u_{j+1} = u_{j+1} u_j$.

Proof. By Lemma 2.1.2 there exists an integer $r(n)$ with the property that if $(w_1, \ldots, w_m)$ is a sequence of m words, with $m \geq r(n)$, then there exists a derived sequence $(u_1, \ldots, u_n)$ of $(w_1, \ldots, w_m)$ such that either

i. $\forall j \in [1, n-1], \; u_j u_{j+1} < u_{j+1} u_j$, or
ii. $\forall j \in [1, n-1], \; u_j u_{j+1} > u_{j+1} u_j$, or
iii. $\forall j \in [1, n-1], \; u_j u_{j+1} = u_{j+1} u_j$.

Moreover, if $(w_1, \ldots, w_m)$ is an m-sequence, then the derived sequence $(u_1, \ldots, u_n)$ is an n-sequence. Hence, if $u_j u_{j+1} > u_{j+1} u_j$ ($u_j u_{j+1} < u_{j+1} u_j$) for all $j \in [1, n-1]$, then by Proposition 2.7.2, the sequence $(u_1, \ldots, u_n)$ is an n-division (inverse n-division), otherwise for any $j \in [1, n-1]$ one has $u_j u_{j+1} = u_{j+1} u_j$.

2.7.2 Bi-ideal sequences and recurrence

We consider infinite bi-ideal sequences. Any such sequence converges to an infinite word called *infinite bi-ideal word.* An infinite word is bi-ideal if and only if it is recurrent.

Uniformly recurrent words are infinite bi-ideal words which have a special interest since any factor of them occurs an infinite number of times but with bounded gaps. In this case one can "localize" in any sufficiently large factor (whose length depends on the recurrency function) the occurrence of the last term of a bi-ideal sequence of any order.

A case of particular interest is when the uniformly recurrent word x is non-periodic. This implies, by Lemma 2.6.2, that x is ω-power-free. In this case on one hand any such word has a factor which is the last term of a bi-ideal sequence of any order n whose canonical factorization is an n-division (see Proposition 2.7.9). On the other hand one can "localize" in any sufficiently large factor (whose length depends on the recurrency and power-free functions) the occurrence of the last term of a bi-ideal sequence of any fixed order whose canonical factorization is either an n-division or an inverse n-division (see Proposition 2.7.8).

Let $f_1, \ldots, f_n, \ldots$ be an infinite bi-ideal sequence, where $f_{i+1} = f_i g_i f_i$ for all $i > 0$. Since for all $i > 0$, f_i is a prefix of the next term f_{i+1} one has that the sequence (f_n) converges, with respect to the prefix metric (see Sect. 1.3), to the infinite word $x \in A^\omega$

$$x = f_1(g_1 f_1)(g_2 f_2) \cdots (g_n f_n) \cdots .$$

However, since for all $i > 0$, f_i is also a suffix of f_{i+1}, one has that the sequence (f_n) converges, with respect to the suffix metric, to the infinite word $x' \in A^{-\omega}$ given by

$$x' = \cdots (f_n g_n) \cdots (f_2 g_2)(f_1 g_1) f_1.$$

Let us observe that one can rewrite x as

$$x = w_1 w_2 \cdots w_n \cdots,$$

where $w_1 = f_1$, $w_{i+1} = g_i f_i$, $i > 0$. For all $n > 0$, $(w_1, \ldots, w_n)$ is the inverse canonical factorization of f_n. Since $(w_1, \ldots, w_n)$ is an inverse n-sequence, one has for all $n > 0$

$$w_{n+1} \in A^* w_1 \cdots w_n,$$

so that, as one easily verifies, the sequence (w_n) converges, with respect to the suffix metric, to the word x'. In a similar way x' can be rewritten as

$$x' = \cdots w_n \cdots w_2 w_1,$$

where $w_1 = f_1$ and $w_{i+1} = f_i g_i$, $i > 0$. For all $n > 0$, $(w_n, \ldots, w_1)$ is the canonical factorization of f_n. Since $(w_n, \ldots, w_1)$ is an n-sequence, then (w_n) converges, with respect to the prefix metric, to the word x.

Definition 2.7.3. *A word $x \in A^\omega$, or $x \in A^{-\omega}$, is* bi-ideal *if it is the limit of a bi-ideal sequence.*

Proposition 2.7.6. *A word $x \in A^\omega$ is recurrent if and only if it is bi-ideal.*

Proof. Let $x \in A^\omega$ be recurrent. We construct, inductively, a bi-ideal sequence $(f_n)_{n>0}$ such that f_n is a prefix of x, for any $n > 0$. We set $f_1 = x_1$. Suppose, by induction, that we have constructed the bi-ideal sequence up to the i th element f_i, with $i > 0$. Since f_i is a prefix of x and x is recurrent, f_i will occur in x infinitely many times, so that there will exist a word $g \in A^*$ such that $f_{i+1} = f_i g f_i$ is still a prefix of x. Thus there exists a bi-ideal sequence $(f_n)_{n>0}$ whose elements are prefixes of x. This implies that $x = \lim_{n\to\infty} f_n$ and, therefore, x is bi-ideal.

Conversely, let $x \in A^\omega$ be an infinite bi-ideal word, that is, $x = \lim_{n\to\infty} f_n$, where (f_n) is a bi-ideal sequence. Let $w \in F(x)$. There exists $\lambda \in A^*$ such that λw is a prefix of x. Thus $w \in F(f_k)$ for a suitable positive integer k. Now for any $p > 0$, f_k will occur at least 2^p times in f_{k+p}. This shows that the number of occurrences of w in x has not an upper bound.

A uniformly recurrent word x is a special infinite bi-ideal word, since any factor of x occurs in x syndetically, that is, with bounded gaps. The following proposition gives a way to localize the last term of a bi-ideal sequence of any order as a suffix of any sufficiently long factor of a uniformly recurrent word.

Proposition 2.7.7. *Let $t \in A^\omega$ be a uniformly recurrent word. For any $n > 0$ there exists a positive integer $D(n)$ such that for any $w \in A^*$, $a \in A$, with $wa \in F(t)$, and $|w| \geq D(n)$ one has that*

$$w = \lambda f_n,$$

where $\lambda \in A^*$, *and* f_n *is the* n *th term of a bi-ideal sequence* $f_{i+1} = f_i g_i f_i$, *with* $g_i \in aA^*$, $i = 1, \ldots, n-1$, *such that* $f_1 \in aA^*$ *and* $|f_i| \leq D(i)$ *for* $i = 1, \ldots, n$.

Proof. The proof is by induction on n. For $n = 1$ we set $D(1) = K(1)$, where K is the recurrency function of the word t. Let $w \in A^*$, $|w| \geq D(1)$, and $wa \in F(t)$. Then in w the letter a has to occur, so that we can factorize w as $w = xay$ with $x, y \in A^*$ and $|ay| \leq D(1)$. The statement follows if we set $f_1 = ay$. Now let $n > 1$. By induction we may suppose that there exists an integer $D(n-1)$ that satisfies the statement for $n-1$. Then we set

$$D(n) = K(D(n-1)+1) + D(n-1).$$

Let $w \in A^*$, $a \in A$ such that $|w| \geq D(n)$ and $wa \in F(t)$. We can write $w = xv$, with $|x| \geq K(D(n-1)+1)$ and $|v| = D(n-1)$. Since $va \in F(t)$, by the induction hypothesis one has

$$v = \lambda' f_{n-1},$$

with $\lambda' \in A^*$, and f_{n-1} is the $(n-1)$ th term of a bi-ideal sequence $f_{i+1} = f_i g_i f_i$, with $f_1 \in aA^*$, $g_i \in aA^*$, $i \in \{1, \ldots, n-2\}$, and $|f_i| \leq D(i)$ for $i \in \{1, \ldots, n-1\}$. By the properties of the function K, one has that x contains va as a factor and, therefore, also $f_{n-1}a$. Hence, one can write $x = \lambda f_{n-1} a \mu$, with $\lambda, \mu \in A^*$, so that

$$w = \lambda f_{n-1} a \mu \lambda' f_{n-1};$$

thus, if we set $g_{n-1} = a\mu\lambda'$, then one has $f_n = f_{n-1} g_{n-1} f_{n-1}$ with $g_{n-1} \in aA^*$. Since $|f_{n-1} a \mu| \leq K(D(n-1)+1)$ and $|\lambda' f_{n-1}| = D(n-1)$, it follows that $|f_n| \leq D(n)$.

We consider now non-periodic uniformly recurrent words. From Lemma 2.6.2 these words are ω-power free. Let $x \in A^\omega$ be a uniformly recurrent and ω-power free word and K, P be the recurrency and the power-free functions of x, respectively. We define the function $q : \mathbb{N}_+ \to \mathbb{N}$ recursively as follows: $q(1) = 1$ and for any $m > 1$,

$$q(m) = 2 \max\{q(m-1)P(q(m-1)), K(q(m-1))\} + q(m-1). \quad (2.16)$$

The following lemma holds [45]:

Lemma 2.7.1. *Let* $x \in A^\omega$ *be a uniformly recurrent and* ω*-power free word. For every pair of integers* m *and* i *such that* $i > q(m)$, $x[i]$ *has as a suffix the* m *th term of a bi-ideal sequence* $(f_s)_{s=1,\ldots,m}$, *where* $f_1 = x_i$ *and* $|f_s| \leq q(s)$, *for* $s = 1, \ldots, m$. *Moreover, if* $(w_1, \ldots, w_m)$ *is the canonical factorization of* f_m, *then for any* h, j, k *such that* $1 \leq h < j < k \leq m$, *one has* $w_h \cdots w_{j-1} w_j \cdots w_{k-1} \neq w_j \cdots w_{k-1} w_h \cdots w_{j-1}$.

Proof. We construct, inductively, a bi-ideal sequence $(f_s)_{s=1,\ldots,m}$ of order m, satisfying the conditions of the lemma. We set $f_1 = x_i$. Suppose now, by induction, that we have already constructed the $(j-1)$ th term f_{j-1}, $1 < j \leq m$, suffix of $x[i]$ and such that $|f_{j-1}| \leq q(j-1)$. Since $|x[i]| = i > q(m) \geq q(j)$, there exist $\lambda, \mu_1, \mu_2 \in A^*$, with

$$|\mu_1| = |\mu_2| = \max\{q(j-1)P(q(j-1)), K(q(j-1))\}$$

such that $x[i] = \lambda\mu_1\mu_2 f_{j-1}$. Being $|\mu_1| \geq K(q(j-1))$ and $|f_{j-1}| \leq q(j-1)$, the word f_{j-1} is a factor of μ_1 so that

$$x[i] = \lambda' f_{j-1} g_{j-1} f_{j-1}, \ \lambda', \ g_{j-1} \in A^*$$

and

$$|f_{j-1} g_{j-1} f_{j-1}| \leq |\mu_1\mu_2 f_{j-1}| \leq q(j).$$

Moreover, one has

$$|g_{j-1}| \geq |\mu_2| \geq q(j-1)P(q(j-1)).$$

Then, setting $f_j = f_{j-1} g_{j-1} f_{j-1}$, the first part of the statement is satisfied. Let us observe that for all $j = 1, \ldots, m$ one has that $|g_j| \geq q(j)P(q(j))$. Now let $(w_1, \ldots, w_m)$ be the canonical factorization of f_m and suppose, by contradiction, that there exist h, j, k with $1 \leq h < j < k \leq m$, such that

$$w_h \cdots w_{j-1} w_j \cdots w_{k-1} = w_j \cdots w_{k-1} w_h \cdots w_{j-1}.$$

Then the two words $u = w_h \cdots w_{j-1}$ and $v = w_j \cdots w_{k-1}$ satisfy the equation $uv = vu$. Then, by Theorem 1.4.2, there exist $z \in A^*$ and $r, s > 0$ such that $u = z^r, v = z^s$. Moreover, since $f_{m-h+1} = w_h \cdots w_m$ and $f_{m-j+1} = w_j \cdots w_m$, one has

$$|w_h \cdots w_{j-1}| = |f_{m-h+1}| - |f_{m-j+1}| \geq |f_{m-j+2}| - |f_{m-j+1}| \geq$$
$$q(m-j+1)P(q(m-j+1));$$

hence

$$|u| = |w_h \cdots w_{j-1}| \geq q(m-j+1)P(q(m-j+1)). \tag{2.17}$$

On the other hand z is a factor of v and, therefore, of f_{m-j+1}. Thus

$$|z| \leq q(m-j+1);$$

moreover, $u = z^r$ implies, by equation (2.17), that $r \geq P(q(m-j+1))$ which is in contradiction with the property of the function P.

Combining Proposition 2.7.5 and Lemma 2.7.1 one obtains the following:

Proposition 2.7.8. *Let $x \in A^\omega$ be a uniformly recurrent and ω-power free word. There exists a map $h : \mathbb{N} \to \mathbb{N}$ such that for every pair of integers n and i such that $i > h(n)$, $x[i]$ has as a suffix the m th term of a bi-ideal sequence $f_{s+1} = f_s g_s f_s$, $s = 1, \ldots, m-1$, with $m \geq n$, such that its canonical factorization has a derived sequence of order n which is either an n-division or an inverse n-division. Moreover, $f_1 \in x_{i+1}A^*$ and $g_s \in x_{i+1}A^*$ for $s = 1, \ldots, m-1$.*

Proof. Let us define for all $n > 0$, $h(n) = q(r(n)+1)$, where r and q are the functions of Proposition 2.7.5 and Lemma 2.7.1, respectively. Let $n \geq 1$ and $m = r(n)$. Let us set $a = x_{i+1}$. By Lemma 2.7.1, if $i > q(r(n)+1)$, then $x[i+1]$ has a suffix f'_{m+1} which is the $(m+1)$ th term of a bi-ideal sequence

$$f'_{s+1} = f'_s g'_s f'_s, \; s = 1, \ldots, m, \text{ with } f'_1 = a.$$

We consider then the bi-ideal sequence $f_{s+1} = f_s g_s f_s$, $s = 1, \ldots, m-1$, defined by $f_s = f'_{s+1}a^{-1}$, for $s = 1, \ldots, m$. By construction, since $f'_1 = a$, one has $f_1 \in aA^*$. Moreover, for $s = 1, \ldots, m-1$

$$f_{s+1} = f'_{s+2}a^{-1} = f'_{s+1}g'_{s+1}f'_{s+1}a^{-1} = f_s a g'_{s+1} f_s,$$

so that $g_s = a g'_{s+1}$.

Let $(w_1, \ldots, w_m)$ and $(w'_1, \ldots, w'_{m+1})$ be the canonical factorizations of f_m and f'_{m+1}. One has that $w_i = w'_i$, for $1 \leq i \leq m$. Indeed, for $i = 1, \ldots, m$

$$w'_i = f'_{m+1-i}g'_{m+1-i} = f_{m-i} a g'_{m+1-i} = f_{m-i}g_{m-i} = w_i.$$

Thus for any h, j, k such that $1 \leq h < j < k \leq m$, one has

$$w_h \cdots w_{j-1} w_j \cdots w_{k-1} \neq w_j \cdots w_{k-1} w_h \cdots w_{j-1}. \tag{2.18}$$

Since $m = r(n)$, by Proposition 2.7.5 there exists a derived sequence $(u_1, \ldots, u_n)$ of $(w_1, \ldots, w_m)$ which is either an n-division or an inverse n-division. In fact, by equation (2.18), condition *iii* of Proposition 2.7.5 cannot be satisfied.

The following proposition shows the existence of some unavoidable regularities which appear in any one sided infinite ω-power free word and then, in particular, in a uniformly recurrent non-periodic word. These regularities include, as special cases, the bi-ideal sequences of Coudrain and Schützenberger and the n-divided words considered by Shirshov.

Proposition 2.7.9. *Let x be an ω-power free one-sided infinite word over a finite totally ordered alphabet A. For any $n > 1$, x has a factor which is the n th term of a strict bi-ideal sequence whose canonical factorization is an n-division.*

Proof. For any one-sided infinite word x there exists, by Theorem 2.3.1, a uniformly recurrent two-sided infinite word w such that $F(w) \subseteq F(x)$. If x is ω-power free, then so will be w. Hence, if we prove the assertion of the proposition for w, then it will be proved also for x. Let $a = \min(\text{alph}(w))$, i.e., a is the letter occurring in w which is minimal with respect to the total order in A. We denote by L_a and M_a the sets defined as

$$L_a = F(w) \cap (A \setminus \{a\})^*, \ M_a = F(w) \cap aa^*.$$

L_a is the set of all factors of w in which the letter a does not occur; it is finite because of the uniform recurrence of w. The set M_a is also finite because w is ω-power-free.

Let $X = a^+(A \setminus \{a\})^+$ and consider the set $Y = F(w) \cap X$; Y is finite since $Y \subseteq M_a L_a$. Let B be a finite alphabet having the same cardinality of Y and consider a bijection $\delta : B \to Y$, which can be extended to a morphism of B^* in Y^*. The alphabet B can be totally ordered by $<_B$ as follows: if $b_1, b_2 \in B$ then

$$b_1 <_B b_2 \Leftrightarrow \delta(b_1) <_A \delta(b_2). \tag{2.19}$$

This order can be extended to the lexicographic order of B^*. Moreover, by Lemma 1.5.1, for all $u, v \in B^*$

$$u <_B v \Rightarrow \delta(u) <_A \delta(v).$$

The set Y is a code (that is, the basis of a free submonoid of A^*) having a synchronization delay equal to 1 (see [8]), i.e., for any pair $(y_1, y_2) \in Y \times Y$ and $\alpha, \beta \in A^*$

$$\alpha y_1 y_2 \beta \in Y^* \Longrightarrow \alpha y_1, y_2 \beta \in Y^*.$$

Since Y is a code, δ is an isomorphism of B^* in Y^*. Moreover, $F(w) \subseteq F(Y^*)$, so that from the synchronization property, one has that any sufficiently large factor of w can, uniquely, be "parsed" in terms of the elements of Y. One can say also that w can, uniquely, be factorized by the elements of Y. Thus there exists a word $\mu \in B^{\pm\omega}$ on the alphabet B such that $w = \delta(\mu)$.

The word μ is ω-power-free. Indeed, let $u \in F(\mu)$. One has that $\delta(u) \in F(w)$. Since w is ω-power free, from the definition of the map p_w (see Def. 2.6.3), it follows

$$(\delta(u))^{p_w(\delta(u))} = \delta(u^{p_w(\delta(u))}) \notin F(w).$$

This implies that $u^{p_w(\delta(u))} \notin F(\mu)$. The word μ is recurrent. In fact let $u \in F(\mu)$; one has that $\delta(u) \in F(w)$; moreover, there exists a letter $x \in A \setminus \{a\}$ such that $x\delta(u)a \in F(w)$. Hence, the factor $x\delta(u)a$ is recurrent in the unique factorization of w in terms of the elements of Y. This implies that u will be recurrent in μ. Moreover, μ is uniformly recurrent; in fact one easily derives that for any n an upper bound to $K_\mu(n)$ is given by $K_w((n+2)l_M)$ where $l_M = \max\{|y| \mid y \in Y\}$.

The proof of the proposition is by induction on the integer n. We begin by proving the basis of the induction, that is, the case $n = 2$. Let us consider in w any factor $u = ya$, with $y \in Y$. We can write u as $u = a^h va$, with $h > 0$, and $v \in L_a$. We can set $w_2 = a$ and $w_1 = a^h v$. One has that $w_2 w_1 = aa^h v <_A w_1 w_2 = a^h va$. Hence, (w_1, w_2) is the canonical factorization of the 2 nd term $w_1 w_2$ of a strict bi-ideal sequence. Let us now suppose that we have proved the result up to $n-1$ for any uniformly recurrent and ω-power-free two-sided infinite word. Thus, by induction, the word μ has a factor

$$u = u_1 \cdots u_{n-1}, \; u_i \in B^+ \; (i = 1, \dots, n-1),$$

such that for any nontrivial permutation $\sigma \in \mathcal{S}_{n-1}$,

$$u_{\sigma(1)} u_{\sigma(2)} \dots u_{\sigma(n-1)} <_B u_1 u_2 \dots u_{n-1} \tag{2.20}$$

and, moreover, $u_i \in u_{i+1} \cdots u_{n-1} B^+$ for all $i \in [1, n-2]$. Let us set $\delta(u_i) = w_i$, $i \in [1, n-2]$. By equations (2.19) and (2.20) for any nontrivial permutation $\sigma \in \mathcal{S}_{n-1}$, one has

$$w_{\sigma(1)} w_{\sigma(2)} \cdots w_{\sigma(n-1)} <_A w_1 w_2 \cdots w_{n-1}$$

and, by the fact that $\delta(B^+) \subseteq aA^*$,

$$w_i \in w_{i+1} \cdots w_{n-1} aA^*, \quad \text{for all} \quad i \in [1, n-2]. \tag{2.21}$$

Let us then set $w_n = a$. One has that $w_1 \cdots w_{n-1} a \in F(w)$. Moreover,

$$w_n w_{n-1} = a w_{n-1} <_A w_{n-1} a = w_{n-1} w_n.$$

By equation (2.21) $(w_1, \dots, w_n)$ is an n-sequence so that the sequence of words $f_i = w_{n-i+1} \cdots w_n$, $1 \le i \le n$, is a bi-ideal sequence of order n whose last term $f_n = w_1 \cdots w_n$ has the canonical factorization $(w_1, \dots, w_n)$. Since for any $i \in [1, n-1]$, $w_{i+1} w_i <_A w_i w_{i+1}$, by Proposition 2.7.2 $(w_1, \dots, w_n)$ is an n-division. Moreover, the bi-ideal sequence is strict since it can never occur that $w_i = w_{i+1} \cdots w_{n-1} a$, for all $i \in [1, n-2]$.

2.7.3 Some extensions of the Shirshov theorem

A more constructive proof of Proposition 2.7.9 and a noteworthy improvement of the result is given by the following:

Proposition 2.7.10. *Let $x \in A^\omega$ be an ω-power free word over a finite and totally ordered alphabet A. For any $n > 1$, x has a factor s which is the n th term of a bi-ideal sequence whose canonical factorization $(w_1, \dots, w_n)$ is an n-division of s. Moreover, the words w_i $(i = 1, \dots, n)$ are Lyndon and such that*

$$w_1 > w_2 > \cdots > w_n.$$

Proof. We can suppose, as before, that x is a uniformly recurrent and ω-power free bi-infinite word on the alphabet A. We give now a procedure in order to find for any integer $n > 0$ a sequence of totally ordered finite alphabets Σ_i $(i = 1, \ldots, n+1)$ and a sequence of bi-infinite uniformly recurrent words $x_i \in \Sigma_i^{\pm\omega}$, with $\Sigma_i = \mathrm{alph}(x_i)$, such that $x_1 = x$ and for any i, $2 \leq i \leq n+1$, there exists a monomorphism

$$\delta_i : \Sigma_i^* \to \Sigma_{i-1}^*,$$

with the property

$$\delta_i(x_i) = x_{i-1}.$$

The construction is given inductively. We totally order $\Sigma_1 = \mathrm{alph}(x) \subseteq A$ and extend this order to the lexicographic order of Σ_1^*. Let us suppose that we have constructed the sequence until the i th step. Then let $x_i \in \Sigma_i^{\pm\omega}$, where $\Sigma_i = \mathrm{alph}(x_i)$ is supposed to be totally ordered. This order can be extended to the lexicographic order $<_{\Sigma_i}$ of Σ_i^*. Let $b_i = \min(\Sigma_i)$ and

$$X_i = b_i^+(\Sigma_i \setminus \{b_i\})^+.$$

Then consider the set

$$\Lambda_{i+1} = F(x_i) \cap X_i.$$

Since x_i is uniformly recurrent and ω-power-free, there are not enough long subwords in x_i which are either a power of b_i or do not contain b_i as a factor, so that Λ_{i+1} is finite.

One can then consider an alphabet Σ_{i+1} with $\mathrm{card}(\Sigma_{i+1}) = \mathrm{card}(\Lambda_{i+1})$ and a bijection $\gamma_{i+1} : \Sigma_{i+1} \to \Lambda_{i+1}$. Let $\delta_{i+1} : \Sigma_{i+1}^* \to \Sigma_i^*$ be the morphism which extends γ_{i+1}. Since Λ_{i+1} is a code with bounded synchronization delay (the synchronization delay is 1), δ_{i+1} is injective and x_i can be uniquely factorized in terms of the elements of Λ_{i+1}. Then there exists a bi-infinite word $x_{i+1} \in \Sigma_{i+1}^{\pm\omega}$ such that

$$\delta_{i+1}(x_{i+1}) = x_i.$$

Since, by the inductive hypothesis, x_i is uniformly recurrent and ω-power free, then one derives, as in the proof of Proposition 2.7.9, that so will be x_{i+1}.

Now we define in Σ_{i+1} a total ordering by setting for $x, y \in \Sigma_{i+1}$

$$x <_{\Sigma_{i+1}} y \Longleftrightarrow \delta_{i+1}(x) <_{\Sigma_i} \delta_{i+1}(y).$$

This total order of Σ_{i+1} can be extended to the lexicographic order $<_{\Sigma_{i+1}}$ of Σ_{i+1}^*. Moreover, for any $u, v \in \Sigma_{i+1}^*$ one has by Lemma 1.5.1

$$u <_{\Sigma_{i+1}} v \Longrightarrow \delta_{i+1}(u) <_{\Sigma_i} \delta_{i+1}(v).$$

Let $w_0, w_1, w_2, \ldots, w_n \in \Sigma_1^+ \subseteq A^*$ be defined as

$$w_n = b_1 = \min(\Sigma_1)$$

and, for $i = 2, \ldots, n+1$,

$$w_{n-i+1} = \delta_2(\delta_3(\cdots \delta_i(b_i) \cdots)),$$

with $b_i = \min(\Sigma_i)$.

We prove that $s = w_1 w_2 \cdots w_n \in F(x)$. Moreover, s is the n th term of a bi-ideal sequence whose canonical factorization is $(w_1, w_2, \ldots, w_n)$. By construction for any $i > 1$ and for any $b \in \Sigma_i$ one has

$$\delta_i(b) = b_{i-1} u_1, \quad \text{with} \quad u_1 \in \Sigma_{i-1}^+,$$

thus, applying the same relation to the first letter of u_1, we obtain

$$\delta_{i-1}(\delta_i(b)) = \delta_{i-1}(b_{i-1}) b_{i-2} u_2, \quad \text{with} \quad u_2 \in \Sigma_{i-2}^+.$$

Iterating this procedure one has

$$\delta_2(\cdots \delta_{i-1}(\delta_i(b)) \cdots) = \delta_2(\cdots \delta_{i-1}(b_{i-1}) \cdots) \cdots \delta_2(b_2) b_1 u_{i-1},$$

with $u_{i-1} \in \Sigma_1^+ \subseteq A^+$. Thus, for $b = b_i$, we obtain for any i, $2 \le i \le n+1$,

$$w_{n-i+1} = w_{n-i+2} \cdots w_n u_{i-1}. \tag{2.22}$$

Thus, in particular, one has

$$w_0 = w_1 \cdots w_n u_n.$$

Since $w_0 \in F(x)$, then $s = w_1 \cdots w_n \in F(x)$. Moreover, equation (2.22) implies that $(w_1, w_2, \ldots, w_n)$ is an n-sequence, so that $s = w_1 \cdots w_n$ is the last term of a bi-ideal sequence of order n whose canonical factorization is $(w_1, \ldots, w_n)$.

Let us now prove that $(w_1, \ldots, w_n)$ is an n-division of s. Indeed, since $\delta_i(b_i) \in b_{i-1}^+(\Sigma_{i-1} \setminus \{b_{i-1}\})^+$, we can write

$$\delta_i(b_i) = b_{i-1}^{r_1} u_1, \quad \text{with} \quad r_1 > 0 \text{ and } u_1 \in (\Sigma_{i-1} \setminus \{b_{i-1}\})^+$$

and, since $b_{i-1} = \min(\Sigma_{i-1})$, one has

$$b_{i-1}\delta_i(b_i) = b_{i-1} b_{i-1}^{r_1} u_1 = b_{i-1}^{r_1} b_{i-1} u_1 <_{\Sigma_{i-1}} b_{i-1}^{r_1} u_1 b_{i-1} = \delta_i(b_i) b_{i-1},$$

so that

$$b_{i-1}\delta_i(b_i) <_{\Sigma_{i-1}} \delta_i(b_i) b_{i-1}. \tag{2.23}$$

For $i = 2$ the above equation becomes

$$w_n w_{n-1} <_A w_{n-1} w_n.$$

Moreover, since the lexicographic ordering is preserved by δ_i, for any i, one obtains from equation (2.23) for $2 < i \leq n$

$$w_{n-i+2}w_{n-i+1} = \delta_2(\cdots\delta_{i-1}(b_{i-1}\delta_i(b_i))\cdots) <_A \delta_2(\cdots\delta_{i-1}(\delta_i(b_i)b_{i-1}))\cdots)$$
$$= w_{n-i+1}w_{n-i+2},$$

i.e.,

$$w_{i+1}w_i <_A w_iw_{i+1}$$

for all $i = 1,\ldots,n-1$. By Proposition 2.7.2, this implies that $(w_1, w_2, \ldots, w_n)$ is an n-division of s.

Let us denote by $\mathcal{L}_i$ the set of the Lyndon words on the alphabet Σ_i. For each $i = 2,\ldots,n$ the word $b_i = \min(\Sigma_i)$ is a Lyndon word on the alphabet Σ_i. Moreover, by Proposition 1.5.4 the monomorphism δ_i preserves Lyndon words. Thus, $\delta_i(b_i) \in \mathcal{L}_{i-1}$ and

$$w_{n-i+1} = \delta_2(\delta_3(\cdots\delta_i(b_i)\cdots)) \in \mathcal{L}_1.$$

All the words w_i $(i = 1,\ldots,n)$ are then Lyndon on the alphabet A. Moreover, since from (2.22) w_{n-i+2} is a proper prefix of w_{n-i+1}, one has

$$w_1 > w_2 > \cdots > w_n.$$

This is, in fact, a further proof, by Lemma 1.5.2, that $(w_1,\ldots,w_n)$ is an n-division.

As a consequence of the preceding proposition, we give the following improvement of Shirshov's theorem.

Theorem 2.7.2. *For all k, p, n positive integers there exists a positive integer $N(k,p,n)$ such that for any totally ordered alphabet A of cardinality k any word $w \in A^*$ whose length is at least $N(k,p,n)$ is such that*

i. there exists $u \neq \epsilon$ such that $u^p \in F(w)$ or
ii. there exists $s \in F(w)$ which is the n th term of a bi-ideal sequence whose canonical factorization $(w_1,\ldots,w_n)$ is an n-division of s. Moreover, the words w_i, $i = 1,\ldots,n$, are Lyndon words such that

$$w_1 > w_2 > \cdots > w_n.$$

Proof. Let A be a totally ordered alphabet of cardinality k. The set of all words of A^* which satisfy either *i* or *ii* is a two-sided ideal $J_{k,n,p}$, or simply J, of A^*. Let $C = A^* \setminus J$. The set C is closed by factors, so that if we suppose that C is infinite, then by Lemma 1.2.2 there exists a one-sided infinite word $x \in A^\omega$ such that $F(x) \subseteq C$. Now either x has a factor which is a p-power and then $F(x) \cap J \neq \emptyset$ or x is p-power free. In this latter case from Proposition 2.7.10 it follows again that $F(x) \cap J \neq \emptyset$. Hence, in both cases we reach a contradiction. Thus C has to be finite and this proves the assertion.

In conclusion of this chapter we mention that a proof of Shirshov's theorem (see Theorem 2.4.3) based on an unavoidable regularity related to Lyndon words was given by Reutenauer in [141]. A proof which uses the uniform recurrence is in [91]. Two different proofs of an improvement of Shirshov's theorem in which the n-divided factor is the canonical factorization of a bi-ideal sequence were given in [45] and [49]. Theorem 2.7.2 is a further generalization which combines in a suitable way the previous unavoidable regularities.

3. Finiteness Conditions for Semigroups

The study of finiteness conditions for semigroups consists in giving some conditions which are satisfied by finite semigroups and which are such as to assure the finiteness of them. In this study one of the properties which is generally required of a semigroup is that of being *finitely generated.*

These conditions are very important both in algebra and automata theory. Indeed, if one supposes that the semigroup is also *periodic* (or *torsion*), the study of these finiteness conditions for semigroups (and groups) is called the (general) *Burnside problem for semigroups* (and *groups*) (see [101]). Several finiteness conditions for finitely generated semigroups have been given in recent years based on different concepts such as: *permutation property, chain conditions on ideals, strong periodicity, iteration properties, and repetitivity* .

As we shall see in more detail in the following chapters, there exists an important relationship between finiteness conditions for semigroups and the *theory of automata* and *regular languages* based on the following fact. To any language L one can associate a semigroup $S(L)$, called the *syntactic semigroup* of L, which satisfies the property that L is acceptable by a finite automaton, or is regular, if and only if $S(L)$ is finite. Hence, in principle, any finiteness condition for finitely generated semigroups can be translated into a regularity condition for languages. The study of regularity conditions for periodic languages (i.e., languages whose syntactic semigroup is periodic) has been called the *Burnside problem for languages* [139].

3.1 Preliminaries on semigroups

In this section we give some notations, definitions and we recall some theorems on semigroups that will be used in the following.

A semigroup S is a set in which a binary associative operation, called *product*, is defined. If $s, t \in S$, then we denote by st the product of s by t. From the associativity one has that for all $s, t, r \in S$ one has

$$(st)r = s(tr).$$

Sometimes the operation defined in S is called *sum* and the sum of $s, t \in S$ is denoted by $s + t$; in such a case one says that S is an *additive semigroup*. A semigroup S is called *monoid* if there exists in S an identity element 1, i.e., such that for all $s \in S$

$$s1 = 1s = s.$$

One easily verifies that the identity element 1 is *unique*. An element s of a monoid is called *invertible* if there exists an element s' such that

$$ss' = s's = 1.$$

If $s \in S$ is invertible, then the element s' satisfying the previous relation is *unique* and denoted by s^{-1}. If any element of a monoid is invertible, then the monoid is called *group*. A semigroup S is *commutative* (*or abelian*) if the product defined in S is commutative, that is, for all $s, t \in S$

$$st = ts.$$

Let S be a semigroup. One can naturally embed S in a monoid S^1 as follows. If S is a monoid, then $S^1 = S$. If S has no identity, then S^1 is obtained from S by adjoining an extra element 1 satisfying the property $s1 = 1s = s$ for all $s \in S^1$.

If $s, t \in S$ we say that s is a *factor* (or *divisor*) of t if $t \in S^1 s S^1$. If $t \in sS^1$ ($t \in S^1 s$), then s is called *left factor* (*right factor*) of t. For any $t \in S$ we denote by $F(t)$ the set of the factors of t. For any subset X of S,

$$F(X) = \bigcup_{t \in X} F(t).$$

One says that X is *factorial* or *closed by factors* if $F(X) = X$.

Let us observe, in order to avoid ambiguities, that in the case of the free semigroup A^+ the factors of a finite, as well as infinite, word w are elements of A^+. Hence, the set $F(w)$ of the factors of w does not contain the empty word.

An element of a semigroup S is called *zero* and denoted by 0 if for all $s \in S$

$$s0 = 0s = 0.$$

One easily verifies that a zero element of a semigroup is *unique*. We denote by S^0 the semigroup obtained from S as follows. If S has a zero element, then $S^0 = S$. If S has no zero element, then S^0 is obtained from S by adjoining an extra element 0 satisfying the property $s0 = 0s = 0$ for all $s \in S^0$.

Let S be a semigroup and $s \in S$. For all $n > 0$ we define s^n inductively as

$$s^1 = s, \quad s^n = ss^{n-1}, \quad \text{for } n > 1. \tag{3.1}$$

An element $e \in S$ is called *idempotent* if

$$e^2 = e.$$

As usual $E(S)$ denotes the set of all idempotents of S.

A subsemigroup S' of a semigroup S is any subset of S closed under the product operation defined in S, that is, for all $s'_1, s'_2 \in S'$ one has

$$s'_1 s'_2 \in S'.$$

A subsemigroup S' of a semigroup S can be a monoid or a group. In such a case the identity of S' is an idempotent of S. If S' is a group with identity $e \in E(S)$, then for any $s \in S'$, s^{-1} will denote the inverse of s in S'; thus one has

$$ss^{-1} = s^{-1}s = e.$$

Let S be a semigroup and $\mathcal{P}(S)$ denote the set of its parts. One can extend the product operation of S to $\mathcal{P}(S)$ as follows: for $X, Y \in \mathcal{P}(S)$ the product $X \cdot Y$, or simply XY, is defined as

$$XY = \{xy \in S \mid x \in X \text{ and } y \in Y\}.$$

This product is associative, so that $\mathcal{P}(S)$ is by itself a semigroup called *semigroup of the parts* of S. For any $n > 0$, X^n is recursively defined, according to equation (3.1), as

$$X^1 = X, \ X^n = XX^{n-1}, \text{ for } n > 1.$$

In $\mathcal{P}(S)$ one can introduce the *Boolean operations* of *union* ($\cup$), *intersection* ($\cap$), and *difference* ($\setminus$). A further important operation is the unary operation *cross* ($^+$) defined for any $X \in \mathcal{P}(S)$ as

$$X^+ = X \cup X^2 \cup \cdots X^n \cup \cdots = \bigcup_{n>0} X^n.$$

The operations ($\cup$), ($\cdot$), and ($^+$) are usually called the *rational operations.* If X is a singleton $\{x\}$, then $\{x\}Y$, $Y\{x\}$ and $\{x\}^+$ are simply denoted by xY, Yx, and x^+, respectively.

Let S be a semigroup and θ be an equivalence relation in S. The equivalence θ is *right invariant (or is a right congruence)* if for all $s_1, s_2, s \in S$

$$s_1 \theta s_2 \Longrightarrow s_1 s \theta s_2 s.$$

In a symmetric way an equivalence θ is *left invariant (or is a left congruence)* if for all $s_1, s_2, s \in S$

$$s_1 \theta s_2 \Longrightarrow s s_1 \theta s s_2.$$

It is well known that an equivalence θ in S is a *congruence* if and only if θ is right and left invariant. In this case the following property, called *substitution property,* holds: for all $s_1, s_2, s_3, s_4 \in S$

$$s_1 \theta s_2, \ s_3 \theta s_4 \Longrightarrow s_1 s_3 \theta s_2 s_4.$$

If θ is a congruence relation in S, then one can consider the *quotient semigroup* S/θ where the product operation ($\circ$) in S/θ is defined as follows: for $s, t \in S$

$$\theta(s) \circ \theta(t) = \theta(st).$$

One easily verifies that, since θ is a congruence, the above product operation is well defined and that it is associative.

If S and T are semigroups a *morphism* of S in T is a map $\phi : S \to T$ such that for all $s_1, s_2 \in S$

$$\phi(s_1 s_2) = \phi(s_1)\phi(s_2).$$

If the map ϕ is surjective (injective, bijective), then the morphism is called *epimorphism* (*monomorphism, isomorphism*). If θ is a congruence in the semigroup S, then the map $\phi : S \to S/\theta$ defined for any $s \in S$ as

$$\phi(s) = \theta(s),$$

is an epimorphism called the *natural epimorphism* associated with θ. Conversely, if $\phi : S \to T$ is a morphism, then the relation $\phi\phi^{-1}$ is a congruence in S, called the *nuclear congruence* associated with ϕ. If ϕ is an epimorphism, then one has that $T \cong S/\phi\phi^{-1}$, that is, T and $S/\phi\phi^{-1}$ are isomorphic.

Let S be a semigroup. A nonempty subset R (subset L) of S is a *right ideal* (*left ideal*) if

$$RS \subseteq R, \quad (SL \subseteq L).$$

A nonempty subset J of S is a *two-sided ideal* if it is both a right and left ideal.

One easily derives with a proof similar to that of Lemma 1.6.1, that *a subset $X \subset S$ is closed by factors if and only if $X = S \setminus J$, where J is a two-sided ideal of S.*

Let X be a subset of a semigroup S. The right ideal *generated* by X is the smallest ideal of S containing X. It is easy to prove that the right ideal generated by X is given by

$$XS^1.$$

With the analogous definitions one has that the left (two-sided) ideal generated by X is given by

$$S^1X, \ (S^1XS^1).$$

If X is a singleton $X = \{a\}$, then the right (left, two-sided) ideal generated by $\{a\}$ is simply denoted by

$$aS^1, \ (S^1a, \ S^1aS^1)$$

and called the *principal* right (left, two-sided) ideal generated by a.

A two-sided (right, left) ideal I of a semigroup S is called *minimal* if it does not properly contain any two-sided (right, left) ideal of S. If S contains a zero 0, then $\{0\}$ is, trivially, the unique minimal two-sided (right, left) ideal of S. Hence, in this case one gives the nontrivial notion of 0-minimality. A two-sided (right, left) ideal I of S is called 0-*minimal* if *i.* $I \neq \{0\}$, *ii.* $\{0\}$ is the unique two-sided (right, left) ideal of S properly contained in I.

A semigroup S is called *right (left) simple* if S does not contain proper right (left) ideals. This is, trivially, equivalent to the condition that for any $s \in S$, $sS = S$ ($Ss = S$). Similarly, S is called *simple* if S does not contain proper two-sided ideals. It is well known (see [22], Chap. 1, p. 6) that a *semigroup S is a group if and only if S is right and left simple.* If a semigroup S has an element 0, then S is called 0-*simple* if $S^2 \neq \{0\}$ and $\{0\}$ is the only proper two-sided ideal of S. This is also equivalent to the condition that $S \neq \{0\}$ and for any $s \in S \setminus \{0\}$, $SsS = S$.

The notion of ideal naturally leads to consider some equivalence relations in a semigroup, called Greeen's relations which have played a fundamental role in the development of the theory of semigroups [22]. We give here the basic definitions and recall the main results that we shall use in this and in the next chapter.

In a semigroup S one can introduce the following equivalence relations $\mathcal{R}$, $\mathcal{L}$, and $\mathcal{J}$ defined as: for $s, t \in S$

$$s \,\mathcal{R}\, t \Longleftrightarrow sS^1 = tS^1,$$

$$s \,\mathcal{L}\, t \Longleftrightarrow S^1 s = S^1 t,$$

$$s \,\mathcal{J}\, t \Longleftrightarrow S^1 s S^1 = S^1 t S^1.$$

From the definition one has that the relation $\mathcal{R}$ (relation $\mathcal{L}$) is a left (right) congruence in S. Moreover, $\mathcal{R}, \mathcal{L} \subseteq \mathcal{J}$. Two further equivalence relations are the relations $\mathcal{H}$ and $\mathcal{D}$, defined as

$$\mathcal{H} = \mathcal{R} \cap \mathcal{L},$$

$$\mathcal{D} = \mathcal{R} \vee \mathcal{L},$$

i.e., $\mathcal{H}$ is the greatest equivalence contained in $\mathcal{R}$ and $\mathcal{L}$, whereas $\mathcal{D}$ is the smallest equivalence in S containing $\mathcal{R}$ and $\mathcal{L}$. Thus one has that $\mathcal{H} \subseteq \mathcal{R}$, $\mathcal{H} \subseteq \mathcal{L}$, and $\mathcal{D} \subseteq \mathcal{J}$. The relations $\mathcal{R}$, $\mathcal{L}$, $\mathcal{J}$, $\mathcal{H}$, and $\mathcal{D}$ are called the *Green relations.* For any $s \in S$,

$$R_s,\ L_s,\ J_s,\ H_s,\ \text{and } D_s$$

denote the $\mathcal{R}$-class, $\mathcal{L}$-class, $\mathcal{J}$-class, $\mathcal{H}$-class, and $\mathcal{D}$-class containing s, respectively.

Lemma 3.1.1. *The relations $\mathcal{R}$ and $\mathcal{L}$ commute, and so*

$$\mathcal{D} = \mathcal{R}\mathcal{L} = \mathcal{L}\mathcal{R}.$$

From this one has that

$$s \mathcal{D} t \Longleftrightarrow \exists r \in S \text{ such that } s \mathcal{R} r \text{ and } r \mathcal{L} t.$$

This implies that any $\mathcal{R}$-class and any $\mathcal{L}$-class in the same $\mathcal{D}$-class have a nonempty intersection.

Theorem 3.1.1. *Let a and c be $\mathcal{D}$-equivalent elements of a semigroup S. Then there exists b in S such that $a \mathcal{R} b$ and $b \mathcal{L} c$, and hence $as = b$, $bs' = a$, $tb = c$, $t'c = b$, for some $s, s', t, t' \in S^1$. The maps $x \to txs$ $(x \in H_a)$ and $z \to t'zs'$ $(z \in H_c)$ are mutually inverse, one-to-one maps of H_a and H_c upon each other.*

From this it follows that the $\mathcal{H}$-classes in the same $\mathcal{D}$-class have the same cardinal number.

Theorem 3.1.2. *(Green) If a, b, and ab all belong to the same $\mathcal{H}$-class H of a semigroup S, then H is a subgroup of S. In particular, any $\mathcal{H}$-class containing an idempotent is a subgroup of S.*

A subgroup G of a semigroup S is called *maximal subgroup* of S if it is not properly contained in any other subgroup of S. One easily derives that *the maximal subgroups of S are precisely the $\mathcal{H}$-classes of S containing idempotents.*

An element s of a semigroup S is called *regular* if there exists $x \in S$ such that

$$s = sxs.$$

Trivially, any idempotent of S is a regular element of S. A $\mathcal{D}$-class D of a semigroup is called *regular* if all its elements are regular.

Theorem 3.1.3. *Let D be a $\mathcal{D}$-class of a semigroup S. The following conditions are equivalent:*

i. D is regular.
ii. D contains a regular element.
iii. D contains an idempotent.

Theorem 3.1.4. *Any two maximal subgroups in the same (regular) $\mathcal{D}$-class are isomorphic.*

Thus if D is a regular $\mathcal{D}$-class, then all the $\mathcal{H}$-classes in D which are groups are isomorphic to the same abstract group, that we call *group of the $\mathcal{D}$-class D* and denote by G_D.

3.2 Finitely generated semigroups

A semigroup (group) S is ***finitely generated*** if there exists a finite subset X of S such that the subsemigroup (subgroup) $\langle X \rangle$ generated by X is S. A semigroup (group) S is called ***locally finite*** if every finitely generated subsemigroup (subgroup) of S is finite.

As is well known, by definition, the subsemigroup of a semigroup S generated by $X \subseteq S$ is the smallest subsemigroup of S containing X. It is given by

$$\langle X \rangle = X \cup X^2 \cup \cdots \cup X^n \cup \cdots = X^+.$$

Let G_e be a subgroup of S having as identity the idempotent e of S and X be a subset of G_e. Let us denote by X^{-1} the set $X^{-1} = \{x^{-1} \mid x \in X\}$, where for any $x \in X$, x^{-1} is the inverse of x in G_e. The subgroup of G_e generated by X, that is the smallest subgroup of G_e containing X, is then given by $(X \cup X^{-1})^+$, i.e., it coincides with the subsemigroup of S generated by $X \cup X^{-1}$.

When a semigroup S is generated by a set X, i.e., $X^+ = S$, one may regard X as an alphabet and consider a surjective morphism from the free semigroup generated by X in S. To avoid a notational ambiguity it is often convenient to denote X, regarded as an alphabet, with a different letter, say A. The surjective morphism

$$\phi : A^+ \to S,$$

from the free semigroup A^+ in S, is usually called the ***canonical epimorphism***. It depends, of course, on the generating set X. Moreover, one has that $S \cong A^+/\phi\phi^{-1}$. Hence, one can identify, up to an isomorphism, S with $A^+/\phi\phi^{-1}$. When S is finitely generated, then the alphabet A is finite.

Let us suppose that A is totally ordered. We can totally order A^+ by the relation $<_a$, called ***alphabetic order***, defined as: for all $u, v \in A^+$

$$u <_a v \iff (|u| < |v|) \text{ or } (|u| = |v| \text{ and } u < v),$$

where $<$ is the lexicographic order. From the definition it follows that $<_a$ is a well-order.

A word w is said to be ***reducible***, with respect to the morphism ϕ and to the order $<_a$, or simply ***reducible***, if there exists $u \in A^+$ such that

$$u <_a w \quad \text{and} \quad \phi(u) = \phi(w).$$

A word which is not reducible will be called ***irreducible***. Let $s \in S$. In the set $\phi^{-1}(s)$ there is a unique minimal element with respect to $<_a$ usually called the ***canonical representative*** of s. Hence, the set of all canonical representatives of the elements of S is the set of all irreducible elements of S. For any set $T \subseteq S$, we denote by C_T the set of the canonical representatives of the elements of T.

An infinite (bi-infinite) word x is *reducible*, relative to the morphism $\phi : A^+ \to S$ and to the order $<_a$, if there exists $w \in F(x)$ which is reducible. An infinite word x which is not reducible, i.e., any $w \in F(x)$ is irreducible, is called *irreducible*.

We recall now some lemmas and propositions on canonical representatives (see [45]).

Lemma 3.2.1. *Let S be a finitely generated semigroup and T be any subset of S closed by factors. Then the set C_T is closed by factors.*

Proof. Let $x \in C_T$ and u be a factor of x. We write $x = \lambda u \mu$, with $\lambda, \mu \in A^*$. Since $\phi(x) = \phi(\lambda)\phi(u)\phi(\mu)$ and T is closed by factors, one has $\phi(u) \in T$ and then $u \in \phi^{-1}(T)$. Suppose now that $u' \in A^+$ exists such that $u' <_a u$ and $\phi(u') = \phi(u)$. If $|u'| < |u|$, then $x' = \lambda u' \mu$ is such that $|x'| < |x|$ and $\phi(x') = \phi(x)$ which is a contradiction. Let us then suppose $|u'| = |u|$ and $u' < u$. Thus $x' = \lambda u' \mu < \lambda u \mu$ and $\phi(x') = \phi(x)$ which is again a contradiction. Hence, $u \in C_T$ which concludes the proof.

Lemma 3.2.2. *Let S be a finitely generated semigroup. If T is an infinite subset of S closed by factors, then there exists a uniformly recurrent irreducible word x such that $F(x) \subseteq C_T$.*

Proof. Since C_T is closed by factors, by Theorem 2.3.1 there exists a uniformly recurrent word $x \in A^{\pm\omega}$ such that $F(x) \subseteq C_T$, so that x is irreducible.

Proposition 3.2.1. *Let S be a finitely generated semigroup. The following conditions are equivalent:*

i. S is infinite.
ii. C_S is infinite.
iii. There exists a uniformly recurrent irreducible word.
iv. There exists an infinite irreducible word.
v. There exist infinitely many finite irreducible words.

Proof. Conditions *i* and *ii* are trivially equivalent since the canonical epimorphism ϕ restricted to C_S is a bijection of S and C_S. *ii* $\Rightarrow$ *iii*: since C_S is infinite and, by Lemma 3.2.1, is closed by factors, one has, by Lemma 3.2.2, that there exists a uniformly recurrent irreducible word x. *iii* $\Rightarrow$ *iv*: trivial. *iv* $\Rightarrow$ *v*: trivial. *v* $\Rightarrow$ *ii*: obvious by the fact that C_S is the set of all irreducible finite words.

As we shall see in the next sections some finiteness conditions for a finitely generated semigroup S will be based on the following argument. One supposes that S is infinite, so that by the preceding proposition there exists a uniformly recurrent word x which is irreducible. In Sect. 2.7 we have seen that a uniformly recurrent word x has many structural properties. Indeed, x is either periodic or ω-power-free. Moreover, in this latter case x has bi-ideal factors of any order whose canonical factorizations satisfy a suitable division property. If one is able to prove, on the base of the given conditions, that there exists one of these factors which is reducible, then one will reach a contradiction and prove the finiteness of S.

Let us observe that the notion of a finite, as well as infinite, irreducible word can be given with respect to any partial order $\leq$ in A^+. If the partial order is a well partial order (see Chap. 6), then for any $s \in S$ the set $\phi^{-1}(s)$ of the representatives of s has at least one and at most a finite number of irreducible elements. Moreover, when in addition $\leq$ is monotone (that is, invariant with respect to the product in A^+), one can derive that the set C_T of all irreducible representatives of any factorial set $T \subseteq S$ is closed by factors. Thus one can easily extend Proposition 3.2.1 to the case of any monotone well partial order of A^+.

As in the case of free monoids one can introduce for any semigroup S the notions of bi-ideal sequence and n-sequence.

Definition 3.2.1. *A sequence $s_1, \ldots, s_n, \ldots$ of elements of a semigroup S is a* bi-ideal sequence *if for any $i > 0$*

$$s_{i+1} \in s_i S^1 s_i.$$

When the sequence is finite and of length n, then $(s_1, \ldots, s_n)$ is called a bi-ideal sequence of order n.

From Lemma 3.2.2 one easily derives the following:

Proposition 3.2.2. *Let S be a finitely generated semigroup. If T is an infinite subset of S closed by factors, then there exists a bi-ideal sequence $(s_n)_{n>0}$ such that for all $n > 0$, $s_n \in T$, and for all positive integers i, j, with $i \neq j$, one has $s_i \neq s_j$.*

Proof. By Lemma 3.2.2 there exists an irreducible uniformly recurrent word $x \in A^\omega$ such that $F(x) \subseteq C_T$. By Proposition 2.7.6, the word x is bi-ideal, so that there exists a bi-ideal sequence $(f_n)_{n>0}$, such that $x = \lim_{n\to\infty} f_n$. Since for every $n > 0$, $f_n \in F(x) \subseteq C_T$ then $\phi(f_n) \in T$, where ϕ is the canonical epimorphism. Moreover, since x is irreducible it follows that for all positive integers i, j, with $i \neq j$, one has $\phi(f_i) \neq \phi(f_j)$. The image by ϕ of the bi-ideal sequence $(f_n)_{n>0}$ is then a bi-ideal sequence $(s_n)_{n>0}$, with $s_n = \phi(f_n) \in T$ for all $n > 0$, such that $s_i \neq s_j$ for $i \neq j$.

Definition 3.2.2. *A sequence $t_1, \ldots, t_n$ of n elements of a semigroup S is called* n-sequence *if for all $i = 1, \ldots, n-1$,*

$$t_i \in t_{i+1} \cdots t_n S^1. \tag{3.2}$$

In a symmetric way one can also introduce the notion of *inverse n-sequence*, where the condition in equation (3.2) is replaced by

$$t_{i+1} \in S^1 t_1 \cdots t_i, \quad i = 1, \ldots, n-1.$$

As in the case of free monoids the notions of bi-ideal sequence of order n and of n-sequence are related. In fact, let $s_1, \ldots, s_n$ be a bi-ideal sequence of S where

$$s_{i+1} = s_i g_i s_i,$$

with $g_i \in S^1$, $i = 1, \ldots, n-1$. As one easily verifies, setting $t_n = s_1$ and $t_{n-i} = s_i g_i$, $i = 1, \ldots, n-1$, the sequence $(t_1, \ldots, t_n)$ is an n-sequence. Conversely, if $(t_1, \ldots, t_n)$ is an n-sequence, then the sequence $s_i = t_{n-i+1} \cdots t_n$, $i = 1, \ldots, n$, is a bi-ideal sequence of order n.

In a similar way if one sets $t_1 = s_1$ and $t_{i+1} = g_i s_i$, $i = 1, \ldots, n-1$, then the sequence $(t_1, \ldots, t_n)$ is an inverse n-sequence. Conversely, if $(t_1, \ldots, t_n)$ is an inverse n-sequence, then the sequence of words $s_i = t_1 \cdots t_i$, $i = 1, \ldots, n$, is a bi-ideal sequence of order n.

Let us, finally, observe that in the case of a group G any sequence $g_1, \ldots, g_n$ of n elements of G is an n-sequence (inverse n-sequence). Indeed, for instance, one can write for all $i = 1, \ldots, n-1$,

$$g_i = g_{i+1} \cdots g_n((g_{i+1} \cdots g_n)^{-1} g_i).$$

Let S be a finitely generated semigroup and $\phi : A^+ \to S$ be the canonical epimorphism. The *growth-function* of S can be defined for all $n > 0$, as

$$g_S(n) = \text{card}\{s \in S \mid \phi^{-1}(s) \cap A^{[n]} \neq \emptyset\}.$$

As one easily verifies, the growth function of S coincides with the growth function of the set C_S of the canonical representatives of S.

Lemma 3.2.3. *Let S be a finitely generated semigroup and S' be a finitely generated subsemigroup of S. If the growth function of S is polynomially upper-bounded, then so will be the growth function of S'.*

Proof. Let $X = \{x_1, x_2, \ldots, x_h\}$ be a set of generators for S and Y be a finite set of generators for S'. Since the growth function g_S is polynomially upper-bounded then a positive integer q and a constant $c > 0$ exist such that for all $n > 0$

$$g_S(n) \leq cn^q.$$

Let us now express the elements of the set Y in terms of the elements of X. One has that there exists an integer p such that for any $y \in Y$ one has

$$y \in \bigcup_{0<j\leq p} X^j.$$

Since

$$g_{S'}(n) = \text{card}\{s \in S' \mid s = y_1 y_2 \cdots y_r \text{ with } y_i \in Y, i = 1, \ldots, r \text{ and } r \leq n\},$$

one has that for all $n > 0$,

$$g_{S'}(n) \leq g_S(np) \leq c(np)^q \leq cp^q n^q.$$

For any $s \in S$ the *order* of s is the order (cardinality) of the subsemigroup $\langle s \rangle$ generated by $\{s\}$. The order of s is finite if and only if there exist positive integers i and j, $i < j$, depending on the element s, such that

$$s^i = s^j. \tag{3.3}$$

Let j be the minimal integer for which the above relation is satisfied for some $i < j$. The integer i, which is unique, is called the *index* and $p = j - i$ the *period* of s. The order of $\langle s \rangle$ is then given by $i + p - 1$. In the case of a group, due to cancellativity, equation (3.3) simply becomes $s^p = 1$, where 1 denotes the identity of the group and the period p depends on the element s.

The following proposition is due to J. Justin [86].

Proposition 3.2.3. *Let S be a finitely generated semigroup such that there exists an integer $n > 0$ for which*

$$g_S(n) < \frac{n(n+3)}{2}.$$

If S is infinite, then S contains an element of infinite order.

Proof. Since S is infinite, then so will be the set C of its canonical representatives. Moreover, $g_S(n) = g_C(n)$ for all $n > 0$. By Proposition 3.2.1 there exists an irreducible infinite word $x \in A^\omega$ such that $F(x) \subseteq C$. This implies

$$g_x(n) \leq g_C(n) = g_S(n).$$

By hypothesis an integer n_0 exists for which $g_S(n_0) < n_0(n_0+3)/2$. Hence, $g_x(n_0) < n_0(n_0+3)/2$ so that by Corollary 1.6.1 the word x is ultimately periodic, $x = uv^\omega$. Let $\phi : A^+ \to S$ be the canonical epimorphism and let $\phi(v) = s \in S$. The order of s is infinite. Indeed, suppose, by contradiction, that there exist integers i and j such that $i < j$ and $s^i = s^j$. One would have

$$\phi(v^i) = \phi(v)^i = \phi(v)^j = \phi(v^j).$$

Hence, v^j is a reducible factor of x which is a contradiction.

Many papers have been devoted to the study of growth functions for semigroups and groups. In particular we recall the following important theorem due to M. Gromov [73]:

Theorem 3.2.1. *Let G be a finitely generated group. If the growth function of G is polynomially upper bounded, then G is nilpotent-by-finite, i.e., it has a (normal) nilpotent subgroup of finite index.*

In concluding this section we recall the following important proposition (see [[101], Lemma 3.4, p. 301]) that we shall use in the following, concerning the groups of the regular $\mathcal{D}$-classes of finitely generated semigroups.

Proposition 3.2.4. *Let S be a finitely generated semigroup and D a regular $\mathcal{D}$-class of S containing a finite number of $\mathcal{R}$-classes ($\mathcal{L}$-classes). Then the group G_D of D is finitely generated.*

3.3 The Burnside problem

A semigroup (group) S is periodic (or torsion) if any element $s \in S$ generates a subsemigroup $\langle s \rangle$ (subgroup) of finite order.

The problem of whether a finitely generated and periodic group is finite was posed by W. Burnside in 1902 and, subsequently, extended to the case of semigroups. However, the conditions of a finite generation and the periodicity are not sufficient to assure the finiteness of a semigroup or a group.

In the case of semigroups this has been shown in 1944 by Morse and Hedlund [120] as a consequence of the fact that on an alphabet with at least three distinct symbols there exist arbitrarily long words without repetitions of consecutive blocks (square-free words) (see Proposition 2.6.2). Indeed these authors, starting with a problem related to "chess games" (that is whether there could exist "unending games"), rediscovered a result shown by Thue in 1906 (see [153, 154]). By means of this result they were able to give an answer to a problem of semigroup theory posed by Dilworth showing the existence of a semigroup S generated by three elements such that $s^2 = 0$ for any element s of S, without S being nilpotent (a semigroup S with a zero element 0 is nilpotent if there exists a positive integer k such that $S^k = 0$). This result gave, obviously, a negative answer to the Burnside problem for semigroups.

Theorem 3.3.1. *(Morse and Hedlund) There exists an infinite 3-generated semigroup that satisfies the identity $x^2 = 0$ and an infinite 2-generated semigroup that satisfies the identity $x^3 = 0$.*

Proof. From Propositions 2.6.1 and 2.6.2 the infinite words t and m are, respectively, cube-free and square-free and $\text{card}(\text{alph}(t)) = 2$ and $\text{card}(\text{alph}(m)) = 3$, so that there exist infinitely many cube-free finite words in a two-letter alphabet and infinitely many square-free finite words in a three-letter alphabet. Let us consider the free semigroup A^+ on the alphabet $A = \{a, b, c\}$ and the semigroup $(A^+)^0$ obtained by adjoining to A^+ a zero element 0. We consider then in $(A^+)^0$ the congruence θ generated by the relation $\{(u^2, 0) \mid u \in A^+\}$ and the quotient semigroup $M = (A^+)^0/\theta$. One has that M is 3-generated and is infinite since any square-free word lies in a single congruence class. Moreover, M satisfies the identity $x^2 = 0$.

In a similar way one derives that the semigroup $T = (B^+)^0/\delta$, where $B = \{a, b\}$ and δ is the congruence in $(B^+)^0$ generated by the relation $\{(u^3, 0) \mid u \in A^+\}$, is a 3-generated semigroup satisfying the identity $x^3 = 0$.

In the case of groups a negative answer to the Burnside problem was given by E. S. Golod in 1964 [69]. This author, using a technique of proof discovered with I. R. Shafarevich based on the non-finiteness of a dimension of a suitable algebra associated with a field (see [94]), was able to show the existence of an infinite 3-generated p-group. Other examples of finitely generated and torsion groups which are infinite were given more recently by S. V. Aliochine and A. I. Olchanski (see [94]), R. I. Grigorchuk [72], and N. Gupta and S. Sidki [75].

If a finitely generated semigroup S is such that for any s of S, $s^m = s^n$, where m and n are fixed exponents such that $0 \leq m < n$, then the problem of the finiteness of S is called the ***bounded Burnside problem***. Also in this case there exist semigroups and groups which are infinite. A particular case which is very interesting and surprising is that of finitely generated and idempotent semigroups (that is, such that for any s of S, $s = s^2$). In this case one can show that S is finite. This result is a consequence of a theorem of J. A. Green and D. Rees (1952) [71] which relates the bounded Burnside problems for groups and for semigroups. More precisely let us denote by $S(r,n)$ ($G(r,n)$) the free semigroup (the free group) in the class of all semigroups (groups) having r generators, satisfying the identity $x = x^{n+1}$ ($x^n = 1$).

Theorem 3.3.2. *(Green and Rees) The following two conditions are equivalent:*

i. $S(r,n)$ *is finite for each positive integer* r.
ii. $G(r,n)$ *is finite for each positive integer* r.

Two more general versions of this theorem will be given later (see Corollary 3.6.5 and Theorem 3.7.2) as a consequence of some finiteness conditions for semigroups based on chain conditions and on iteration conditions, respectively.

In the case $n = 1$ the group $G(r,1)$ is trivial and thus the idempotent semigroup $S(r,1)$ is finite. We recall that it has been shown that $G(r,n)$ is finite for every positive integer r, for $n = 2,3,4$ and 6 (see [94]). The problem is still open for $n = 5$. Moreover, as a consequence of some works of P. S. Novikov and S. I. Adjan (1968) and Adjan (see [1]), it is known that if n is an *odd* integer ≥ 665, then $G(r,n)$ is infinite for any $r > 1$. Recently I. G. Lysionok [2] proved that $G(r,n)$ is infinite for any $r > 1$ and n *odd* integer ≥ 115. Moreover, S. V. Ivanov [82] proved that $G(r,n)$ is infinite for any $r > 1$ and n integer $\geq 2^{48}$.

The Burnside problem has a positive answer for some classes of semigroups and groups as, for instance, commutative semigroups and solvable groups. Moreover, we mention the classical results of I. Schur (1911) (see [144]) and I. Kaplansky (1965) (see [93]) which give a positive answer to Burnside's problem in the case of groups of matrices over the field of complex numbers and over an arbitrary field, respectively. As regards semigroups of matrices with elements in a field we recall the following result [113]:

Theorem 3.3.3. *(McNaughton and Zalcstein) A torsion semigroup of* $n \times n$ *matrices over a field is locally finite.*

In Sect. 3.6.4 we shall give a proof of the preceding theorem based on a suitable finiteness condition for semigroups which generalizes a theorem of Coudrain and Schützenberger. Other proofs of this theorem have been given by G. Jacob [83] and, in a more restricted form, by A. Mandel and I. Simon [110].

A further positive answer to the Burnside problem was obtained in the case of unitary rings satisfying a polynomial identity [152]. We recall that a unitary ring T satisfies a polynomial identity (PI-ring) if there exists a polynomial $p \in \mathbb{Z}\langle x_1, \ldots, x_r \rangle$, $p \neq 0$, in non-commutative indeterminates $x_1, \ldots, x_r$ and with integral coefficients such that

$$p(t_1, \ldots, t_r) = 0 \text{ for all } t_1, \ldots, t_r \in T.$$

Theorem 3.3.4. *(Straubing) A torsion subsemigroup of a unitary ring which satisfies a polynomial identity is locally finite.*

The proof, that will be given in Sect. 3.6.4, uses Theorem 3.3.3 and is based on Shirshov's theorem utilized in the case of rings satisfying a polynomial identity. A similar result in the case of groups was proved by C. Procesi [131].

From a classical result (Amitsur-Levitzki's theorem, see [132]) the algebra of all $n \times n$ matrices over a unitary commutative ring satisfies a polynomial identity, so that one derives from Theorem 3.3.4 the following extension of Theorem 3.3.3:

Corollary 3.3.1. *A torsion semigroup of $n \times n$ matrices over a unitary commutative ring is locally finite.*

3.4 Permutation property

A finitely generated, torsion, and commutative semigroup is obviously finite. A. Restivo and C. Reutenauer introduced in 1984 [137] a property of semigroups, called *permutation property*, which generalizes commutativity and is such that a finitely generated and torsion semigroup is finite if and only if it is permutable. Let us give the following:

Definition 3.4.1. *Let S be a semigroup and n be an integer > 1. A sequence $s_1, \ldots, s_n$ of n elements of S is called* permutable *if the product $s_1 \cdots s_n$ remains invariant under some nontrivial permutation of its factors, that is, there exists a permutation $\sigma \in S_n$, different from the identity, such that $s_1 s_2 \cdots s_n = s_{\sigma(1)} s_{\sigma(2)} \cdots s_{\sigma(n)}$.*

We say that a semigroup S is *n-permutable*, or satisfies the property $\mathcal{P}_n$, if any sequence of n elements of S is permutable. The property $\mathcal{P}_2$ is obviously equivalent to commutativity. We say that S is *permutable*, or satisfies the property $\mathcal{P}$, if there exists an integer $n > 1$ such that S is n-permutable.

Proposition 3.4.1. *Let S be a finitely generated semigroup. If S is permutable, then C_S is a bounded language and the growth function of S is polynomially upper-bounded.*

Proof. Suppose that S is n-permutable, $n \geq 2$. Let A be a finite alphabet and $\phi : A^+ \to S$ be the canonical epimorphism. The growth function of S coincides with the growth function of the set C_S of the canonical representatives of S. Let $w \in A^+$. Suppose that $w = xv_1v_2 \cdots v_ny$ and $v_1v_2 \cdots v_n$ is n-divided. Then there exists a nontrivial permutation $\sigma \in S_n$ such that $\phi(w) = \phi(w')$ and $w' = xv_{\sigma(1)}v_{\sigma(2)} \cdots v_{\sigma(n)}y$. Moreover, w' precedes w in the lexicographic ordering and so w is reducible. Therefore, the words of C_S do not contain n-divided factors, since they are irreducible. By Theorem 2.5.1, C_S is a bounded language and, by Proposition 2.5.1, the growth function of C_S, as well as that of S, is polynomially upper-bounded.

Theorem 3.4.1. *(Restivo and Reutenauer) Let S be a finitely generated and periodic semigroup. S is finite if and only if it is permutable.*

Proof. A first proof is based on the theorem of Shirshov (see Theorem 2.4.3). It is clear that if S is finite, then it is $(n+1)$-permutable with $n = \text{card}(S)$. Conversely, suppose that S is n-permutable and that $\phi : A^+ \to S$ is the canonical epimorphism. Since S is periodic and A is finite, with $\text{card}(A) = k$, one can take p such that $p \geq 2n$ and for each word w of length $< n$,

$$\phi(w)^p = \phi(w)^q,$$

for some $q < p$. Let us now prove by means of the theorem of Shirshov, that for any word w such that $|w| \geq N(k,p,n)$ there exists a word w' such that $w' <_a w$ and $\phi(w) = \phi(w')$. Indeed, from Corollary 2.4.3 one has either

$$w = \lambda u^p \mu, \quad \text{with } \lambda, \mu, u \in A^* \text{ and } 0 < |u| < n,$$

or

$$w = \lambda x_1 \cdots x_n \mu, \quad \text{with } \lambda, \mu, u \in A^*, x_i \in A^+ \; i = 1, \ldots, n$$

and

$$x_1 \cdots x_n > x_{\sigma(1)} \cdots x_{\sigma(n)}$$

for any nontrivial permutation σ of $\{1, \ldots, n\}$. In the first case since $\phi(u)^p = \phi(u)^q$ with $q < p$, if one sets $w' = \lambda u^q \mu$, then one has $w' <_a w$ and $\phi(w) = \phi(w')$.

In the second case since S is n-permutable, there exists a nontrivial permutation $\sigma \in S_n$ such that

$$\phi(x_1) \cdots \phi(x_n) = \phi(x_{\sigma(1)}) \cdots \phi(x_{\sigma(n)}).$$

Hence, setting $w' = \lambda x_{\sigma(1)} \cdots x_{\sigma(n)} \mu$, one has again $w' <_a w$ and $\phi(w) = \phi(w')$. By Proposition 3.2.1, the result follows.

The following different proof of the theorem uses Proposition 3.4.1. Indeed, since S is permutable, the set C_S of the canonical representatives of S is a bounded language, so that there exist words $u_1, \ldots, u_k$ such that

$$C_S \subseteq u_1^* \cdots u_k^*.$$

Any element $c \in C_S$ can be represented as

$$c = u_1^{r_1} \cdots u_k^{r_k}.$$

Thus

$$|c| = \sum_{i=1}^{k} r_i |u_i| \leq r \sum_{i=1}^{k} |u_i|,$$

with $r = \max\{r_i \mid i = 1, \ldots, k\}$. Let us set $s_i = \phi(u_i)$, $i = 1, \ldots, k$. Since S is periodic there exists an integer p such that for all $i = 1, \ldots, k$ one has $s_i^p = s_i^{q_i}$ with $q_i < p$. If S is infinite, then by Proposition 3.2.1, C_S is infinite too. Thus the length of a canonical representative c, as well as the integer r associated with it, becomes arbitrarily large. When $r > p$ we have that c is reducible which is a contradiction.

In the following we will prove more general theorems (see Theorem 3.4.2, Theorem 3.4.10) from which Theorem 3.4.1 is derived as a corollary. A further proof of Theorem 3.4.1 based on *well quasi-orders*, will be given in Chap. 6. Now we give a first generalization of Theorem 3.4.1. We consider a weak notion of permutability of a semigroup S, requiring that for some integer $n > 1$ not all the sequences of n elements are permutable but only the n-sequences of S.

Theorem 3.4.2. *Let S be a finitely generated and periodic semigroup. S is finite if and only if there exists an integer $n \geq 2$ such that any n-sequence of S is permutable.*

Proof. The proof is similar to that of Theorem 3.4.1, the only difference is in using Proposition 2.7.9, instead of Theorem 2.4.3. The "only if" part is trivial, so we prove the "if" part. Let $n \geq 2$ be an integer such that any n-sequence of S is permutable. Let $\phi : A^+ \to S$ be the canonical epimorphism and suppose by contradiction that S is infinite.

By Proposition 3.2.1 there exists an irreducible and uniformly recurrent word t relative to ϕ. Since S is periodic, t is ω-power-free; indeed, otherwise, by Lemma 2.6.2, t is periodic so that it contains a factor u^p, such that $\phi(u^p) = \phi(u)^p = \phi(u)^q = \phi(u^q)$, with $1 \leq q < p$, and this contradicts the irreducibility of t. By Proposition 2.7.9, t contains a factor x which is the n th term of a bi-ideal sequence whose canonical factorization is an n-division. We can write $x = w_1 w_2 \cdots w_n$, where $(w_1, w_2, \ldots, w_n)$ is the canonical factorization of x. Since $(\phi(w_1), \phi(w_2), \ldots, \phi(w_n))$ is an n-sequence of S, it is permutable; then for a nontrivial permutation $\sigma \in S_n$ one has

$$\phi(x) = \phi(w_1)\phi(w_2) \cdots \phi(w_n) = \phi(w_{\sigma(1)})\phi(w_{\sigma(2)}) \cdots \phi(w_{\sigma(n)}).$$

On the other hand one has $x > w_{\sigma(1)} w_{\sigma(2)} \cdots w_{\sigma(n)}$ and this contradicts the irreducibility of t.

Permutation property introduced for semigroups in connection with Burnside's problem, was extensively studied and many papers have been devoted to this subject (see [123]). In the case of groups we recall the following important characterization of finitely generated permutable groups [26]:

Theorem 3.4.3. *A finitely generated group G is permutable if and only if it is abelian-by-finite, i.e., G has an abelian (normal) subgroup of finite index.*

The proof of sufficiency of the preceding theorem is a straightforward consequence of the following:

Lemma 3.4.1. *Let H be an abelian subgroup of a group G such that the index of H in G is finite. Then G is n-permutable with $n = 2[G : H]$.*

Proof. Let $m = [G : H]$ be the index of H in G and $g_1, \ldots, g_n$ be any sequence of $n = 2m$ elements of G. We denote by g_0 the identity of G and consider the sequence of $n+1$ left cosets of H in G

$$g_0H = H,\ g_0g_1H,\ \ldots,\ g_0g_1 \cdots g_nH.$$

Since the number of elements of the above sequence is $2m + 1$, from the pigeon-hole principle there must exist three equal elements in the sequence, i.e., there exist integers i, j, k such that $0 \le i < j < k \le n$ and

$$g_0g_1 \cdots g_iH = g_0g_1 \cdots g_jH = g_0g_1 \cdots g_kH.$$

By the left invariance of the left-cosets partition, one derives then

$$g_{i+1} \cdots g_jH = g_{j+1} \cdots g_kH = H,$$

so that

$$g_{i+1} \cdots g_j,\ \ g_{j+1} \cdots g_k \in H.$$

If we set $u = g_{i+1} \cdots g_j$, $v = g_{j+1} \cdots g_k$, then, in view of the commutativity of H, one has $uv = vu$. Hence, one derives

$$g_1 \cdots g_n = g_1 \cdots g_iuvg_{k+1} \cdots g_n = g_1 \cdots g_ivug_{k+1} \cdots g_n,$$

that proves our assertion.

The theorem of Restivo and Reutenauer can then be easily derived in the case of groups as follows. Let G be a finitely generated and torsion group. If G is permutable, then from the preceding theorem, G has an abelian subgroup H of finite index. Since any subgroup of finite index of a finitely generated group is finitely generated (see [76], Chap. 7), one has that H is finitely generated, periodic, and commutative. Hence, H is finite. This implies that G is finite.

Remark 3.4.1. Another way of proving Theorem 3.4.1 in the case of groups is obtained by Gromov's theorem (see Theorem 3.2.1) and Proposition 3.4.1 (which is actually based on Shirshov's theorem). In fact, a finitely generated and permutable group G has, in view of Proposition 3.4.1, a growth function which is polynomially upper-bounded. By Theorem 3.2.1, G has a nilpotent subgroup H of a finite index. The subgroup H is finitely generated. If G is periodic, then so will be H. This implies that H is finite and then that G is finite.

Let us recall (see [76]) that if G is a group, then the ***derivate*** G' of G is the subgroup of G generated by all commutators $x^{-1}y^{-1}xy$ with $x, y \in G$.

The following theorem characterizes the structure of any permutable group [27].

Theorem 3.4.4. *A group G satisfies the permutation property if and only if it has a normal subgroup N of finite index such that its derivate group N' is finite.*

In the case of a ***division ring*** the following has been recently proven [28], as a consequence of a more general result.

Proposition 3.4.2. *Let D be a division ring. If the multiplicative group of D is permutable, then D is a field.*

This proposition is a nontrivial extension of the famous theorem of J. H. M. Wedderburn (see [76]) stating that ***a finite division ring is a field.***

Let K be a commutative ring with unit element 1. A K-algebra M is a K-module such that for any $k \in K$ and $m_1, m_2 \in M$, one has $k(m_1m_2) = (km_1)m_2 = m_1(km_2)$. We shall also make the hypothesis that for any $m \in M$ one has $1m = m$.

The K-algebra M satisfies a ***polynomial identity*** (PI-algebra) if there exists an alphabet $X = \{x_1, \ldots, x_q\}$ and a polynomial $p \in K\langle x_1, \ldots, x_q\rangle$, $p \neq 0$, in the non-commutative indeterminates $x_1, \ldots, x_q$ and coefficients in K, such that

$$p(m_1, \ldots, m_q) = 0$$

for any $m_1, \ldots, m_q \in M$. The degree of p is the maximal length of the words of p. An identity is ***admissible*** if there exists in p a word having a length equal to the degree of p and such that its coefficient is an invertible element of K. This condition is obviously verified when K is a field. Let us recall the following classical result on PI-algebras (compare [104, 131, 142]):

Proposition 3.4.3. *If a K-algebra satisfies an admissible polynomial identity of degree n, then it satisfies a multilinear identity of the form*

$$x_1 \cdots x_n = \sum_{\sigma \in S_n \setminus \{id\}} k_\sigma x_{\sigma(1)} \cdots x_{\sigma(n)},$$

for some $k_\sigma \in K$.

Let S be a semigroup. The *semigroup algebra* $K[S]$ *of* S *over* K, can be defined as the set of all formal sums

$$f = \sum_{s \in S} \alpha_s s,$$

with $\alpha_s \in K$ and such that the set $\{s \in S \mid \alpha_s \neq 0\}$ is finite. One can introduce in $K[S]$ the operation of sum (+), product (·), and multiplication by an element of K as follows. Let $\lambda \in K$ and $f, g \in K[S]$ be

$$f = \sum_{s \in S} \alpha_s s, \quad g = \sum_{s \in S} \beta_s s,$$

then $f + g$, fg, and λf are defined as

$$f + g = \sum_{s \in S} (\alpha_s + \beta_s) s, \quad fg = \sum_{s \in S} \Big(\sum_{uv = s} \alpha_u \beta_v \Big) s,$$

$$\lambda f = \sum_{s \in S} (\lambda \alpha_s) s.$$

$K[S]$ is then a K-algebra under the above operations.

The following lemma shows that if the semigroup algebra $K[S]$ satisfies an admissible polynomial identity, then the semigroup S has to be permutable.

Lemma 3.4.2. *If* $K[S]$ *satisfies an admissible polynomial identity of degree* n*, then* S *is* n*-permutable.*

Proof. From Proposition 3.4.3, $K[S]$ satisfies a multilinear identity of degree n. If we take as elements of $K[S]$ the elements of S, then one has for all $s_1, \ldots, s_n \in S$

$$s_1 \cdots s_n = \sum_{\sigma \in S_n \setminus \{id\}} k_\sigma s_{\sigma(1)} \cdots s_{\sigma(n)},$$

for some $k_\sigma \in K$. This equality in $K[S]$ implies that $t = s_1 \cdots s_n$ has to appear in the right hand side of the above equation with a coefficient $\neq 0$. Hence, there exists at least a permutation $\sigma \in S_n \setminus \{id\}$ such that

$$s_1 \cdots s_n = s_{\sigma(1)} \cdots s_{\sigma(n)}.$$

In their paper [137] Restivo and Reutenauer asked if the converse of the statement of the preceding lemma is true.

Problem 3.4.1. For each semigroup S which satisfies the permutation property, does $K[S]$ satisfy a polynomial identity?

If the multiplicative semigroup M of a K-algebra is n-permutable, then M satisfies a nontrivial polynomial identity, namely

$$\prod_{\sigma \in S_n \setminus \{id\}} (x_1 \cdots x_n - x_{\sigma(1)} \cdots x_{\sigma(n)}) = 0.$$

Thus an interesting problem is to analyze the relationship between the permutation property of S and the property that $K[S]$ satisfies a polynomial identity. This problem originated many investigations.

A negative answer to problem 3.4.1 was given by M. Garzon and Y. Zalcstein [66]. Indeed, they constructed a $\mathcal{P}_3$-group G such that $K[G]$, where K is a field, does not satisfy a polynomial identity. Their construction is based on the following:

Proposition 3.4.4. *The group algebra $K[G]$ of a $\mathcal{P}_3$-group G satisfies a polynomial identity if and only if G is abelian-by-finite or the field K has characteristic* 2.

The example of Garzon and Zalcstein is then obtained by constructing a suitable $\mathcal{P}_3$-group G which is not abelian-by-finite. From Theorem 3.4.3 it follows that G cannot be finitely generated.

A positive answer was obtained by the same authors [66] in the case of finitely generated groups when the field has characteristic 0.

Theorem 3.4.5. *Let K be a field of characteristic* 0. *A finitely generated group G satisfies the permutation property if and only if its group algebra $K[G]$ satisfies a polynomial identity.*

Some extensions of this result to some classes of finitely generated semigroups such as *periodic, cancellative, and inverse* and with K being any field, were also given (see [67, 124]). However, there exist examples of finitely generated, regular semigroups which are permutable and such that $K[S]$ does not satisfy a polynomial identity.

3.4.1 The weak permutability

The are many generalizations of the permutation property. Here we recall the *weak-permutation property.*

Definition 3.4.2. *Let S be a semigroup. For each $n > 1$ one can consider the following property $\mathcal{P'}_n$: for any sequence $s_1, s_2, \ldots, s_n$ of n elements of S there exist two permutations $\sigma, \tau \in S_n$, $\sigma \neq \tau$, such that $s_{\sigma(1)} s_{\sigma(2)} \cdots s_{\sigma(n)} = s_{\tau(1)} s_{\tau(2)} \cdots s_{\tau(n)}$.*

A semigroup S which satisfies $\mathcal{P'}_n$ is called *n-weakly permutable.* One says that S is *weakly permutable*, or satisfies $\mathcal{P'}$, if there exists an $n > 1$ such that S is n-weakly permutable. It is obvious that if a semigroup S is permutable, then it is weakly permutable. One can ask the question whether the converse is true. It has been proved by R. D. Blyth (1987) that in the case of groups the weak permutability is equivalent to permutability [10].

Theorem 3.4.6. *Let G be a group. G is permutable if and only if it is weakly permutable.*

The result of Blyth cannot be extended to the case of semigroups. In fact G. Pirillo [127] gave an example of a finitely generated semigroup which belongs to the class $\mathcal{P}'$ and not to the class $\mathcal{P}$. Moreover, one can ask the question whether in the Theorem 3.4.1 one can make the weaker hypothesis that the semigroup S belongs to the class $\mathcal{P}'$. Also in this case the answer is negative; in fact, as we shall prove in the following, a large class of counterexamples can be obtained by considering Rees quotient monoids of finitely generated free monoids by two-sided ideals, whose growth-functions are quadratically upper-bounded (see [37, 38, 42]). To illustrate this result we make the following considerations.

Let J be a two-sided ideal of A^* and set $L = A^* \setminus J$. The language L, by Lemma 1.6.1, is closed by factors so that any factor of a word of L is still an element of L. We can consider the Rees quotient monoid of A^* by the ideal J (see [101]). This monoid, usually denoted by A^*/J, is defined as the quotient A^*/ρ_J, where $\rho_J = \{id\} \cup (J \times J)$ is a congruence relation, called the Rees congruence relative to the ideal J, whose equivalence classes consist of the ideal J and singletons. Thus the monoid A^*/J coincides, up to an isomorphism, with the monoid $M(L)$ whose support is the set $L \cup \{0\}$, where 0 is a new element not in L and the product $\circ$ is defined as follows: for any $f_1, f_2 \in M(L)$,

$$f_1 \circ f_2 = \begin{cases} f_1 f_2 & \text{if } f_1,\ f_2,\ f_1 f_2 \in L \\ 0 & \text{otherwise.} \end{cases}$$

The monoid $M(L)$, which can be defined for any factorial language L, will be called *the factor monoid of* L. We observe that the growth functions of L and $M(L)$ coincide.

Theorem 3.4.7. *Let M be a Rees quotient monoid of A^* by a two-sided ideal J. If the growth function of M is quadratically upper-bounded (i.e., $g_M(n) \le cn^2$, for all $n > 0$), then M is weakly permutable.*

Proof. Let L be the factorial language $A^* \setminus J$ and $M(L)$ be the factor monoid of L. By hypothesis $g_L(n) \le cn^2$, for all $n > 0$. We prove that $M(L)$ is $5r$-weakly permutable for any $r > 3$, such that $r! > 12d + 1$, where $d = 22c$. Let $x_1, \ldots, x_{2r}, y_1, \ldots, y_r, z_1, \ldots, z_{2r}$ be $5r$ elements of $M(L)$. If at least one of these elements is 0, then the property $\mathcal{P}'_{5r}$ is, trivially, satisfied. Let us then consider the case when all these elements belong to L. We may always suppose that

$$|x_1| \ge \cdots \ge |x_{2r}| \ge |y_1| \ge \cdots \ge |y_r| \tag{3.4}$$

and

$$|z_i| \ge |x_1| \ \ (i = 1, \ldots, 2r). \tag{3.5}$$

We set $v = z_1 \cdots z_{2r}$ and $n = \sum_{i \in [1,2r]} |x_i|$. From equations (3.4) and (3.5) one has $|v| \ge n$. Let us consider the set

$$V = \{x_{\sigma(1)} \cdots x_{\sigma(2r)} \mid \sigma \in \mathcal{S}_{2r}\}.$$

Now either there exist two permutations $\sigma, \tau \in \mathcal{S}_{2r}$, $\sigma \neq \tau$, such that

$$x_{\sigma(1)} \cdots x_{\sigma(2r)} = x_{\tau(1)} \cdots x_{\tau(2r)}$$

or $\mathrm{card}(V) = (2r)!$. In the first case the property $\mathcal{P}'_{5r}$ is satisfied. Let us then suppose that $\mathrm{card}(V) = (2r)!$. Since $r > 3$, by Lemma 1.4.3, there exists $\sigma \in \mathcal{S}_{2r}$ such that $w_\sigma = x_{\sigma(1)} \cdots x_{\sigma(2r)}$ is not a sesquipower of order > 3. Let us then consider the set

$$T = \{w_\sigma y_{\tau(1)} \cdots y_{\tau(r)} v \mid \tau \in \mathcal{S}_r\}.$$

One has that

$$\mathrm{card}(T \cap L) \leq \mathrm{card}(L_{w_\sigma, v, m}),$$

where

$$L_{w_\sigma, v, m} = w_\sigma A^m v \cap L,$$

and $m = \sum_{i \in [1,r]} |y_i| \leq (1/2)n$. Let us first suppose that $\mathrm{card}(L_{w_\sigma, v, m}) \leq 12d$. Now either there exist two permutations $\eta, \tau \in \mathcal{S}_r$, $\eta \neq \tau$, such that

$$y_{\eta(1)} \cdots y_{\eta(r)} = y_{\tau(1)} \cdots y_{\tau(r)}$$

or $\mathrm{card}(T) = r! > 12d + 1$, so that there exist two permutations $\eta, \tau \in \mathcal{S}_r$, $\eta \neq \tau$, such that

$$w_\sigma y_{\eta(1)} \cdots y_{\eta(r)} v, \; w_\sigma y_{\tau(1)} \cdots y_{\tau(r)} v \notin L.$$

In both cases the property $\mathcal{P}'_{5r}$ is satisfied. Let us then suppose that $\mathrm{card}(L_{w_\sigma, v, m}) > 12d$. By Proposition 1.6.4, w_σ has a suffix which is a strong sesquipower of order $h \geq 6$, period $p < (1/8)n$, and is such that $hp > (5/8)n$. Let us suppose that this is the case and consider the set

$$Q = \{v y_{\tau(1)} \cdots y_{\tau(r)} w_\sigma \mid \tau \in \mathcal{S}_r\}.$$

One has that

$$\mathrm{card}(Q \cap L) \leq \mathrm{card}(L_{v, w_\sigma, m}),$$

where $L_{v, w_\sigma, m} = v A^m w_\sigma \cap L$. If $\mathrm{card}(L_{v, w_\sigma, m}) \leq 12d$ then, following an argument similar to the above one, the property $\mathcal{P}'_{5r}$ is satisfied. Let us then suppose that $\mathrm{card}(L_{v, w_\sigma, m}) > 12d$. By Proposition 1.6.4, w_σ has a prefix which is a strong sesquipower of order $h' \geq 6$, period $q < (1/8)n$, and is such that $h'q > (5/8)n$. Since

$$hp + h'q - (p + q) > n,$$

by Lemma 1.4.2 it follows that w_σ is a strong sesquipower of order ≥ 6 which is a contradiction.

A way of constructing a Rees quotient monoid of A^* is the following: let w be an infinite word over the alphabet A. We consider the factor monoid $M(F(w))$, where $F(w)$ is the set of the factors of w. We simply denote $M(F(w))$ by $M(w)$. The monoid $M(w)$ is finitely generated by the set alph(w) and is infinite because so is $F(w)$. Let us now suppose that the infinite word w is p-power-free, with $p > 1$, i.e., w does not contain factors like u^p with $u \neq \epsilon$. In this case the monoid $M(w)$ is periodic since for any $u \neq \epsilon$, $u^p = 0$.

We have seen in the previous chapter that the Fibonacci word f and the Thue–Morse words t and m in two and three symbols are respectively 4, 3, and 2 power-free. The following holds:

Proposition 3.4.5. *The Fibonacci word f and the Thue Morse words t and m have a growth function which is quadratically upper-bounded.*

Proof. The subword complexities of the words f, t, and m are linearly upper-bounded. In the case of f, since f is a Sturmian word, the subword complexity is given for all n by $n+1$. The subword complexities f_t and f_m have been exactly calculated in [14, 37] and [36], respectively. One can derive that for all $n \geq 2$

$$f_m(n) = f_t(n+1) \leq 10n/3.$$

From this one has that the growth functions of f, t, and m are quadratically upper-bounded (see Proposition 1.6.2).

From the preceding proposition one has that the growth functions of the monoids $M(f)$, $M(t)$, and $M(m)$ are quadratically upper-bounded. By Theorem 3.4.7 it follows that these monoids are weakly permutable. Moreover, since they are finitely generated, periodic, and infinite, by Theorem 3.4.1 one has that $M(f)$, $M(t)$, and $M(m)$ are not permutable. These results are of some interest for the Burnside problem for semigroups since they provide examples of ***finitely generated and torsion semigroups which are weakly permutable and infinite.***

We recall that Restivo proved first in [135] that $M(f)$ is 8-weakly permutable. A proof that $M(t)$ and $M(m)$ are 5-weakly permutable was given in [37]. Subsequently, these results were widely generalized by proving that ***$M(w)$ is weakly-permutable in the case of any infinite p-power-free word w whose subword complexity is linearly upper-bounded*** [42].

The above results were extended later on by F. Mignosi [115] to the case of any infinite word whose subword complexity is ***linearly upper-bounded.*** However, it should be remarked that in the case of an infinite word w the considered language $L = F(w)$ is ***right prolongable,*** whereas in the statement of Theorem 3.4.7 one supposes only that the language is ***factorial.*** Moreover, Theorem 3.4.7 provides a further generalization since one considers a quadratic upper bound to the growth-function; in fact, if the subword complexity of a factorial language is linearly upper-bounded, then the growth-function is quadratically upper-bounded. However, the converse of this statement is true only under suitable hypotheses on the language as, for instance, that it is prolongable in at least one direction (see Proposition 1.6.2).

Let us remark that as a consequence of a result of Mignosi [115] the quadratic upper bound in Theorem 3.4.7 is ***optimal*** in the sense that if $g : \mathbb{N} \to \mathbb{N}$ is a function such that

$$\limsup_{n\to\infty}(g(n)/n^2) = \infty,$$

then one can construct a Rees quotient monoid M which is not weakly permutable and such that $g_M = g$.

One can ask the question whether there exists a weakly permutable factorial language whose growth function is more than quadratic. A first positive answer was obtained by considering the factor monoid $M(L)$ of the language L of all overlap-free words in a two-letter alphabet. The subword complexity f_L of L satisfies the following inequality [98]:

$$c_1 n^{1.53} \leq f_L(n) \leq c_2 n^{1.57},$$

where c_1 and c_2 are two positive constants. Hence, the growth function of L, and then of $M(L)$, is more than quadratic. However, it has been proved in [117] that the monoid $M(L)$ is weakly permutable.

A further example is given by the factor monoid $M(St)$ of the language St of finite factors of all Sturmian words. From (1.7) one easily derives that the growth function of St is more than quadratic. However, it was proved in [92] that $M(St)$ is weakly permutable. In [39] the following was posed:

Conjecture 3.4.1. Let S be a finitely generated semigroup such that its growth function is quadratically upper-bounded. Then S is weakly permutable.

Theorem 3.4.7 shows that the conjecture has a positive answer in the case of semigroups which are Rees quotient monoids of A^*. However, this is a very special class of semigroups. In the case of groups the following result holds [108]:

Theorem 3.4.8. *If the growth function of a group G is upper-bounded by a polynomial of degree ≤ 3, then G is permutable.*

From this it follows that the conjecture is true for groups. By using the preceding theorem one can also derive that the conjecture is true for completely 0-simple semigroups. Let us recall (see [22]) that a semigroup S is ***completely simple (0-simple)*** if S is simple (0-simple) and contains a ***primitive idempotent*** that is an idempotent which is minimal (minimal among the nonzero idempotents) in the set $E(S)$ of the idempotents, with respect to the partial order $\leq$ defined in $E(S)$ as: for $e, f \in E(S)$

$$e \leq f \text{ if and only if } ef = fe = e.$$

Proposition 3.4.6. *A completely* 0*-simple semigroup whose growth function is upper-bounded by a polynomial of degree* ≤ 3 *is permutable.*

Proof. If S is a completely 0-simple semigroup, then $S \setminus \{0\}$ is a regular $\mathcal{D}$-class D of S. Moreover, for any $s \in D$ the $\mathcal{R}$-class R_s ($\mathcal{L}$-class L_s) is such that $R_s \cup \{0\}$ ($L_s \cup \{0\}$) is a 0-minimal right (left) ideal (see [22]). Since S is finitely generated it follows that S has finitely many $\mathcal{R}$-classes and $\mathcal{L}$-classes and then $\mathcal{H}$-classes, so that S is a finite union of $\mathcal{H}$-classes. From Proposition 3.2.4 any $\mathcal{H}$-class which is a group is finitely generated. By Theorem 3.1.4 all these groups are isomorphic to the same group G. From the proof of Lemma 3.2.3 one derives that the growth function of G is upper-bounded by a polynomial of degree ≤ 3, so that by Theorem 3.4.8, G is permutable. An $\mathcal{H}$-class H of S which is not a group is such that $H^2 = \{0\}$, so that for any two elements $h_1, h_2 \in H$ one has $h_1 h_2 = h_2 h_1 = 0$. By using a simple argument based on Ramsey's theorem one derives that S is permutable. Indeed, let us suppose that G is k-permutable and that the number of $\mathcal{H}$-classes is p. We set $m = R(2, k+1, p) - 1$ where R is the function of Ramsey's theorem and consider any sequence $s_1, \ldots, s_m$ of m elements of S. Let H_s ($s = 1, \ldots, p$) denote the $\mathcal{H}$-classes and define the sets

$$B_s = \{\{i, j\} \mid 1 \leq i < j \leq m + 1 \text{ and } s_i \cdots s_{j-1} \in H_s\},$$

with $s = 1, \ldots, p$. By Ramsey's theorem there exist integers $i_1, i_2, \ldots, i_{k+1}$ such that $1 \leq i_1 < i_2 < \cdots < i_{k+1} \leq m + 1$ and all the pairs $\{i_h, i_{h'}\}$ such that $1 \leq h < h' \leq k + 1$ are in the same class B. We can then decompose $s_1 \cdots s_m$ as

$$s_1 \cdots s_m = \lambda t_1 \cdots t_k \mu,$$

where

$$t_1 = s_{i_1} \cdots s_{i_2 - 1},$$

$$\vdots$$

$$t_k = s_{i_k} \cdots s_{i_{k+1} - 1}$$

with $\lambda = s_1 \cdots s_{i_1 - 1}$ and $\mu = s_{i_{k+1}} \cdots s_n$. Now all the elements t_i, $i = 1, \ldots, k$ belong to the same $\mathcal{H}$-class H which is a group isomorphic to G, because $t_1 t_2 \in H$ (see Theorem 3.1.2). Since G is k-permutable there exists a permutation $\sigma \in S_k \setminus \{id\}$ such that

$$t_1 \cdots t_k = t_{\sigma(1)} \cdots t_{\sigma(k)}.$$

From this one has that the product $s_1 \cdots s_k$ can be rewritten in a nontrivial way. This implies that S is m-permutable.

3.4.2 The ω-permutability

In this section we shall introduce the following concept of *ω-permutable* (*ω-weakly permutable*) semigroup which is, as we shall see, more general than that of permutable (weakly permutable).

Definition 3.4.3. *A semigroup S is called* ω-permutable (ω-weakly permutable) *if for any infinite sequence $s_1, s_2, \ldots, s_k, \ldots$ of elements of S there exist an integer $n > 1$ and a permutation $\sigma \in S_n$, $\sigma \neq id$ (two permutations $\sigma, \tau \in S_n$ with $\sigma \neq \tau$) such that $s_1 s_2 \cdots s_n = s_{\sigma(1)} s_{\sigma(2)} \cdots s_{\sigma(n)}$, ($s_{\sigma(1)} s_{\sigma(2)} \cdots s_{\sigma(n)} = s_{\tau(1)} s_{\tau(2)} \cdots s_{\tau(n)}$).*

One easily verifies that if a semigroup is permutable (weakly permutable), then it is also ω-permutable (ω-weakly permutable). We shall denote by $\mathcal{P}_\omega$ (by $\mathcal{P}'_\omega$) the class of all ω-permutable (weakly ω-permutable semigroups). The ω-permutation property, as well as the weakly ω-permutation property, has been introduced in the case of groups by Blyth and Rhemtulla [11]. The main result of these authors is the following:

Theorem 3.4.9. *A group G is ω-permutable if and only if it is permutable.*

Moreover, Blyth and Rhemtulla proved also that if a group is ω-weakly permutable, then in general it is not weakly permutable. In the case of semigroups the situation is quite different since an ω-permutable semigroup can be non-permutable.

Example 3.4.1. Let $A_\omega = \{a_1, a_2, \ldots, a_n, \ldots\}$ be a denumerable infinite alphabet. For any $w = a_{i_1} \cdots a_{i_r} \in A^*$ we set $\inf(w) = \min\{i_1, \ldots, i_r\}$. Let us define the language $L = \{w \in A_\omega^* \mid |w| \leq \inf(w)\}$. The language L is closed by factors, so that $J = A_\omega^* \setminus L$ is a two-sided ideal. We consider then the factor monoid $M(L)$ of L, that is, the Rees quotient monoid of A_ω^* by the ideal J. One easily verifies that $M(L)$ is not permutable but it is ω-permutable. Indeed, for any infinite sequence $f_1, f_2, \ldots, f_n, \ldots$ an integer n_0 exists, depending on the sequence, for which $f_1 f_2 \cdots f_{n_0} = 0$ and this, obviously, implies that the preceding elements are permutable. If we consider the sequence $a_{M+1}, \ldots, a_{M+n}, \ldots$, with M a sufficiently large positive integer, then the integer n_0 becomes arbitrarily large and, moreover, the sequence $a_{M+1}, \ldots, a_{M+n}$ is not permutable for $n < n_0$.

The following holds [41]:

Theorem 3.4.10. *Let S be a finitely generated and periodic semigroup. S is finite if and only if S is ω-permutable.*

Proof. The "only if" part of the theorem is trivial. Let us then prove the "if" part. Let S be a finitely generated semigroup and $\phi : A^+ \to S$ be the canonical epimorphism. If S is infinite, then by Proposition 3.2.1 there exists an infinite irreducible word x relative to ϕ. From Corollary 2.4.2 there exists a word $s \in A^\omega$ such that $F(s) \subseteq F(x)$ and s is ultimately periodic or ω-divided. The word s is obviously irreducible. Let us first suppose that s is ultimately periodic. In this case there exist words $u \in A^*$, $v \in A^+$ such that $s = uv^\omega$. Let $\phi(v) = \xi$. Since S is periodic, there exist i and j, $i \neq j$, such that $\xi^i = \xi^j$. Thus uv^i and uv^j would have the same image by ϕ which is a contradiction with the irreducibility of s. Let us now suppose that s is ω-divided, i.e., s can be factorized as $s = u_1 \cdots u_n \cdots$ with $u_i \in A^+$, $i > 0$, and for any $k > 0$ and $\sigma \in \mathcal{S}_k$, $\sigma \neq id$,

$$u_1 \cdots u_k > u_{\sigma(1)} \cdots u_{\sigma(k)}.$$

Let us set for any $n > 0$, $t_n = \phi(u_n)$ and consider the infinite sequence $t_1, \ldots, t_n, \ldots$. Since S is ω-permutable there exist an $n > 1$ and a permutation $\sigma \in \mathcal{S}_n$, $\sigma \neq id$, such that

$$t_1 t_2 \cdots t_n = t_{\sigma(1)} t_{\sigma(2)} \cdots t_{\sigma(n)}.$$

Now, by construction, the sequence $u_1 \cdots u_n$ is the canonical representative of $t_1 t_2 \cdots t_n$, but $u_1 \cdots u_n > u_{\sigma(1)} \cdots u_{\sigma(n)}$, which is again in contradiction with the irreducibility of s.

3.5 Partial commutations

In this section we will be concerned with square-free elements of partially commutative free monoids. Roughly speaking, partially commutative free monoids are obtained by considering words in which only some pairs of letters can commute. Partially commutative free monoids have been considered first by P. Cartier and D. Foata [21] for problems related to rearrangements of words. Moreover, it has been shown that these objects are of some interest in the study of *concurrency processes* and *parallel computation* (see [30] and references therein).

Let A be a finite alphabet and denote by θ a reflexive and symmetric relation on A. We say that the letters $a, b \in A$ *commute* if $(a, b) \in \theta$. The relation θ is also called *commutation relation.*

We denote by $\sim_\theta$ the congruence relation in A^* generated by the set

$$\{(ab, ba) \mid (a, b) \in \theta\}.$$

In other terms for any $u, v \in A^*$, one has $u \sim_\theta v$ if and only if $u = v$ or there exists an integer $h > 0$ and a sequence of words $w_0, w_1, \ldots, w_h$ such that $u = w_0$, $v = w_h$, and for $i = 1, \ldots, h$ one has

$$w_{i-1} = f_i ab g_i, \; w_i = f_i ba g_i, \quad f_i, g_i \in A^*$$

with $(a, b) \in \theta$.

The quotient monoid $M(A,\theta) = A^*/\sim_\theta$ is also called the ***partially commutative free monoid*** over A relative to the commutation relation θ. An element of $M(A,\theta)$ is also called ***trace.*** If $\theta = \emptyset$, then $M(A,\emptyset)$ is equal to the ***free monoid*** A^*; if $\theta = A \times A$, then $M(A, A \times A)$ is equal to the ***free commutative monoid.***

A trace $m \in M(A,\theta)$ is called ***square-free*** if $m \neq rs^2t$ with $r,t \in M(A,\theta)$ and $s \not\sim_\theta \epsilon$; in other terms a square-free element of $M(A,\theta)$ is a congruence class of A^* mod $\sim_\theta$ which contains only square-free words of A^*. We shall denote by $L_2(M(A,\theta))$ the set of all square-free traces of $M(A,\theta)$.

Example 3.5.1. Let $A = \{a,b,c\}$ and the commutation relation θ be such that a commutes with b and c, whereas b does not commute with c. The congruence class containing the word $abcb$ is a square-free trace since the words in the class are

$$abcb, bacb, bcab, bcba$$

and these are all square-free. On the contrary, the congruence class of the word $abca$ is a trace which is not square-free. Indeed, it contains the word ba^2c which is not square-free.

The following theorem [20], whose proof we omit, gives an effective procedure to decide whether a given partially commutative free monoid contains infinitely many square-free elements.

Theorem 3.5.1. *Let $M(A,\theta)$ be a partially commutative free monoid. The set $L_2(M(A,\theta))$ of square-free traces is infinite if and only if at least one of the following conditions is satisfied:*

i. There exist three distinct letters $a,b,c \in A$ such that

$$(a,b),(b,c),(c,a) \notin \theta.$$

ii. There exist four distinct letters $a,b,c,d \in A$ such that

$$(a,b),(a,c),(a,d) \notin \theta.$$

iii. There exist four distinct letters $a,b,c,d \in A$ such that

$$(a,b),(b,d),(d,c),(c,a) \notin \theta.$$

iv. There exist five distinct letters $a,b,c,d,e \in A$ such that

$$(a,b),(b,c),(c,d),(d,e) \notin \theta.$$

Given the alphabet A and the commutation relation θ one can consider the graph $G = (A, E_\theta)$ where each node represents a letter and an edge joins two nodes if and only if the corresponding letters commute. However, for our purposes it is more convenient to refer ourselves to the complementary graph $G^c = (A, E_{\theta^c})$ where θ^c is the complementary relation of θ called the *non-commutation relation.* In terms of the non-commutation relation the conditions *i–iv* of Theorem 3.5.1 correspond to the existence in the graph G^c of at least one of the following subgraphs:

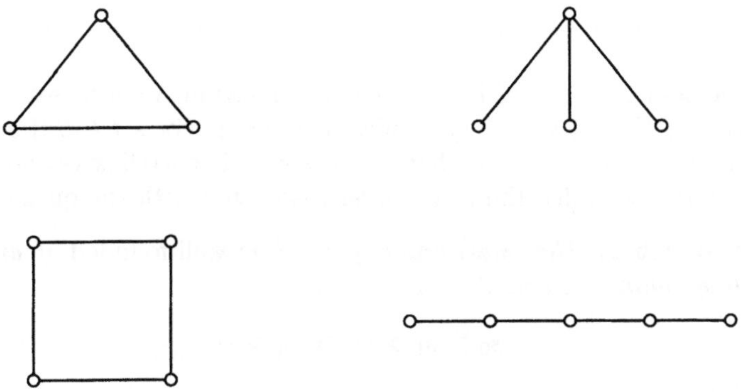

Fig. 1

Moreover, Theorem 3.5.1 can be restated as follows: $L_2(M(A,\theta))$ *is infinite if and only if in the non-commutation graph* G^c *there exists at least one node of degree* > 2 *or one node of degree* 2 *having two adjacent nodes of degree* 2. Hence, to decide whether $L_2(M(A,\theta))$ is infinite is reduced to a simple property of the graph G^c.

Let us give now an application of Theorem 3.5.1 to the Burnside problem. Let $[M(A,\theta)]^0$ be the monoid obtained by adjoining a zero element 0 to $M(A,\theta)$ and by δ the congruence in $[M(A,\theta)]^0$ generated by the relation

$$\{(u^2,0) \mid u \in M(A,\theta)\}.$$

Let $M'(A,\theta)$ be the quotient $[M(A,\theta)]^0/\delta$. This monoid is finitely generated and torsion. Since

$$M'(A,\theta) = L_2(M(A,\theta)) \cup \{0\},$$

one derives:

Proposition 3.5.1. *The finitely generated and torsion monoid* $M'(A,\theta)$ *is infinite if and only if at least one of the conditions i–iv of Theorem 3.5.1 is satisfied.*

In the case $\theta = \emptyset$, $L_2(M(A,\emptyset))$ is the set of square-free words in the free monoid A^*. As proved by Morse and Hedlund (see Theorem 3.3.1) $M'(A,\emptyset)$ is infinite if and only if $\mathrm{card}(A) \geq 3$.

3.6 Chain conditions

A binary relation $\leq$ on a set S is a *quasi-order* if it is reflexive and transitive. If for all $s,t \in S$, $s \leq t \leq s$ implies $s = t$, then $\leq$ is a *partial order.* If $s \leq t$ implies $t \leq s$, then $\leq$ is an equivalence relation. The inverse relation $\leq^{-1}$

of $\leq$ is a quasi-order in S usually denoted by $\geq$. The meet $\leq \cap \leq^{-1}$ is an equivalence relation $\sim$ and the quotient of S by $\sim$ is a *poset* (partially ordered set). It is clear that any quasi-order induces a partial order over the quotient set $S/\sim$.

An element $s \in X \subseteq S$ is *minimal (maximal)* in X with respect to $\leq$ if, for every $x \in X$, $x \leq s$ ($s \leq x$) implies that $x \sim s$. For $s, t \in S$ if $s \leq t$ and s is not equivalent to t mod $\sim$, then we set $s < t$. The irreflexive and transitive relation $<$ is also called the *strict order* associated with the quasi-order $\leq$.

Definition 3.6.1. *The quasi-order $\leq$ in S is* well-founded *if any strictly descending chain of elements of S*

$$s_0 > s_1 > \cdots > s_n > \cdots$$

has a finite length.

Now let us consider the following relations in a semigroup S defined as: for $s, t \in S$ we set

$$s \leq_{\mathcal{L}} t \Longleftrightarrow S^1 s \subseteq S^1 t,$$
$$s \leq_{\mathcal{R}} t \Longleftrightarrow sS^1 \subseteq tS^1,$$
$$s \leq_{\mathcal{J}} t \Longleftrightarrow S^1 sS^1 \subseteq S^1 tS^1.$$

One can easily see that $\leq_{\mathcal{L}}$, $\leq_{\mathcal{R}}$, and $\leq_{\mathcal{J}}$ are quasi-order relations of S. The equivalence relations $\sim_{\mathcal{L}}$, $\sim_{\mathcal{R}}$, and $\sim_{\mathcal{J}}$ are the Green relations $\mathcal{L}$, $\mathcal{R}$, and $\mathcal{J}$, respectively (see Sect. 3.1). One can then partially order the $\mathcal{L}$-classes ($\mathcal{R}$-classes, $\mathcal{J}$-classes) by setting for $s, t \in S$

$$L_s \leq L_t \Longleftrightarrow s \leq_{\mathcal{L}} t,$$

$$R_s \leq R_t \Longleftrightarrow s \leq_{\mathcal{R}} t,$$

$$J_s \leq J_t \Longleftrightarrow s \leq_{\mathcal{J}} t.$$

Definition 3.6.2. *A semigroup satisfies the* minimal condition on principal right (left, two-sided) *ideals if the quasi-order $\leq_{\mathcal{R}}$ ($\leq_{\mathcal{L}}$, $\leq_{\mathcal{J}}$) is well-founded. We denote by* $\min_R$ ($\min_L, \min_J$) *this minimal condition.*

Equivalently S satisfies $\min_R$ ($\min_L, \min_J$) if and only if any nonempty subset of $\mathcal{R}$-classes ($\mathcal{L}$-classes, $\mathcal{J}$-classes) contains a minimal element (see [22]).

The following lemma (*chain lemma*) gives an interesting sufficient condition under which a finitely generated semigroup satisfies $\min_R$ (see [34]).

Lemma 3.6.1. *Let S be a finitely generated semigroup satisfying the following condition: there exists an integer $m > 0$ such that for any sequence $s_1, s_2, \ldots, s_m$ of m elements of S there exist integers i, j, $1 \leq i, j \leq m$, such that*

$$s_1 \cdots s_j \in s_1 \cdots s_j s_i S^1.$$

Then S satisfies $\min_R$.

Proof. Let X be a finite set of generators for S and let $k = \text{card}(X)$. We show that any strictly descending chain of principal right ideals of S has a length at most equal to mk (that is, it contains at most $n = 1 + mk$ elements). In fact, let $s_0, s_1, \ldots, s_n$ be $n+1$ elements of S such that

$$s_0S^1 \supset s_1S^1 \supset \cdots \supset s_nS^1.$$

For any integer i, $1 \leq i \leq n$, there exists then $t_i \in S^1, t_i \neq 1$, such that

$$s_i = s_{i-1}t_i. \tag{3.6}$$

Let us express each t_i, $1 \leq i \leq n$ in terms of the elements of the generating set X. We can then write

$$t_i = u_i x_i v_i, \tag{3.7}$$

where $u_i, v_i \in S^1$ and x_i is the left-most generator in the preceding factorization such that

$$s_{i-1}u_iS^1 \supset s_{i-1}u_ix_iS^1.$$

Hence, one has

$$s_{i-1}S^1 = s_{i-1}u_iS^1 \supset s_{i-1}u_ix_iS^1 \supseteq s_{i-1}u_ix_iv_iS^1 = s_iS^1. \tag{3.8}$$

Since $n = 1 + mk$, by the pigeon-hole principle, it follows that there exist an $x \in X$ and integers $i_1, i_2, \ldots, i_{m+1}$ such that

$$1 \leq i_1 < i_2 < \cdots < i_{m+1} \leq n \quad \text{and} \quad x_{i_1} = x_{i_2} = \cdots = x_{i_{m+1}} = x.$$

Let i, j be any two integers such that $0 \leq i < j \leq n$. From (3.6) one gets

$$s_j = s_{j-1}t_j = \cdots = s_it_{i+1}\cdots t_j.$$

Hence, by equations (3.7) and (3.8), one derives that

$$s_iu_{i+1}S^1 \supset s_iu_{i+1}x_{i+1}S^1$$

and for $j > i+1$,

$$s_it_{i+1}\cdots t_{j-1}u_jS^1 \supset s_it_{i+1}\cdots t_{j-1}u_jx_jS^1.$$

Thus if we set

$$w_0 = s_{i_1-1}u_{i_1}$$

and for $1 \leq r \leq m$,

$$w_r = v_{i_r}t_{i_r+1}\cdots t_{i_{r+1}-1}u_{i_{r+1}},$$

then one obtains for all p, $1 \leq p \leq m$

$$w_0xw_1x\cdots xw_pS^1 \supset w_0xw_1x\cdots xw_pxS^1. \tag{3.9}$$

Let us then consider the sequence $xw_1, xw_2, \ldots, xw_m$. By the hypothesis made on S there exist i, j, $1 \le i, j \le m$, and $t \in S^1$ for which

$$xw_1\cdots xw_j = xw_1\cdots xw_jxw_it.$$

Thus

$$xw_1\cdots xw_jS^1 = xw_1\cdots xw_jxw_itS^1 \subseteq xw_1\cdots xw_jxS^1.$$

Multiplying on the left both the sides of the above equation by w_0, we reach a contradiction with (3.9) in the case $p = j$.

Definition 3.6.3. *An element s of a semigroup S is called* right stable (left stable) *if for any $t \in J_s$, $tS^1 \subseteq sS^1$ ($S^1t \subseteq S^1s$) implies $s \mathcal{R} t$ ($s \mathcal{L} t$). An element of S is called* stable *if it is both right and left stable. A subset $X \subseteq S$ is called* right stable (left stable, stable) *if any element of X is right stable (left stable, stable).*

The following lemma is well known (see [22]).

Lemma 3.6.2. *Let S be a periodic semigroup, then S is stable.*

Proof. Let $s, t \in S$ be such that $s \mathcal{J} t$. Suppose that $sS^1 \subseteq tS^1$. This implies $s = tx$, $x \in S^1$. Since $s \mathcal{J} t$, there exist $\lambda, \mu \in S^1$ such that

$$t = \lambda s\mu = \lambda tx\mu = \lambda^n t(x\mu)^n, \ n \ge 0.$$

Since S is periodic, there exist integers i, j such that $i < j$ and $\lambda^i = \lambda^j$. Thus, setting $p = j - i$, one derives

$$t = \lambda^{i+p}t(x\mu)^{i+p} = \lambda^i t(x\mu)^{i+p} = \lambda^i t(x\mu)^i(x\mu)^p = tx\mu(x\mu)^{p-1} = s\mu(x\mu)^{p-1}.$$

Hence, $s \mathcal{R} t$ so that S is right stable. In a symmetric way one proves that S is left stable.

Definition 3.6.4. *A semigroup S satisfies the condition* $\min_R^*$ ($\min_L^*$) *if for each $\mathcal{J}$-class the set of all $\mathcal{R}$-classes ($\mathcal{L}$-classes) contained in the $\mathcal{J}$-class has a minimal element.*

It is clear that if the semigroup S satisfies $\min_R$ ($\min_L$), then it satisfies $\min_R^*$ ($\min_L^*$), whereas the converse is not generally true. The following proposition (see [22]) relates the condition $\min_R^*$ ($\min_L^*$) with the right (left) stability.

Proposition 3.6.1. *A semigroup S is right (left) stable if and only if S satisfies* $\min_R^*$ ($\min_L^*$).

Proof. We give a proof in the "right" case. A symmetric proof can be done in the "left" case. If S is a right stable semigroup, then trivially it satisfies $\min_R^*$. Conversely, suppose that S satisfies $\min_R^*$ and let $a, b \in S$ be such that

$$a \mathcal{J} b \text{ and } R_a \leq R_b.$$

This latter condition implies that there exists $x \in S^1$ such that $a = bx$. Moreover, by hypothesis there exists an $\mathcal{R}$-class, say R_c, which is minimal in the set of all $\mathcal{R}$-classes contained in $J_a = J_b = J_c$. Since $b \mathcal{J} c$, there exist $y, z \in S^1$ such that $b = ycz$; hence, one has

$$a = yczx.$$

Let us set $d = czx$. One has $dS^1 \subseteq cS^1$ and then $R_d \leq R_c$ and $J_d \leq J_c$. We prove that $R_d = R_c$. If $J_c = J_d$, then the result is obvious from the minimality of R_c. Let us then suppose $J_d < J_c$. In this case

$$J_a = J_{yd} \leq J_d < J_c,$$

which is a contradiction since $J_a = J_c$. Hence, one has $R_d = R_c$. We can then write

$$c = dw = czxw, \; w \in S^1,$$

and

$$b = ycz = yczxwz = awz.$$

Thus $bS^1 \subseteq aS^1$ and, therefore, $R_b = R_a$.

3.6.1 The J-depth decomposition theorem

In this section we prove an important theorem of structure, called $\mathcal{J}$-depth decomposition theorem, for finitely generated semigroups which are right stable and whose subgroups are locally finite. This theorem, as we shall see in the following sections, has several applications to finiteness conditions for finitely generated semigroups whose subgroups are locally finite.

Definition 3.6.5. *Let s be an element of a semigroup S. The* $\mathcal{J}$-depth *of s is the length of the longest strictly ascending chain of two-sided principal ideals starting with s. The $\mathcal{J}$-depth of s can be infinite. A semigroup S admits a* $\mathcal{J}$-depth function $d_{\mathcal{J}}$ *if for every $s \in S$ the $\mathcal{J}$-depth $d_{\mathcal{J}}(s)$ of s is finite.*

We remark that the existence of the $\mathcal{J}$-depth function is stronger than the ascending chain condition on the order $\leq_{\mathcal{J}}$ and weaker than the ascending chain condition on the ideals.

Example 3.6.1. Let S be the semigroup $S = \{x_n \mid n \in \mathbb{N}\} \cup \{0\}$ where $x_i x_j = 0$ for all $i, j \in \mathbb{N}$. For each $n \geq 0$ the set $I_n = \{x_1, x_2, \ldots, x_n\} \cup \{0\}$ is a two-sided ideal of S since $I_n S = S I_n = \{0\}$. One has

$$I_1 \subset I_2 \subset \cdots \subset I_n \subset \cdots,$$

so that the ascending chain condition is not satisfied. However, S has a $\mathcal{J}$-depth function. Indeed, for any $s \in S$, $S^1 s S^1 = \{0\}$ that implies $d_{\mathcal{J}}(s) = 0$.

For $s, t \in S$, if $J_s < J_t$, then one says that the $\mathcal{J}$-class J_t is *above* J_s.

Definition 3.6.6. *A semigroup S is* weakly finite $\mathcal{J}$-above *if each $\mathcal{J}$-class of S has only finitely many $\mathcal{J}$-classes above it.*

One easily verifies that a semigroup S is weakly finite $\mathcal{J}$-above if and only if for any $s \in S$ the factors of s can lie only in a finite number of $\mathcal{J}$-classes. It is clear from the definition that if S is weakly finite $\mathcal{J}$-above, then it has a $\mathcal{J}$-depth function.

Definition 3.6.7. *A semigroup S is* finite $\mathcal{J}$-above *if for each $s \in S$ the set $\{t \in S \mid t \geq_{\mathcal{J}} s\}$ is finite, i.e., S is weakly finite $\mathcal{J}$-above and every $\mathcal{J}$-class is finite.*

Let S be a semigroup. We define inductively a sequence $(K_n)_{n \geq 0}$ of sets as follows: $K_0 = \emptyset$ and, for all $n > 0$,

$$K_n = \bigcup_{j=1,\ldots,n} C_j,$$

where for $j > 0, C_j$ is the set of the elements of $S \setminus K_{j-1}$ which are maximal with respect to $\leq_{\mathcal{J}}$ in $S \setminus K_{j-1}$. Moreover, we set $K_S = \cup_{j>0} K_j$.

Lemma 3.6.3. *Let S be a semigroup. For all $j > 0$, K_j is closed by factors and is a union of $\mathcal{J}$-classes.*

Proof. The proof is by induction (i.e., by supposing that K_{j-1} is closed by factors). Let $w \in K_j$ and m be a factor of w. We write $w = \lambda m \mu$, with $\lambda, \mu \in S^1$. We prove that $m \in K_j$. If $w \in K_{j-1}$, then the conclusion would be trivially true by the induction hypothesis. Let us then suppose that $C_j \neq \emptyset$ and that $w \in C_j$. If $S^1 w S^1 \subset S^1 m S^1$, then by the maximality of w in $S \setminus K_{j-1}$ one derives $m \in K_{j-1}$ and, therefore, $m \in K_j$ (since $K_{j-1} \subseteq K_j$). Let us then suppose that $S^1 w S^1 = S^1 m S^1$, that is, m and w are in the same $\mathcal{J}$-class.

If, by contradiction, $m \notin K_j$, then there exists $n \in S \setminus K_{j-1}$ such that $n >_{\mathcal{J}} m$, i.e., $S^1 n S^1 \supset S^1 m S^1 = S^1 w S^1$, so that we have that $n >_{\mathcal{J}} w$ which is in contrast with the maximality of w in $S \setminus K_{j-1}$. The proof of the base of the induction is trivial.

K_j is a union of $\mathcal{J}$-classes. Let $s \in K_j$ and $t \ \mathcal{J} \ s$, i.e., $S^1 t S^1 = S^1 s S^1$. This implies $s = xty$ for suitable $x, y \in S^1$. Since K_j is closed by factors it follows that $t \in K_j$.

Definition 3.6.8. *A semigroup S has a* weak $\mathcal{J}$-depth decomposition *if for all $j > 0$ the sets K_j are finite. Moreover, if S is infinite then K_S has to be infinite. A semigroup S has a* $\mathcal{J}$-depth decomposition *if it has a weak $\mathcal{J}$-depth decomposition and $S = K_S$.*

We remark that if S is finite, then $S = K_S$. Indeed, since $K_n \subseteq K_{n+1}$ for all $n \geq 0$, there must exist a positive integer h such that $K_n = K_h$ for all $n \geq h$. Moreover, $S = K_h$, otherwise, one would have $C_{h+1} \neq \emptyset$ and $K_h \subset K_{h+1}$.

Proposition 3.6.2. *Let S be a semigroup. The following conditions are equivalent*

i. S has a $\mathcal{J}$-depth function and a weak $\mathcal{J}$-depth decomposition.
ii. S has a $\mathcal{J}$-depth decomposition.

Proof. $i \Rightarrow ii$: we have to prove that $S = K_S$. If S is finite, then the result is trivial so that we assume that S is infinite. Suppose to the contrary that $S \supset K_S$. If $s \in S \setminus K_S$, then for all $j > 0$, $s \notin K_j$. Let $n > d_{\mathcal{J}}(s)$, where $d_{\mathcal{J}}(s)$ denotes the $\mathcal{J}$-depth of s. Since $s \notin K_n$, it follows that there exists $t_{n-1} \in S \setminus K_{n-1}$ such that $s <_{\mathcal{J}} t_{n-1}$. By iteration one derives that there exists a sequence $t_0, t_1, t_2, \ldots, t_{n-1}$ of n elements of S such that $t_i \in S \setminus K_i$ $(i = 0, 1, \ldots, n-1)$, and

$$s <_{\mathcal{J}} t_{n-1} <_{\mathcal{J}} \cdots <_{\mathcal{J}} t_1 <_{\mathcal{J}} t_0.$$

Hence, we have a strictly ascending $\leq_{\mathcal{J}}$-chain starting with s and having a length greater than $d_{\mathcal{J}}(s)$ which is a contradiction. Hence, $S = K_S$.

$ii \Rightarrow i$: we have to prove that S has a $\mathcal{J}$-depth function. Let $s \in S$ and $h \geq 0$ be the integer such that $s \in C_{h+1}$. We prove that $d_{\mathcal{J}}(s) = h$. Since $s \in C_{h+1}$, there exists a strictly ascending chain

$$s <_{\mathcal{J}} t_h <_{\mathcal{J}} \cdots <_{\mathcal{J}} t_2 <_{\mathcal{J}} t_1,$$

with $t_i \in C_i$ $(i = 1, \ldots, h)$. Thus $h \leq d_{\mathcal{J}}(s)$. We prove that $d_{\mathcal{J}}(s) = h$. Suppose that there exists a chain $s <_{\mathcal{J}} s_k <_{\mathcal{J}} \cdots <_{\mathcal{J}} s_1$, with $k > h$. Let $s_1 \in C_p$, $p \geq 1$; one easily derives that $s \in C_q$ with $q \geq p + k$. Hence, in any case $q > h + 1$, which is a contradiction.

Proposition 3.6.3. *If a semigroup S has a $\mathcal{J}$-depth decomposition, then S is finite $\mathcal{J}$-above.*

Proof. Let $s \in C_h \subseteq K_h$ for a suitable $h > 0$ and t be an element of S such that $s \leq_{\mathcal{J}} t$. This implies that t is a factor of s. By Lemma 3.6.3, K_h is closed by factors so that one derives $t \in K_h$. Since for any h the set K_h is finite, the result follows.

A partial converse of this proposition under the hypothesis that S is finitely generated will be proved in Chap. 4 (see Corollary 4.3.1). The following basic theorem on the $\mathcal{J}$-depth decomposition holds [47]:

Theorem 3.6.1. *($\mathcal{J}$-depth decomposition theorem) Let S be a finitely generated semigroup, which is right stable and whose subgroups are locally finite. Then S has a weak $\mathcal{J}$-depth decomposition.*

Proof. We begin by proving, following an inductive argument, that for all $j > 0$ if K_{j-1} is finite and $S \setminus K_{j-1} \neq \emptyset$, then $C_j \neq \emptyset$ and K_j, as well as C_j, is a finite set. Let m be an element of $S \setminus K_{j-1}$. Since S is finitely generated one can write $m = x_1 \cdots x_n$, with $x_i \in X$ $(i = 1, \ldots, n)$, where X is a finite set of generators for S. It follows that either

$$(i) \quad m = ay, \quad \text{with} \quad a \in S \setminus K_{j-1} \cap X \quad \text{and} \quad y \in S^1,$$

or

$$(ii) \quad m = m'ay, \quad \text{with} \quad m' \in K_{j-1},\ m'a \in S \setminus K_{j-1},\ a \in X \quad \text{and} \quad y \in S^1.$$

In the first case one has $S^1mS^1 \subseteq S^1aS^1$; in the second case $S^1mS^1 \subseteq S^1m'aS^1$. Since X and K_{j-1} are finite sets, one derives that there is a finite subset Z of $S \setminus K_{j-1}$ with the property that for any $m \in S \setminus K_{j-1}$ there exists $z \in Z$ such that $m \leq_{\mathcal{J}} z$. Since Z is finite, one has that in Z, and then in $S \setminus K_{j-1}$, there are maximal elements with respect to $\leq_{\mathcal{J}}$. Hence, C_j is not empty. Thus C_1 is not empty. From Lemma 3.6.3, for all $j > 0$, K_j is closed by factors and union of $\mathcal{J}$-classes.

We prove now that K_j *contains only finitely many* $\mathcal{R}$*-classes.* In fact, if m is an element of C_j, then either $m = ay$, with $a \in X \cap C_j$ and $y \in S^1$, or $m = m'ay$, with $m' \in K_{j-1}$, $y \in S^1$, $m'a \in C_j$, and $a \in X$. In the first case $mS^1 \subseteq aS^1$ and then $S^1mS^1 \subseteq S^1aS^1$. Since $m \in C_j$, one has $m \mathcal{J} a$ and by the fact that S is right stable it follows that $m \mathcal{R} a$. In the latter case $S^1mS^1 \subseteq S^1m'aS^1$ and $mS^1 \subseteq m'aS^1$. Since $m \in C_j$, it follows that $m \mathcal{J} m'a$ and then, by the stability condition, $m \mathcal{R} m'a$. Thus in any case the number of $\mathcal{R}$-classes in C_j is less than or equal to $\mathrm{card}(X) + \mathrm{card}(X)\,\mathrm{card}(K_{j-1})$. By the inductive hypothesis K_{j-1} is a finite set so that the result follows.

K_j *contains only finitely many* $\mathcal{L}$*-classes.* Since the number of $\mathcal{J}$-classes in K_j is finite, it is sufficient to prove that each $\mathcal{J}$-class J in K_j contains only a finite number of $\mathcal{L}$-classes. Let us first prove that if J is a $\mathcal{J}$-class in C_j, then there exist in J elements which are maximal with respect to $\leq_{\mathcal{L}}$. In fact let $m \in J \subseteq C_j \subseteq S \setminus K_{j-1}$. One can write $m = x_1 \cdots x_n$, with $x_i \in X$ $(i = 1, \ldots, n)$. It follows that either

$$(iii) \quad m = ya, \quad \text{with} \quad a \in S \setminus K_{j-1} \cap X \quad \text{and} \quad y \in S^1,$$

or

$$(iv) \quad m = yam', \quad \text{with} \quad m' \in K_{j-1},\ am' \in S \setminus K_{j-1},\ a \in X \quad \text{and} \quad y \in S^1.$$

Thus $S^1mS^1 \subseteq S^1aS^1$ or $S^1mS^1 \subseteq S^1am'S^1$; since $m \in C_j$, it follows that $S^1mS^1 = S^1aS^1$ or $S^1mS^1 = S^1am'S^1$, that is, $m \mathcal{J} a$ or $m \mathcal{J} am'$. Moreover, from (*iii*) and (*iv*) one has also $S^1m \subseteq S^1a$ or $S^1m \subseteq S^1am'$, i.e., $m \leq_{\mathcal{L}} a$ or $m \leq_{\mathcal{L}} am'$. Since X and K_{j-1} are finite sets, one derives that there is a finite subset Z of J with the property that for any $m \in C_j$ there exists $z \in Z$ such that $m \leq_{\mathcal{L}} z$. Since Z is finite one has that in Z, and then in J, there are maximal elements with respect to $\leq_{\mathcal{L}}$.

Now let us prove that *all the elements in J are left stable.* Let $a, b \in J$ and suppose that $S^1 a \subset S^1 b$. One has $a = xb$, with $x \in S^1$. Let w be an element of J which is maximal with respect to $\leq_{\mathcal{L}}$. Since $w \mathcal{J} a$, one has

$$w = \lambda a \mu = \lambda x b \mu, \quad \lambda, \mu \in S^1.$$

Moreover, one has $S^1 w S^1 = S^1 \lambda x b \mu S^1 \subseteq S^1 b \mu S^1 \subseteq S^1 b S^1$ so that, since $w \mathcal{J} b$, there follows $b\mu \mathcal{J} b$. Now $b\mu S^1 \subseteq bS^1$; since S is right stable, it follows that $b\mu S^1 = bS^1$. This implies that there exists $\gamma \in S^1$ such that $b = b\mu\gamma$. Hence, one has

$$S^1 a\mu = S^1 x b \mu \subseteq S^1 b \mu.$$

If $S^1 a\mu = S^1 b\mu$, then $S^1 a\mu\gamma = S^1 b\mu\gamma = S^1 b$. Moreover, $S^1 a = S^1 xb = S^1 x b\mu\gamma = S^1 a\mu\gamma = S^1 b$ which is a contradiction. Let us then assume $S^1 a\mu \subset S^1 b\mu$. One has $S^1 w = S^1 \lambda a \mu \subseteq S^1 a\mu \subset S^1 b\mu$ which is a contradiction since $w \mathcal{J} b\mu$ and w is maximal with respect to $\leq_{\mathcal{L}}$.

Turning back to (*iii*) and (*iv*) we have seen that if $m \in J \subseteq C_j$, then $m \leq_{\mathcal{L}} a$ or $m \leq_{\mathcal{L}} am'$ with $a \in X$, $m' \in K_{j-1}$, and $m \mathcal{J} a$ or $m \mathcal{J} am'$. From the stability on the left of the elements of C_j, it follows that $m \mathcal{L} a$ or $m \mathcal{L} am'$. This implies that the number of $\mathcal{L}$-classes in J is upper-bounded by $\mathrm{card}(X) + \mathrm{card}(X)\,\mathrm{card}(K_{j-1})$.

We can prove now that K_j is finite. Suppose by contradiction that K_j is infinite. Since K_j is closed by factors, by Proposition 3.2.2 there exists a bi-ideal sequence $(f_n)_{n>0}$ of elements of K_j such that $f_1 \in X$, $f_n = f_{n-1} g_{n-1} f_{n-1}$, $g_{n-1} \in S^1$, $n > 1$, and

$$f_n \neq f_m, \quad \text{for } n \neq m.$$

Since the number of $\mathcal{H}$-classes in K_j is finite, an integer k exists such that

$$f_k \,\mathcal{H}\, f_n, \quad \text{for infinitely many } n \geq k.$$

Moreover, from $f_n S^1 \subseteq f_{n-1} S^1$ and $S^1 f_n \subseteq S^1 f_{n-1}$, for $n > 1$, one derives that

$$f_k \,\mathcal{H}\, f_n, \quad \text{for all } n \geq k.$$

The elements f_n for $n > k$ are all regular. Indeed, by $f_n \mathcal{R} f_{n+1}$ and $f_n \mathcal{L} f_{n+1}$, it follows that there exist $\lambda, \mu \in S^1$ such that

$$f_n = f_{n+1}\lambda \quad \text{and} \quad f_n = \mu f_{n+1}.$$

Thus

$$f_{n+1} = f_n g_n f_n = f_{n+1} \lambda g_n \mu f_{n+1}$$

and f_{n+1} is regular. The elements of the sequence $(f_n)_{n\geq k}$ lie in the same $\mathcal{H}$-class H of a regular $\mathcal{D}$-class D which is contained in K_j. Hence, D has only finitely many $\mathcal{R}$-classes. In such a case by Proposition 3.2.4 the maximal subgroups of D are finitely generated and by our hypothesis they are finite.

As we have seen in Sect. 3.1 the maximal subgroups of S coincide with the $\mathcal{H}$-classes containing idempotents; moreover, by Theorem 3.1.1 any two $\mathcal{H}$-classes contained in the same $\mathcal{D}$-class have the same cardinality. This implies that H is finite in contradiction with the fact that the elements of the sequence (f_n) are infinitely many. The proof of the base of the induction is trivial. Since for every $n \geq 0$, K_n is finite and $K_n \subseteq K_{n+1}$, one derives that if S is infinite, then for every $n \geq 0$, $K_n \subset K_{n+1}$ so that K_S is infinite.

We remark that the preceding theorem can be generalized, with a similar proof, under the weaker hypotheses that S is a finitely generated semigroup such that K_S is right stable and the subgroups of S in K_S are locally finite.

Corollary 3.6.1. *Let S be a finitely generated and periodic semigroup. Either of the two following conditions*

i. S is weakly permutable,
ii. the growth function of S is polynomially upper-bounded,

implies that the subgroups of S are locally finite and that S has a weak J-depth decomposition.

Proof. Let us first suppose that S is weakly permutable. If G is a finitely generated group, then by Theorem 3.4.6 it is permutable. Moreover, since G is torsion, by Theorem 3.4.1 G is finite.

Suppose now that the growth function of S is polynomially upper-bounded. Let G be a finitely generated group. By Lemma 3.2.3 the growth function of G is polynomially upper-bounded. Since G is periodic by Gromov's theorem (see Theorem 3.2.1 and Remark 3.4.1) it follows that G is finite.

The fact that in both cases S has a weak $\mathcal{J}$-depth decomposition is a consequence of Lemma 3.6.2 and the $\mathcal{J}$-depth decomposition theorem.

3.6.2 Minimal conditions on principal right ideals

In this section we shall give an interesting application of the $\mathcal{J}$-depth decomposition theorem for finiteness conditions for semigroups. The following theorem [43] is a generalization of a theorem of E. Hotzel [81].

Theorem 3.6.2. *Let S be a finitely generated semigroup S whose subgroups are locally finite. If S satisfies $\min_R$ ($\min_L$), then S is finite.*

Proof. If S satisfies $\min_R$, then S is right stable. Suppose that S is infinite; then by Theorem 3.6.1 so will be K_S. Since K_S is closed by factors, from Proposition 3.2.2 one has that there exists a bi-ideal sequence $(f_n)_{n>0}$ of elements of K_S such that

$$f_n = f_{n-1}g_{n-1}f_{n-1},\ g_{n-1} \in S^1,\ n > 1,$$

and $f_n \neq f_m$, for $n \neq m$. Since $f_nS^1 \subseteq f_{n-1}S^1$, from $\min_R$ there exists an integer k such that for all $n \geq k$, $f_n \mathcal{R} f_k$. Let j be such that $f_k \in K_j$. Since K_j is a union of $\mathcal{J}$-classes, it follows that $f_n \in K_j$ for $n \geq k$, which is a contradiction since K_j is a finite set. A symmetric proof can be done in the $\min_L$ case.

The preceding theorem generalizes Hotzel's theorem since one requires only that *finitely generated* subgroups are finite (instead of *all* subgroups as in Hotzel's theorem). This generalization is important since it allows us to derive finiteness conditions for finitely generated semigroups which can be brought back to finiteness conditions on finitely generated groups. In fact, as a consequence of this result one can find important finiteness conditions for semigroups, some of which provide significant answers to the Burnside problem (see [43]).

We shall give some applications of Theorem 3.6.2; in all of them the hypothesis that only *finitely generated subgroups* of a given semigroup are finite is essential.

Theorem 3.6.3. *Let S be a finitely generated, periodic semigroup satisfying $\min_L$ ($\min_R$). If the growth function of S is polynomially upper-bounded, then S is finite.*

Proof. By Corollary 3.6.1 the subgroups of S are locally finite. Since S satisfies $\min_L$ ($\min_R$), then from Theorem 3.6.2 the result follows.

Corollary 3.6.2. *Let S be a finitely generated and periodic semigroup satisfying $\min_L$ ($\min_R$). If S is weakly permutable, then S is finite.*

Proof. By Theorems 3.4.6 and 3.4.1 all the finitely generated subgroups of S are finite, so that by Theorem 3.6.2, S is finite.

Corollary 3.6.3. *Let S be a finitely generated semigroup such that*

$$\forall s \in S\ \exists x \in S^1 \text{ such that } s = s^2x.$$

If all finitely generated subgroups of S are finite, then S is finite.

Proof. Since the hypotheses of chain lemma (see Lemma 3.6.1) are satisfied by taking $m = 1$, one has that S satisfies $\min_R$. Hence, by Theorem 3.6.2, it follows that S is finite.

A semigroup is called *completely regular* (see [126]) if it is the union of its maximal subgroups. A completely regular semigroup S verifies the hypotheses of Corollary 3.6.3. In fact for any $s \in S$ one has

$$s = ss^{-1}s \text{ and } ss^{-1} = s^{-1}s,$$

where s^{-1} is the inverse of s in the maximal subgroup of S containing s. Therefore, $s = s^2s^{-1}$. Hence, the following corollary holds:

Corollary 3.6.4. *Let S be a completely regular semigroup. S is finite if and only if S is finitely generated and all finitely generated subgroups have a finite order.*

Corollary 3.6.5. *Let S be a finitely generated semigroup such that*

i. for any $s \in S$ there exists $k > 1$ such that $s = s^k$,
ii. the subgroups of S are locally finite.

Then S is finite.

Proof. A semigroup S satisfies i if and only if it is a periodic completely regular semigroup. Indeed, a semigroup S satisfying i is trivially periodic. Moreover, let s be any element of S; since $s = s^k$ for a suitable $k > 1$, the subsemigroup $\langle s \rangle$ generated by s is a cyclic group G of order $k-1$. Let e be the identity of G. Since H_e is the maximal subgroup of S having the identity e, one derives $G \subseteq H_e$. Hence, $s \in H_e$. This shows that S is the union of its maximal subgroups. Conversely, suppose that S is completely regular and periodic. For any $s \in S$ there exist integers i, j, depending on s, such that $1 \leq i < j$ and $s^i = s^j$. By hypothesis H_s is a group, so that $s^{j-i+1} = s$ and the condition i is satisfied. The remaining part of the proof follows from Corollary 3.6.4.

We remark that Corollary 3.6.5 (see [157]) gives a generalization of the theorem of Green and Rees (see Theorem 3.3.2). A further generalization of this theorem will be proved in Theorem 3.7.2.

3.6.3 Minimal conditions on principal bi-ideals

We recall (see [22]) that a *bi-ideal* B of S is a subsemigroup of S such that

$$BSB \subseteq B.$$

A bi-ideal is called *principal* if it is of the kind sS^1s, where s is any element of S. A famous theorem of Coudrain and Schützenberger [24] states that *if S is a finitely generated semigroup satisfying the minimal condition on principal bi-ideals and all subgroups of S are finite, then S is finite.*

We prove a remarkable generalization of the theorem of Coudrain and Schützenberger. More precisely let S be a semigroup. For $s, t \in S$ we set $s \leq_B t$ if $s \in \{t\} \cup tS^1t$; we say that S satisfies the condition $\min_B$ if any strictly descending chain with respect to $\leq_B$ of elements of S has a finite length. One can easily prove (see Proposition 3.6.6) that if S satisfies the minimal condition on principal bi-ideals, then S satisfies $\min_B$.

Let T be a semigroup and T' a subsemigroup of T. We say that a subgroup G of T is *locally finite in T'* if any subgroup of G which is generated by a finite subset of T' is finite.

The main result of this section is the following stronger version [53] of the theorem of Coudrain and Schützenberger (see Theorem 3.6.4). *Let T be a semigroup satisfying* $\min_B$. *Let T' be a subsemigroup of T such that all subgroups of T are locally finite in T'. Then T' is locally finite.* The theorem of Coudrain and Schützenberger is then derived when *i.* $T' = T$, *ii.* the condition $\min_B$ is replaced by the stronger minimal condition on principal bi-ideals, *iii.* the local finiteness of subgroups of T in T' is replaced by the finiteness of all subgroups.

A noteworthy application of the main theorem will be a straightforward proof of the theorem of McNaughton and Zalcstein (see Theorem 3.3.3) that will be given in the next section.

If C is any nonempty subset of S, then the bi-ideal $C \cup CS^1C$ is the bi-ideal generated by C, that is, the smallest bi-ideal of S containing C. For any $s \in S$, we denote by $B(s)$ the bi-ideal generated by s,

$$B(s) = sS^1s \cup \{s\}.$$

Definition 3.6.9. *Let S be a semigroup. For $s,t \in S$ we set $s \leq_B t$ if $B(s) \subseteq B(t)$.*

The relation $\leq_B$ is a quasi-order since it is reflexive and transitive; moreover one easily verifies that $s \leq_B t$ if and only if $s = t$ or $s \in tS^1t$. Let $\equiv_B$ be the equivalence relation

$$\equiv_B \, = \, \leq_B \cap (\leq_B)^{-1}.$$

One has then

$$s \equiv_B t \iff s = t \text{ or } \exists u, v \in S^1 \text{ such that } s = tut \text{ and } t = svs.$$

Let us observe that a bi-ideal sequence of S, $(s_i)_{i>0}$ with $s_{i+1} \in s_iS^1s_i$, $i > 0$ (see Def. 3.2.1), determines a descending chain with respect to $\leq_B$,

$$\cdots \leq_B s_n \leq_B \cdots \leq_B s_2 \leq_B s_1.$$

Conversely, if $s_1, s_2, \ldots, s_n, \ldots$ is a strictly descending chain with respect to $\leq_B$,

$$\cdots <_B s_n <_B \cdots <_B s_2 <_B s_1,$$

then $(s_i)_{i>0}$ is a bi-ideal sequence of S.

Definition 3.6.10. *A semigroup S satisfies* $\min_B$ *if the quasi-order $\leq_B$ is well-founded, i.e., any strictly descending chain with respect to $\leq_B$ has a finite length. S satisfies* $\min_B^*$ *if $s \leq_B t$ and $s \mathcal{J} t \Rightarrow s \equiv_B t$.*

Proposition 3.6.4. *If a semigroup S satisfies* $\min_B$, *then S satisfies* $\min_B^*$.

Proof. Let I be a $\mathcal{J}$-class of S. I contains at least one element which is minimal in I with respect to $\leq_B$. In fact, otherwise, there would exist an infinite strictly descending chain with respect to $\leq_B$ of elements of I which is absurd. Now we prove that any element of I is minimal in I with respect to $\leq_B$. Let a be a minimal element of I and let $b \mathcal{J} a$. One has

$$b = xay, \; x, y \in S^1.$$

Let $c \in I$ and $c \leq_B b$. We prove that $c \equiv_B b$. Since $c \leq_B b$, if $c \neq b$, then there exists $\lambda \in S^1$ such that

$$c = b\lambda b = xay\lambda xay.$$

Let $d = ay\lambda xa$. One has $d \mathcal{J} a$. In fact a is a factor of d and d is a factor of c so that

$$S^1aS^1 \supseteq S^1dS^1 \supseteq S^1cS^1.$$

Moreover,

$$S^1cS^1 = S^1bS^1 = S^1aS^1,$$

hence, $S^1aS^1 = S^1dS^1$.

Since $d = ay\lambda xa$, one has $d \leq_B a$ and, by the minimality of a, it follows that $d \equiv_B a$. If $d = a$, then $c = xdy = xay = b$. Let us then suppose $d \neq a$. One has that there exists $\mu \in S^1$, such that

$$a = d\mu d = ay\lambda xa\mu ay\lambda xa.$$

By substitution one derives

$$a = ay\lambda x(ay\lambda xa\mu ay\lambda xa)\mu(ay\lambda xa\mu ay\lambda xa)y\lambda xa$$

$$= (ay\lambda xay)\lambda xa\mu ay\lambda xa\mu ay\lambda xa\mu ay\lambda(xay\lambda xa) = (ay\lambda xay)\mu'(xay\lambda xa),$$

where $\mu' = \lambda xa\mu ay\lambda xa\mu ay\lambda xa\mu ay\lambda$. Hence

$$b = xay = (xay\lambda xay)\mu'(xay\lambda xay) = c\mu' c.$$

Thus one has $b \leq_B c$ and then $b \equiv_B c$.

Proposition 3.6.5. *If a semigroup S is periodic, then S satisfies* $\min_B^*$.

Proof. Let $s \leq_B t$ and $s \mathcal{J} t$. If $s = t$ the result is trivially true; let us then suppose $s \neq t$. One has then $s = txt$, $s = \alpha t\beta$, and $t = \gamma s\delta$, with $x, \alpha, \beta, \gamma, \delta \in S^1$. Let us introduce now the following sequence

$$\begin{aligned}
f_0 &= s,\\
f_1 &= f_0\delta x\gamma f_0,\\
f_2 &= f_1\delta^2 x\gamma^2 f_1,\\
&\vdots\\
f_n &= f_{n-1}\delta^n x\gamma^n f_{n-1},\\
&\vdots
\end{aligned}$$

It is easy to prove by induction that for any $n \geq 0$ one has $s = \gamma^n f_n \delta^n$. Indeed $s = f_0$ and

$$\gamma^n f_n \delta^n = \gamma^n f_{n-1} \delta^n x \gamma^n f_{n-1} \delta^n$$

$$= \gamma\gamma^{n-1} f_{n-1} \delta^{n-1} \delta x \gamma\gamma^{n-1} f_{n-1} \delta^{n-1} \delta = \gamma s \delta x \gamma s \delta = txt = s.$$

On the other hand for any $n \geq 1$ one has also

$$t = \gamma s \delta = \gamma\gamma^n f_n \delta^n \delta = \gamma^{n+1} f_n \delta^{n+1}.$$

Since S is periodic, there exist two integers h, k, $h < k$ such that $\gamma^h = \gamma^k$ and $\delta^h = \delta^k$. Thus

$$t = \gamma^k f_{k-1} \delta^k = \gamma^h f_{k-1} \delta^h.$$

If $k - 1 = h$, then $t = s$ which is in contrast with the hypothesis. Let us then suppose $k - 1 > h$. By construction one has $f_{k-1} = f_h \delta^h \xi \gamma^h f_h$ for a suitable $\xi \in S^1$; so that $t = \gamma^h f_h \delta^h \xi \gamma^h f_h \delta^h$. Now $s = \gamma^h f_h \delta^h$ and then $t = s\xi s$ that implies $s \equiv_B t$.

We give now the following definition [24].

Definition 3.6.11. *A semigroup S satisfies the* minimal condition on principal bi-ideals *if any strictly descending chain*

$$s_1 S^1 s_1 \supset s_2 S^1 s_2 \supset \cdots \supset s_n S^1 s_n \supset \cdots,$$

with $s_1, s_2, \ldots, s_n, \ldots \in S$, has a finite length.

Proposition 3.6.6. *If a semigroup S satisfies the minimal condition on principal bi-ideals, then S satisfies* $\min_B$.

Proof. Suppose, by contradiction, that there exists an infinite strictly descending chain with respect to $\leq_B$ of elements of S

$$s_1 >_B s_2 >_B \cdots >_B s_n >_B s_{n+1} >_B \cdots.$$

One has $s_{n+1} = s_n x_n s_n$, $x_n \in S^1$, $n > 0$, and

$$s_{n+1} S^1 s_{n+1} = s_n x_n s_n S^1 s_n x_n s_n \subseteq s_n S^1 s_n.$$

From the minimal condition on principal bi-ideals there exists m such that for $n \geq m$ one has $s_n S^1 s_n = s_m S^1 s_m$. Thus, for $n \geq m$

$$s_{n+1} = s_n x_n s_n = s_{n+2} y s_{n+2}, \; y \in S^1,$$

that implies $s_{n+1} \leq_B s_{n+2}$. Since $s_{n+2} \leq_B s_{n+1}$, it follows that $s_{n+1} \equiv_B s_{n+2}$ which is a contradiction.

Proposition 3.6.7. *If a semigroup S satisfies* $\min_R$ *and* $\min_L$*, then S satisfies* $\min_B$.

Proof. Suppose, by contradiction, that there exists an infinite strictly descending chain

$$f_1 >_B f_2 >_B \cdots f_n >_B f_{n+1} >_B \cdots,$$

with $f_i \in S$, $i > 0$. For any $n > 1$ one has then

$$f_{n+1} = f_n x_n f_n, \; x_n \in S^1. \tag{3.10}$$

This implies that $f_n S^1 \supseteq f_{n+1} S^1$ and $S^1 f_n \supseteq S^1 f_{n+1}$. Therefore, by $\min_R$ and $\min_L$ it follows that there exists an integer m such that for any $n \geq m$ one has $f_n \; \mathcal{R} \; f_{n+1}$ and $f_n \; \mathcal{L} \; f_{n+1}$; thus $f_n = u f_{n+1}$ and $f_n = f_{n+1} v$, with $u, v \in S^1$. One has

$$f_n S^1 f_n = f_{n+1} v S^1 u f_{n+1} \subseteq f_{n+1} S^1 f_{n+1}.$$

From (3.10),

$$f_n S^1 f_n \supseteq f_{n+1} S^1 f_{n+1},$$

so that for any $n \geq m$

$$f_n S^1 f_n = f_{n+1} S^1 f_{n+1}.$$

Hence,

$$f_{n+1} = f_n x_n f_n = f_{n+2} \lambda f_{n+2},$$

with $\lambda \in S^1$. Since $f_{n+2} = f_{n+1} x_{n+1} f_{n+1}$, one has $f_{n+1} \equiv_B f_{n+2}$. This is a contradiction.

Lemma 3.6.4. *Let S be a semigroup and $s, t \in S$. If $s \equiv_B t$, then $s \; \mathcal{H} \; t$.*

Proof. Let $s \leq_B t$. If $s = t$ the result is trivially true; let us then suppose $s \neq t$. Then there exists $x \in S^1$ such that $s = txt$. This implies $sS^1 = txtS^1 \subseteq tS^1$ and $S^1 s = S^1 txt \subseteq S^1 t$. If $t \leq_B s$, then in a similar way one proves that $tS^1 \subseteq sS^1$ and $S^1 t \subseteq S^1 s$. Hence, $s \equiv_B t \Rightarrow s \; \mathcal{R} \; t$ and $s \; \mathcal{L} \; t$, that is, $s \; \mathcal{H} \; t$.

Lemma 3.6.5. *If a semigroup S satisfies* $\min_J$ *and* $\min_B^*$*, then it satisfies* $\min_B$.

Proof. Suppose by contradiction that there exists in S an infinite strictly descending chain

$$s_1 >_B s_2 >_B \cdots >_B s_n >_B \cdots.$$

Since $s_n >_B s_{n+1}$ implies $s_n \geq_J s_{n+1}$, $n \geq 1$, then by $\min_J$ there exists an integer m such that $s_n \; \mathcal{J} \; s_m$, for all $n \geq m$. Using $\min_B^*$ it follows that $s_n \equiv_B s_m$, for $n \geq m$, which is absurd.

The following theorem gives a finiteness condition for semigroups which is a strong generalization of the theorem of Coudrain and Schützenberger [24].

Theorem 3.6.4. *Let T be a semigroup satisfying* $\min_B$. *Let T' be a subsemigroup of T such that the subgroups of T are locally finite in T'. Then T' is locally finite.*

Proof. Let S be a finitely generated subsemigroup of T'; we prove that S is finite. Suppose that S is infinite. Since S is finitely generated, one can consider the canonical epimorphism $\phi : A^+ \to S$, where A is a finite alphabet. By Proposition 3.2.1 there exists a uniformly recurrent word $m \in A^\omega$ which is irreducible.

We prove now that the following fact holds:

Fact 1. There exists $K > 0$ such that if $\lambda w \mu \in F(m)$ and $|w| \geq K$, then $\phi(w) \; \mathcal{R} \; \phi(w\mu)$ and $\phi(w) \; \mathcal{L} \; \phi(\lambda w)$.

Indeed, consider the set $\phi(F(m))$. Since T satisfies $\min_B$, there exists an element $s_0 \in \phi(F(m))$ which is minimal with respect to the relation $\leq_B$. Let $x \in F(m)$ be such that $\phi(x) = s_0$. We take $K = \delta(x)$ where δ is the word recurrency function of m.

Let $\lambda w \mu \in F(m)$ with $\lambda, \mu \in A^*$ and $|w| \geq K$. Since m is uniformly recurrent, there exist $z', z \in A^*$ such that $w = z'xz$. Moreover, there exists $u \in A^*$ such that

$$\lambda z' x z \mu u \lambda z' x z \mu \in F(m).$$

This implies

$$x z \mu u \lambda z' x \in F(m).$$

Since

$$\phi(x z \mu u \lambda z' x) = s_0 \phi(z \mu u \lambda z') s_0 \leq_B s_0,$$

one has, by the minimality of s_0 in $\phi(F(m))$ with respect to $\leq_B$, that

$$s_0 \phi(z \mu u \lambda z') s_0 \equiv_B s_0.$$

By Lemma 3.6.4 one easily derives

$$\phi(x) \; \mathcal{R} \; \phi(xz) \; \mathcal{R} \; \phi(xz\mu)$$

and

$$\phi(x) \; \mathcal{L} \; \phi(z'x) \; \mathcal{L} \; \phi(\lambda z' x).$$

Since $\mathcal{R}$ is left invariant and $\mathcal{L}$ is right invariant, it follows that

$$\phi(w) = \phi(z'xz) \; \mathcal{R} \; \phi(w\mu)$$

and

$$\phi(w) = \phi(z'xz) \; \mathcal{L} \; \phi(\lambda w)$$

which concludes the proof of Fact 1.

Since m is uniformly recurrent, we can factorize m as

$$m = w\lambda_0 w\lambda_1 w\lambda_2 \cdots w\lambda_n \cdots$$

where $|w| = K$ and for all $i \geq 0$, $|\lambda_i| \leq \delta(w) = D$. We can then consider the alphabet $Y = \{w\lambda_i \mid i \geq 0\}$; Y is trivially finite; one can rewrite m on the alphabet Y as an infinite word $s \in Y^\omega$

$$s = y_0 y_1 y_2 \cdots y_n \cdots$$

where $y_i = w\lambda_i \in Y$ for $i \geq 0$. Moreover, we can suppose that s is uniformly recurrent as an element of Y^ω. Indeed, otherwise, by Theorem 2.3.1 there exists a uniformly recurrent word $t \in Y^\omega$ such that $F(t) \subseteq F(s)$. Then we can write

$$m = y_0 \mu_0 y_0 \mu_1 y_0 \cdots y_0 \mu_n y_0 \cdots$$

where $y_0 = w\lambda_0 = w\lambda$ (we set $\lambda = \lambda_0$) and

$$\begin{aligned}
\mu_0 &= w\lambda_1 \cdots w\lambda_{j_1}, \\
\mu_1 &= w\lambda_{j_1+1} \cdots w\lambda_{j_2}, \\
&\vdots \\
\mu_k &= w\lambda_{j_k+1} \cdots w\lambda_{j_{k+1}}, \\
&\vdots
\end{aligned}$$

In view of the uniform recurrence of s, for any $k \geq 0$ one has $j_{k+1} - j_k < M$ for a suitable positive integer M. We prove now that for any $k \geq 0$

$$\phi(w\lambda) \ \mathcal{H} \ \phi(\mu_k w\lambda) \ \mathcal{H} \ \phi(w\lambda\mu_k w\lambda). \tag{3.11}$$

Indeed, since $w\lambda\mu_k w\lambda \in F(m)$, by Fact 1 one has

$$\phi(w) \ \mathcal{R} \ \phi(w\lambda)\mathcal{R} \ \phi(w\lambda\mu_k w\lambda) \tag{3.12}$$

and

$$\phi(w\lambda) \ \mathcal{L} \ \phi(\mu_k w\lambda) \ \mathcal{L} \ \phi(w\lambda\mu_k w\lambda). \tag{3.13}$$

Moreover, $\mu_k \in wA^*$ so that, since $\mu_k w\lambda \in F(m)$, one has by Fact 1

$$\phi(w) \ \mathcal{R} \ \phi(\mu_k w\lambda). \tag{3.14}$$

Hence, from (3.12), (3.13), and (3.14) one derives equation (3.11).

Let us now set $t_k = \mu_k w\lambda$; from equation (3.11) one has that for all $k \geq 0$

$$\phi(w\lambda t_k) \ \mathcal{H} \ \phi(w\lambda) \ \mathcal{H} \ \phi(t_k).$$

If H is the $\mathcal{H}$-class of $\phi(w\lambda)$, then $\phi(w\lambda t_k), \phi(w\lambda), \phi(t_k) \in H$, so that $H^2 \cap H \neq \emptyset$ and H is a group by Theorem 3.1.2. Hence, m can be rewritten as

$$m = w\lambda t_0 t_1 \cdots t_k \cdots$$

where $\phi(t_j) \in H$, $j \geq 0$, and $|t_j| \leq (M+1)(D+K)$. Thus all the factors of m of the kind $t_r \cdots t_s$ are such that $\phi(t_r \cdots t_s)$ belong to a subgroup G of H generated by a finite subset of T'. By hypothesis G is finite so that there exist two factors $t_0 \cdots t_i$ and $t_0 \cdots t_j$, $i < j$, such that $\phi(t_0 \cdots t_i) = \phi(t_0 \cdots t_j)$; this is absurd since m is irreducible.

We remark that a different proof of Theorem 3.6.4, based on a suitable $\mathcal{J}$-depth decomposition of the semigroup S, can be found in [51].

If in the preceding theorem we identify T' with T, then one obtains, by Proposition 3.6.6, the following:

Corollary 3.6.6. *Let T be a semigroup satisfying the minimal condition on principal bi-ideals. If the subgroups of T are locally finite, then T is locally finite.*

We note that Corollary 3.6.6 is a remarkable generalization of the theorem of Coudrain and Schützenberger, since in the latter theorem one supposes that *all subgroups* of T are finite.

Corollary 3.6.7. *Let T be a semigroup satisfying* $\min_B$. *If T' is a periodic subsemigroup whose subgroups are locally finite, then T' is locally finite.*

Proof. Let G be a subgroup of T generated by a finite subset X of T'. In view of the periodicity of T' one has $G \subseteq T'$; indeed in this case the inverse of any element $x \in X$ is still an element of T'. Since the subgroups of T' are locally finite, then G is finite. Hence, from Theorem 3.6.4 the result follows.

Let us remark that if we drop the hypothesis that T' is periodic, Corollary 3.6.7 does not, in general, hold true. For instance, take T equal to the group of integers and T' equal to the subsemigroup of positive integers.

Corollary 3.6.8. *Let T be a periodic semigroup satisfying* $\min_J$. *If T' is a subsemigroup of T whose subgroups are locally finite, then T' is locally finite.*

Proof. From Proposition 3.6.5, T satisfies $\min_B^*$, so that by Lemma 3.6.5, T satisfies $\min_B$. Since T' is periodic, by Corollary 3.6.7 the result follows.

3.6.4 The McNaughton–Zalcstein and Straubing theorems

In this section we give a proof of the McNaughton and Zalcstein theorem on the local finiteness of periodic semigroups of matrices of a finite dimension on a field (see Theorem 3.3.3). In particular, we will show that the semigroups of all $n \times n$ matrices over a field satisfy some interesting minimal conditions. Using these conditions and Corollary 3.6.7 one can reduce the local finiteness of a periodic subsemigroup of matrices to that of its subgroups. Moreover, using the result of McNaughton and Zalcstein and Shirshov's theorem, we prove Straubing's theorem (see Theorem 3.3.4) showing that a periodic subsemigroup of a unitary ring which satisfies a polynomial identity is locally finite.

Let F denote a field and $\mathcal{M}_n(F)$ the semigroup of $n \times n$ square matrices over F. We shall identify, up to an isomorphism, $\mathcal{M}_n(F)$ with the semigroup $\mathrm{End}_n(V, F)$ of the endomorphisms of a vectorial space V of dimension n over the field F. Let us recall that for $f \in \mathrm{End}_n(V, F)$,

$$\mathrm{Im}(f) = \{v \in V \mid f(w) = v \text{ for some } w \in V\},$$

$$\mathrm{Ker}(f) = \{v \in V \mid f(v) = 0\}$$

and

$$\mathrm{rank}(f) = \dim(\mathrm{Im}(f)) = \dim(V/\mathrm{Ker}(f)).$$

The following theorem (see [101]) gives a characterization of Green's relations in $\mathrm{End}_n(V, F)$.

Theorem 3.6.5. *Let* $f, g \in \mathrm{End}_n(V, F)$. *One has*

i. $f \ \mathcal{J} \ g$ *if and only if* $\mathrm{rank}(f) = \mathrm{rank}(g)$,
ii. $f \ \mathcal{L} \ g$ *if and only if* $\mathrm{Im}(f) = \mathrm{Im}(g)$,
iii. $f \ \mathcal{R} \ g$ *if and only if* $\mathrm{Ker}(f) = \mathrm{Ker}(g)$.

Proposition 3.6.8. *The semigroup* $\mathrm{End}_n(V, F)$ *satisfies* $\min_L$ *and* $\min_R$.

Proof. We prove that $\mathrm{End}_n(V, F)$ satisfies $\min_L$. Let us denote $\mathrm{End}_n(V, F)$ by S. Suppose, by contradiction, that there exists an infinite strictly descending chain of principal left ideals

$$S^1 f_1 \supset S^1 f_2 \supset \cdots \supset S^1 f_i \supset \cdots.$$

For any $i \geq 1$ one has $f_{i+1} = x f_i$, $x \in S$; thus $\mathrm{Im}(f_{i+1}) \subseteq \mathrm{Im}(f_i)$ and so $\mathrm{rank}(f_{i+1}) \leq \mathrm{rank}(f_i) \leq n$. Thus there exists an integer m such that for any $i \geq m$ one has

$$\mathrm{rank}(f_i) = \mathrm{rank}(f_m).$$

On the other hand for any $i \geq m$ one has also

$$\mathrm{Im}(f_m) \supseteq \mathrm{Im}(f_i)$$

and, since $\dim(\mathrm{Im}(f_i)) = \dim(\mathrm{Im}(f_m))$, one obtains $\mathrm{Im}(f_m) = \mathrm{Im}(f_i)$. Thus by Theorem 3.6.5, one derives that for any $i \geq m$, $f_i \ \mathcal{L} \ f_m$ and this is a contradiction.

Let us prove now that $\mathrm{End}_n(V, F)$ satisfies $\min_R$. Suppose by contradiction that there exists an infinite strictly descending chain of principal right ideals

$$f_1S^1 \supset f_2S^1 \supset \cdots \supset f_iS^1 \supset \cdots .$$

As before, for any $i \geq 1$ one has $f_{i+1} = f_i x, x \in S$; thus $\mathrm{Ker}(f_i) \subseteq \mathrm{Ker}(f_{i+1})$ and so $\mathrm{rank}(f_{i+1}) \leq \mathrm{rank}(f_i)$. Thus there exists an integer m such that for any $i \geq m$ one has $\mathrm{rank}(f_i) = \mathrm{rank}(f_m)$. Moreover, for any $i \geq m$ one has $\mathrm{Ker}(f_m) \subseteq \mathrm{Ker}(f_i)$ and, since $\dim(V/\mathrm{Ker}(f_i)) = \dim(V/\mathrm{Ker}(f_m))$, one derives $\mathrm{Ker}(f_m) = \mathrm{Ker}(f_i)$. Thus by Theorem 3.6.5, for any $i \geq m$, $f_i \,\mathcal{R}\, f_m$, and this is absurd.

Let us now give a proof of the McNaughton–Zalcstein theorem (see Theorem 3.3.3) stating: *a torsion semigroup of $n \times n$ matrices over a field is locally finite.*

Proof of Theorem 3.3.3. Let S be a finitely generated and periodic subsemigroup of $\mathcal{M}_n(F)$. Since $\mathcal{M}_n(F)$ is isomorphic to $\mathrm{End}_n(V, F)$, it follows by Proposition 3.6.8 that $\mathcal{M}_n(F)$ satisfies $\min_R$ and $\min_L$ and then, by Proposition 3.6.7, also the condition $\min_B$. It is well known that any finitely generated and periodic subgroup of $\mathcal{M}_n(F)$ is finite (see [93]). Then all finitely generated subgroups of S are finite. From Corollary 3.6.7 it follows that S is finite.

We recall that G. Jacob [83] gave a further proof of this theorem; moreover, he proved that it is possible to decide, under certain rather general assumptions, whether a finitely generated semigroup of matrices over a field is finite.

We are now in a position to give a proof of Straubing's theorem stating: *a torsion subsemigroup of a unitary ring which satisfies a polynomial identity is locally finite.*

Proof of Theorem 3.3.4. Let T be a unitary ring satisfying a polynomial identity and S be a periodic finitely generated subsemigroup of T. We prove that S is finite. By Proposition 3.4.3, T satisfies a multilinear identity

$$x_1 \cdots x_n = \sum_{\sigma \in \mathcal{S}_n \setminus \{id\}} a_\sigma x_{\sigma(1)} \cdots x_{\sigma(n)}, \tag{3.15}$$

where $a_\sigma \in \mathbb{Z}$ for any $\sigma \in \mathcal{S}_n \setminus \{id\}$.

Since S is finitely generated, there exists a finite alphabet $A = \{a_1, \ldots, a_k\}$ and the canonical epimorphism $\phi : A^+ \to S$.

Let n be the positive integer of equation (3.15). Since S is periodic and A is finite, one can take p such that $p \geq 2n$ and for each word w of length $< n$, $\phi(w)^p = \phi(w^p) = \phi(w)^q = \phi(w^q)$ for some $q < p$.

Let P be the additive subgroup of T generated by the elements of S; it coincides with the set of all integral linear combinations of elements of S. Since S is a semigroup, P is a subring of T. In fact the product of two elements of P, as one can easily verify, is an integral linear combination of elements of S. We prove that P is the additive subgroup of T generated by the set $Q = \{\phi(u) \mid |u| < N(k,p,n)\}$, where N is the function of Corollary 2.4.3. We prove that for any $w \in A^+$, $\phi(w)$ is an integral linear combination of elements of Q. In order to do this it is enough to prove that for any $w \in A^+$, with $|w| \geq N(k,p,n)$, $\phi(w)$ is an integral linear combination of elements $\phi(v)$, with $v <_a w$.

Indeed, from Corollary 2.4.3 one has either

$$w = \lambda u^p \mu,$$

with $\lambda, \mu \in A^*$, $0 < |u| < n$ or

$$w = \lambda v_1 \cdots v_n \mu,$$

with $\lambda, \mu \in A^*$, $v_i \in A^+$, $1 \leq i \leq n$, and

$$v_1 \cdots v_n > v_{\sigma(1)} \cdots v_{\sigma(n)}$$

for any nontrivial permutation σ of $\{1, \ldots, n\}$. In the first case, since $\phi(u)^p = \phi(u)^q$, with $q < p$, if one sets $v = \lambda u^q \mu$, then one has $v <_a w$ and $\phi(w) = \phi(v)$.

In the second case, by the multilinear identity of equation (3.15), one has

$$\phi(v_1) \cdots \phi(v_n) = \sum_{\sigma \in S_n \setminus \{id\}} a_\sigma \phi(v_{\sigma(1)}) \cdots \phi(v_{\sigma(n)})$$

so that

$$\phi(w) = \sum_{\sigma \in S_n \setminus \{id\}} a_\sigma \phi(\lambda v_{\sigma(1)} \cdots v_{\sigma(n)} \mu),$$

where $\lambda v_{\sigma(1)} \cdots v_{\sigma(n)} \mu <_a w$ for any $\sigma \in S_n \setminus \{id\}$.

Thus S is contained in a ring P whose additive group is finitely generated, as Q is finite. Let U be the torsion group of P, i.e.,

$$U = \{t \in P \mid jt = 0 \text{ for some } j \in \mathbb{N}_+\}.$$

U is an ideal of P. Indeed $jt = 0$, $j \in \mathbb{N}_+$, $t \in U$, implies $j(pt) = p(jt) = 0$ and $j(tp) = (jt)p = 0$, for any $p \in P$. As is well known, any subgroup of a finitely generated abelian group is finitely generated. Therefore, U is finitely generated and then finite. Thus the finiteness of S is equivalent to the finiteness of its projection $\pi(S)$ in the quotient ring P/U, since any element of $\pi(S)$ has only finitely many counterimages in P. We observe that the additive group P/U is finitely generated and torsion-free, thus it is isomorphic to $\mathbb{Z}^m$ for some positive integer m. On the other hand the right multiplication by an element of P/U determines a linear transformation of $\mathbb{Z}^m$, i.e., a matrix of $\mathbb{Z}^{m \times m}$. We have then a morphism from P/U to $\mathbb{Z}^{m \times m}$. This morphism is injective, since P/U is a unitary ring. Therefore, $\pi(S)$ is isomorphic to a periodic finitely generated subsemigroup of $\mathbb{Z}^{m \times m}$. Since $\mathbb{Z}$ is contained in the field of rational numbers, by Theorem 3.3.3, $\pi(S)$ is finite.

3.7 Iteration property

As we have seen in the previous sections, in relation to Burnside's problem, "periodicity" is not, in general, a finiteness condition for a finitely generated semigroup. In order to obtain a finiteness condition one has to require that the semigroup satisfies some further properties such as, for instance, "permutability".

In this section we introduce some conditions called "iteration properties". These properties are stronger than periodicity, and in some cases (see Theorem 3.7.3) they are finiteness conditions for a finitely generated semigroup S. In the general case the finiteness of S is brought back to that of its finitely generated subgroups (see Theorem 3.7.2). This will be a remarkable generalization of the theorem of Green and Rees.

A semigroup S satisfies an iteration property if in any product $s_1 s_2 \cdots s_m$ of a sufficiently great number m of elements of S, there exists a factor $s_i \cdots s_j$ $(1 \leq i \leq j \leq m)$ which can be iterated, i.e., replaced by a suitable power $(s_i \cdots s_j)^n$, $n \neq 1$, without changing the value of the product.

Iteration conditions can be either *central* or *one-sided.* Moreover, the exponent n, as well as the integer m, are usually fixed. A particular kind of iteration, called ***strong periodicity*** is considered in Sect. 3.7.2. Moreover, a weaker iteration condition called ω-iteration property, is introduced in Sect. 3.7.1.

Iteration properties, as we shall see in Chap. 5, are very important in formal language theory, since they naturally reflect in the case of semigroups the "pumping properties" of regular languages.

Definition 3.7.1. *Let S be a semigroup and m and n two integers such that $m > 0$ and $n \geq 0$. We say that the sequence $s_1, s_2, \ldots, s_m$ of m elements of S is n-*iterable *if there exist i, j such that $1 \leq i \leq j \leq m$ and*

$$s_1 \cdots s_m = s_1 \cdots s_{i-1}(s_i \cdots s_j)^n s_{j+1} \cdots s_m. \tag{3.16}$$

We say that S is (m,n)-iterable, or satisfies the property $C(n,m)$ if all sequences of m elements of S are n-iterable.

Let us observe that the property $C(1,m)$ is always trivially true. Moreover, the property $C(0,m)$ is actually a cancellation property (*m-cancellation property*) which obviously implies the finiteness of any finitely generated semigroup satisfying it, so that a semigroup satisfies properly the iteration property $C(n,m)$ only if $n > 1$. We say that a semigroup S satisfies the *central iteration property* if there exist $m > 0$ and $n > 1$ such that S satisfies $C(n,m)$.

We shall consider in the following also the weaker iteration condition $C(m)$:

Definition 3.7.2. *Let m be a positive integer. A semigroup S satisfies the condition $C(m)$ if for any sequence $s_1, \ldots, s_m$ of m elements of S there exist integers i, j, n such that $1 \leq i \leq j \leq m$, $0 \leq n \neq 1$, and*

$$s_1 \cdots s_m = s_1 \cdots s_{i-1}(s_i \cdots s_j)^n s_{j+1} \cdots s_m.$$

Note that in this case the exponent n may depend on the sequence $s_1, \ldots, s_m$. Obviously, if S satisfies $C(n,m)$ with $n \neq 1$, then it satisfies $C(m)$.

Proposition 3.7.1. *If S is a finite semigroup, then S satisfies $C(n,m)$ with $m = \mathrm{card}(S) + 1$ and any $n \geq 0$. If S satisfies $C(m)$, then S is periodic.*

Proof. Let $s_1, \ldots, s_m$ be a sequence of m elements of S, with $m = \mathrm{card}(S)+1$; we consider then the sequence

$$s_1,\ s_1 s_2, \ldots,\ s_1 s_2 \cdots s_m.$$

Since $m > \mathrm{card}(S)$ there exist integers i, j such that $1 \leq i < j \leq m$ and

$$s_1 \cdots s_i = s_1 \cdots s_i(s_{i+1} \cdots s_j) = s_1 \cdots s_i(s_{i+1} \cdots s_j)^n,$$

for all $n \geq 0$. Thus

$$s_1 \cdots s_i s_{j+1} \cdots s_m = s_1 \cdots s_i(s_{i+1} \cdots s_j)^n s_{j+1} \cdots s_m,$$

so that $C(n,m)$ holds for all $n \geq 0$.

If S satisfies $C(m)$, then for any $s \in S$, consider the sequence $s_1 = s_2 = \cdots = s_m = s$. One has that there exist integers i, j and $n \neq 1$ such that $1 \leq i \leq j \leq m$ and

$$s^m = s^{m+(n-1)(j-i+1)},$$

so that S is periodic.

In [34] the following stronger iteration property $D(n,m)$, $n \geq 0$, $m > 0$, called ***iteration property on the right***, was considered. A semigroup S satisfies $D(n,m)$, if for any sequence $s_1, s_2, \ldots, s_m$ of m elements of S there exist integers i, j such that $1 \leq i \leq j \leq m$ and

$$s_1 \cdots s_j = s_1 \cdots s_{i-1}(s_i \cdots s_j)^n.$$

Let S be a finitely generated semigroup. If S satisfies $D(n,m)$ with $n > 1$, then by Lemma 3.6.1 it satisfies $\min_R$. It was proved in [34], as a consequence of the theorem of Hotzel, that $D(2,m)$ assures the finiteness of S. A more combinatorial proof of this result was obtained by K. Hashiguchi [77]. He was also able to give an upper bound for the number of elements of the semigroup. More recently it has been proved [43], by using Theorem 3.6.2 and Proposition 3.7.2, that $D(3,m)$ also implies the finiteness of S:

Theorem 3.7.1. *Let S be a finitely generated semigroup. S is finite if and only if S satisfies the properties $D(2,m)$ or $D(3,m)$.*

The question was posed in [34] as to whether the central iteration property implies the finiteness of a finitely generated semigroup. A positive answer will be given later by Theorem 3.7.3 which is a remarkable generalization of Theorem 3.7.1.

Proposition 3.7.2. *Let G be a group satisfying the following condition*

(D). A positive integer n exists such that any sequence $g_1, \ldots, g_n$ of n elements of G either:

i. is permutable, or
ii. there exist integers i, j, $1 \le i \le j \le n$, such that

$$(g_i \cdots g_j)^2 = 1, \tag{3.17}$$

where 1 is the identity of G.

Then G is $2n^2$-weakly permutable.

Proof. Let G be a group satisfying the condition (D) and set $r = 2n^2$. We prove that G is r-weakly-permutable. Let $g_1, g_2, \ldots, g_r$ be a sequence of r elements of G. If there exists a subsequence of $g_1, g_2, \ldots, g_r$ which is permutable, then $g_1, g_2, \ldots, g_r$ is trivially weakly-permutable. So we will suppose, without loss of generality, that all subsequences of n elements of $g_1, g_2, \ldots, g_r$ satisfy equation (3.17). We can set

$$\begin{aligned} t_1 &= g_1 \cdots g_n, \\ t_2 &= g_{n+1} \cdots g_{2n}, \\ &\vdots \\ t_{2n} &= g_{r-n+1} \cdots g_r. \end{aligned}$$

In view of condition (D) for each i, $1 \le i \le 2n$, we can factorize t_i as

$$t_i = p_i h_i q_i,$$

where $p_i, h_i, q_i \in G$ and $h_i^2 = 1$, where 1 denotes the identity of G. Thus $h_i = h_i^{-1}$. Up to a permutation the product $g_1 g_2 \cdots g_r$ can be rewritten as

$$u h_1 \cdots h_{2n} v.$$

Let us set now

$$s_j = h_{2j-1} h_{2j}, \text{ for } 1 \le j \le n.$$

Thus the preceding product can be written as

$$u s_1 \cdots s_n v.$$

By hypothesis the sequence $s_1, \ldots, s_n$ of elements of G is either permutable or there exist integers i, j, $1 \le i \le j \le n$, such that

$$(s_i \cdots s_j)^2 = 1.$$

In the first case there exists a permutation $\sigma \in \mathcal{S}_n$, $\sigma \neq id$, such that $s_1 s_2 \cdots s_n = s_{\sigma(1)} s_{\sigma(2)} \cdots s_{\sigma(n)}$, so that one easily derives that the sequence $g_1, g_2, \ldots, g_r$ is weakly-permutable. In the second case one has

$$s_i \cdots s_j = (s_i \cdots s_j)^{-1} = s_j^{-1} \cdots s_i^{-1}.$$

Since $h_k = h_k^{-1}$ for all $k \in [1, 2n]$, one derives

$$h_{2i-1} h_{2i} \cdots h_{2j-1} h_{2j} = h_{2j} h_{2j-1} \cdots h_{2i} h_{2i-1}.$$

This shows that G is r-weakly permutable.

A further condition that we consider is the following: we say that S satisfies the iteration condition $C(r, s; m)$ of order m, with $r, s, m > 0$, if for any sequence $s_1, s_2, \ldots, s_m$ of m elements of S there exist i, j such that $1 \leq i \leq j \leq m$ and

$$s_1 \cdots s_m = s_1 \cdots s_{i-1} (s_i \cdots s_j)^r s_{j+1} \cdots s_m = s_1 \cdots s_{i-1} (s_i \cdots s_j)^s s_{j+1} \cdots s_m.$$

Lemma 3.7.1. *Let S be a semigroup. If S satisfies the property $C(n, n+1; m)$, then all subgroups of S have finite orders. If S satisfies $C(n, n+2; m)$ or $C(n+2, 2n+1; m)$, then all finitely generated subgroups of S have finite orders.*

Proof. Let S be a semigroup satisfying the condition $C(n, n+1; m)$ and let G be any subgroup of S. We prove that card$(G) \leq m$. In fact let g_0 be the identity element of G and $g_0, g_1, \ldots, g_m$ be $m+1$ distinct elements of G. We consider then the sequence $h_1, \ldots, h_m$ of m elements of G, with $h_i = g_{i-1}^{-1} g_i$ $(i = 1, \ldots, m)$. From the condition $C(n, n+1; m)$ there exist integers i, j, $1 \leq i \leq j \leq m$, such that

$$h_1 \cdots h_{i-1} (h_i \cdots h_j)^n h_{j+1} \cdots h_m = h_1 \cdots h_{i-1} (h_i \cdots h_j)^{n+1} h_{j+1} \cdots h_m.$$

By cancellation one derives

$$h_i \cdots h_j = g_0 = g_{i-1}^{-1} g_j;$$

hence, $g_{i-1} = g_j$ which is a contradiction (one reaches the same conclusion also when the condition $C(n+2, 2n+1; m)$ holds with n an even integer).

Suppose now that S satisfies $C(n, n+2; m)$ or $C(n+2, 2n+1; m)$ and let G be a subgroup of S. The preceding conditions imply that G satisfies the following property.

For any sequence $g_1, g_2, \ldots, g_m$ of m elements of G there exist i, j such that $1 \leq i \leq j \leq m$ and

$$(g_i \cdots g_j)^2 = g_0. \tag{3.18}$$

From the preceding proposition one has that G is $2m^2$-weakly permutable and, therefore, permutable by Blyth's theorem (see Theorem 3.4.6); hence, there exists an integer $h > 1$ such that any product of h elements of G can be rewritten in a nontrivial way. Moreover, from (3.18), G is periodic. Hence, if one supposes that G is finitely generated, then by using the theorem of Restivo and Reutenauer (see Theorem 3.4.1) it follows that G is finite.

Theorem 3.7.2. *Let S be a finitely generated semigroup satisfying the iteration condition $C(m)$. If the subgroups of S are locally finite, then S is finite.*

Proof. Let S be a finitely generated semigroup and $\phi : A^+ \to S$ be the canonical epimorphism. By Proposition 3.7.1, S is periodic and then, by Lemma 3.6.2, it is stable. Since the subgroups of S are locally finite then from the $\mathcal{J}$-depth decomposition theorem (see Theorem 3.6.1) one has that S is infinite if and only if $K_S = \cup_{j>0} K_j$ is an infinite subset of S. Moreover, K_S is closed by factors. Suppose, by contradiction, that S is infinite, then also K_S will be so.

By Lemma 3.2.2 there exists an irreducible uniformly recurrent word $x \in A^\omega$ such that $F(x) \subseteq C$, where C denotes the set of the canonical representatives of the elements of K_S. One has then for every $i > 0$,

$$\phi(x[i])S^1 \supseteq \phi(x[i+1])S^1.$$

Suppose now, first, that an integer h exists such that for all $i \geq h$

$$\phi(x[i])S^1 = \phi(x[i+1])S^1.$$

This implies $\phi(x[i]) \mathcal{R} \phi(x[h])$ for all $i \geq h$. Since $\phi(x[h]) \in K_S$, a positive integer j exists such that $\phi(x[h]) \in K_j$. We recall that K_j is a union of $\mathcal{R}$-classes so that

$$\phi(x[i]) \in K_j,$$

for all $i \geq h$. Since K_j is a finite set, there must exist integers i and j, $i < j$, for which $\phi(x[i]) = \phi(x[j])$. This is a contradiction since $x[i]$ and $x[j]$ belong to C. Thus there exist infinitely many integers i for which

$$\phi(x[i])S^1 \supset \phi(x[i+1])S^1. \tag{3.19}$$

Suppose now that S satisfies the iteration condition $C(m)$. We can take i so large that $i \geq D(m)$ (see Proposition 2.7.7) and the preceding relation (3.19) is satisfied. Let $x[i+1] = x[i]a$, $a \in A$. Since x is uniformly recurrent, by Proposition 2.7.7 one has that $x[i]$ has a suffix f_m which is the m th term of a bi-ideal sequence $\sigma = (f_1, \ldots, f_m)$, where $f_{i+1} = f_i g_i f_i$ $(i = 1, \ldots, m-1)$, $f_1 \in aA^*$ and $g_i \in aA^*$. The word f_m can be factorized as

$$f_m = w_1 \cdots w_m,$$

where $(w_1, \ldots, w_m)$ is the canonical factorization of f_m. Let us then consider the sequence $\phi(w_1), \ldots, \phi(w_m)$ of m elements of S. Since S satisfies $C(m)$, then there exist integers n, r, s such that $0 \leq n \neq 1$, $1 \leq r \leq s \leq m$, and

$$\phi(f_m) = \phi(h_{r,s,n}),$$

with $h_{r,s,n} = w_1 \cdots w_{r-1}(w_r \cdots w_s)^n w_{s+1} \cdots w_m$. If $n = 0$, then f_m will be reducible which is a contradiction. Let us then suppose $n > 1$. In such a case one can write $h_{r,s,n}$ as

$$h_{r,s,n} = w_1 \cdots w_{r-1} w_r \cdots w_s w_r h, \tag{3.20}$$

for a suitable $h \in A^*$. Let us first suppose $s < m$. For all $i \in [1, m-1]$ one has by equation (2.8) that

$$w_i = w_{i+1} \cdots w_m g_{m-i},$$

so that, since $r \leq s$, one derives, by iteration of the preceding formula (see (2.9)), that

$$w_r = w_{s+1} \cdots w_m g_{m-s} u,$$

with $u \in A^*$. Hence, one has that

$$h_{r,s,n} \in f_m g_{m-s} A^*.$$

Since $g_{m-s} \in aA^*$, it follows that

$$\phi(f_m) \; \mathcal{R} \; \phi(f_m a). \tag{3.21}$$

This implies, since relation $\mathcal{R}$ is left invariant,

$$\phi(x[i]) \; \mathcal{R} \; \phi(x[i+1]), \tag{3.22}$$

which is a contradiction. If $s = m$, then from (3.20) one has $h_{r,m,n} = f_m w_r h$. Since $w_r \in aA^*$ $(r = 1, \ldots, m)$, one has again that equations (3.21) and (3.22) are satisfied which is a contradiction.

Let us observe that the preceding theorem is a generalization of the Green and Rees theorem. In fact, in the case $m = 1$ one obtains Corollary 3.6.5.

Corollary 3.7.1. *Let S be a finitely generated semigroup and m, n integers such that $m > 0$ and $n \geq 0$. If S satisfies $C(n, n+1; m)$ or $C(n, n+2; m)$ or $C(n+2, 2n+1; m)$, then S is finite.*

Proof. By Lemma 3.7.1 one has that all subgroups of S are locally finite. Moreover, any one of the iteration conditions that, by hypothesis, S satisfies implies $C(m)$, so that from Theorem 3.7.2 the result follows.

From the above results one derives [44]:

Theorem 3.7.3. *A finitely generated semigroup S is finite if and only if S satisfies $C(2, m)$ or $C(3, m)$.*

Proof. If S is finite, then from Proposition 3.7.1 one has that $C(2,m)$ and $C(3,m)$ are satisfied. Conversely, we remark that condition $C(1,n;m)$ is equivalent to condition $C(n,m)$, so that the result follows from Corollary 3.7.1, by considering conditions $C(1,2;m)$ and $C(1,3;m)$.

We consider now a central iteration condition for a semigroup S which is weaker than $C(n,m)$ since one supposes that only the sequences $s_1,\ldots,s_m$ of m elements of S which are m-sequences are n-iterable.

Definition 3.7.3. *A semigroup S satisfies the condition $B(n,m)$, $m>0$, $n>1$, if for any m-sequence $s_1,\ldots,s_m$ of elements of S there exist integers i,j such that $1\le i\le j\le m$ and*

$$s_1\cdots s_m = s_1\cdots s_{i-1}(s_i\cdots s_j)^n s_{j+1}\cdots s_m.$$

S is called iterable on the m-sequences *if there exists an integer $n>1$ such that S satisfies $B(n,m)$.*

It is easy to prove that a semigroup S satisfying $B(n,m)$ is periodic. Indeed, let $f:\mathbb{N}_+\to\mathbb{N}_+$ defined as

$$f(1)=1,\ f(m)=2^{m-2},\ \text{for all } m>1.$$

Then for any $s\in S$, the sequence $s_1,\ldots,s_m$, where for any $j=1,\ldots,m$,

$$s_j = s^{f(m-j+1)},$$

is an m-sequence of S, so that from the condition $B(n,m)$ it follows that there are two distinct exponents p and q such that $s^p=s^q$.

Theorem 3.7.4. *A finitely generated semigroup S is finite if and only if S satisfies the condition $B(2,m)$ or $B(3,m)$.*

Proof. The "only if" part of the theorem is trivial. Let us prove the "if" part. Let S be a finitely generated semigroup satisfying $B(2,m)$ or $B(3,m)$. As we have seen above, S is periodic. Moreover, we observe that, since any sequence $g_1,\ldots,g_m$ of m elements of a group G is in fact an m-sequence, one derives by using Proposition 3.7.2, that the subgroups of S are locally finite. The remaining part of the proof is identical to that of Theorem 3.7.2 in which only the iteration on m-sequences is used.

3.7.1 ω-iteration property

In this section we introduce the ***ω-iteration property.*** This property is weaker than the iteration property $C(2,m)$ and, as we shall see, is a finiteness condition for finitely generated semigroups.

Definition 3.7.4. *A semigroup S has the* ω-iteration property *if for each infinite sequence $s_1, s_2, \ldots, s_i, \ldots$ of elements of S there exist $n \geq 1$ (depending on the sequence chosen) and integers i, j such that $n \geq i \geq j \geq 1$ and*

$$s_n s_{n-1} \cdots s_2 s_1 = s_n s_{n-1} \cdots s_{i+1} (s_i \cdots s_j)^2 s_{j-1} \cdots s_2 s_1 .$$

We observe that one could define a similar *ω-iteration property on the left* (*on the right*) requiring that $s_i s_{i-1} \cdots s_2 s_1 = (s_i \cdots s_j)^2 s_{j-1} \cdots s_2 s_1$, $(s_1 s_2 \cdots s_i = s_1 s_2 \cdots s_{j-1} (s_j \cdots s_i)^2)$.

It is easy to show that the ω-iteration property on the right (on the left) implies $\min_R$ ($\min_L$) and the finiteness of all subgroups. Therefore, by Theorem 3.6.2, these properties are finiteness conditions for finitely generated semigroups.

We prove now that the same holds for the ω-iteration property [130].

Theorem 3.7.5. *A finitely generated semigroup has the ω-iteration property if and only if it is finite.*

Proof. First of all we observe that, for any $m \geq 1$, the iteration property $C(2, m)$ implies the ω-iteration property. If S is a finite semigroup, then, by Proposition 3.7.1, S satisfies $C(2, m)$, for some $m \geq 1$ and, therefore, S satisfies the ω-iteration property. The "if" part is then proved.

We now prove the "only if" part. Let S be a finitely generated semigroup satisfying the ω-iteration property and let $\phi : A^+ \to S$ be the canonical epimorphism. Suppose, by contradiction, that S is infinite. By Proposition 3.2.1 there exists a bi-infinite uniformly recurrent word $m \in A^{\pm\omega}$ which is irreducible.

Claim 1. *There exists a positive integer k such that, for each $u \in F(m)$ and $a \in A$, if $ua \in F(m)$ and $|u| \geq k$, then $\phi(u) \mathcal{R} \phi(ua)$.*

Proof of Claim 1. If k does not exist, then there exist $a \in A$ and an infinite set U of words u such that $ua \in F(m)$ and $\phi(u)$ and $\phi(ua)$ are not in Green's relation $\mathcal{R}$.

By Lemma 1.2.1, there exists a left infinite word m' such that each of its finite right factors is a right factor of some $ua \in Ua$. As the factors of m' are factors of the uniformly recurrent word m, also m' is uniformly recurrent.

Now, by the uniform recurrence of m', with arguments similar to those used in the proof of Proposition 2.7.7, one derives that m' can be factorized as

$$m' = \cdots w_n w_{n-1} \cdots w_2 w_1 a,$$

where

$$w_1 = a f_1,$$

$$w_2 = w_1 a f_2,$$

$$\vdots$$

$$w_h = w_{h-1}w_{h-2}\cdots w_2 w_1 a f_h,$$
$$\vdots$$

for some set of words f_h, $h \geq 1$.

By hypothesis there exists an integer $n \geq 1$ such that the product

$$\phi(w_n)\phi(w_{n-1})\cdots\phi(w_2)\phi(w_1)$$

can be centrally iterated. So, for some integers i, j, $i \geq j \geq 1$, we have

$$w_n \cdots w_2 w_1 \equiv w_n \cdots w_{i+1}(w_i \cdots w_j)^2 w_{j-1} \cdots w_1,$$

where $\equiv$ denotes the congruence $\phi\phi^{-1}$. Thus one has

$$w_n \cdots w_2 w_1 \equiv w_n \cdots w_{i+1} w_i \cdots w_j w_i u \tag{3.23}$$

for a suitable $u \in A^*$. Let us first suppose $j > 1$. As w_i has the prefix $w_{j-1}w_{j-2}\cdots w_2 w_1 a$, we have, for a suitable $v \in A^*$

$$w_n \cdots w_1 \equiv w_n \cdots w_1 a v.$$

From (3.23) one reaches the same result also when $j = 1$ since $w_i \in aA^*$ for all $i \geq 1$. Since there exists $\lambda \in A^*$ such that $\lambda w_n \cdots w_1 = u \in U$, it follows that

$$u \equiv uav$$

and, therefore, $\phi(u)\ \mathcal{R}\ \phi(ua)$ which is a contradiction. This proves Claim 1.

The previous claim can be improved as follows.

Claim 2. *There exists a positive integer k such that, for each $u, v \in F(m)$, if $uv \in F(m)$ and $|u| \geq k$, then $\phi(u)\ \mathcal{R}\ \phi(uv)$.*

Proof of Claim 2. Let $u, v \in F(m)$, $v = a_1 a_2 \cdots a_m$ for some $a_1, a_2, \ldots, a_m \in A$, such that $|u| \geq k$ and $uv \in F(m)$. By Claim 1 one has

$$\phi(u)\ \mathcal{R}\ \phi(u)\phi(a_1)\ \ \mathcal{R}\ \phi(u)\phi(a_1)\phi(a_2)\ \mathcal{R}\ \cdots\ \mathcal{R}\ \phi(u)\phi(a_1)\cdots\phi(a_m).$$

This proves Claim 2.

Since m is uniformly recurrent we can write

$$m = \cdots w_h w_{h-1} \cdots w_2 w_1 t,$$

where $t \in A^\omega$, $|w_1| \geq k$, and

$$w_h \in w_{h-1}w_{h-2}\cdots w_2 w_1 A^*,$$

for any $h > 1$.

Let us consider the infinite sequence $\phi(w_2), \phi(w_3), \ldots, \phi(w_m), \ldots$ of elements of S. By hypothesis there exists an integer $n \geq 2$ such that the product

$$\phi(w_n)\phi(w_{n-1})\cdots\phi(w_2)$$

can be centrally iterated. Therefore, for some integers $i \geq j > 1$ we have

$$w_n \cdots w_2 \equiv w_n \cdots w_{i+1}(w_i \cdots w_j)^2 w_{j-1} \cdots w_2$$

and then

$$w_n \cdots w_1 \equiv w_n \cdots w_{i+1}(w_i \cdots w_j)^2 w_{j-1} \cdots w_1.$$

Since $|w_i| > k$ for all $i = 2, \ldots, n$, thus by Claim 2 one has $\phi(w_i) \ \mathcal{R} \ \phi(w_i \cdots w_1)$. Moreover, w_i has the prefix $w_{j-1}w_{j-2}\cdots w_2w_1$, so that one has $\phi(w_i) \ \mathcal{R} \ \phi(w_{j-1}\cdots w_1)$. Thus $\phi(w_i \cdots w_1) \ \mathcal{R} \ \phi(w_{j-1}\cdots w_1)$. This implies

$$w_{j-1} \cdots w_1 \equiv w_i \cdots w_j \cdots w_1 \zeta,$$

for a suitable $\zeta \in A^*$. One has then

$$w_n \cdots w_i \cdots w_j w_{j-1} \cdots w_1 \equiv w_n \cdots w_{i+1}(w_i \cdots w_j)^2 w_{j-1} \cdots w_1 \zeta$$

$$\equiv w_n \cdots w_1 \zeta \equiv w_n \cdots w_{i+1} w_{j-1} \cdots w_1.$$

Thus $w_n w_{n-1} \cdots w_2 w_1$ is reducible and this is a contradiction.

Remark 3.7.1. The ω-iteration property is weaker than the iteration condition $C(2,m)$. Thus, from Theorem 3.7.5 one has that $C(2,m)$ is a finiteness condition. However, the argument of the proof of Theorem 3.7.5, which is different and more combinatorial than that given for Theorem 3.7.3, cannot be extended to the case of condition $C(3,m)$.

3.7.2 Strong periodicity

Another important property, strictly related to the iteration property is the *strong periodicity*. Let S be a semigroup and $E(S)$ the set of its idempotent elements.

Definition 3.7.5. *Let m be a positive integer. A semigroup S is* strongly m-periodic *if for any sequence $s_1, \ldots, s_m$ of m elements of S there exist integers i and j such that $1 \leq i \leq j \leq m$ and $s_i \cdots s_j \in E(S)$.*

A semigroup S is *strongly periodic* if there exists a positive integer m such that S is strongly m-periodic [149]. The origin of the term strongly m-periodic is due to the fact that if S is strongly m-periodic, then S is certainly periodic and, moreover, the index and the period of any element are less than or equal to m.

Theorem 3.7.6. *(Simon) Let S be a finitely generated semigroup. The following conditions are equivalent:*

i. S is finite.

ii. $S \setminus E(S)$ *is finite.*
iii. S *is strongly periodic.*

Proof. The implication $i \Rightarrow ii$ is trivial. For that which concerns the implication $ii \Rightarrow iii$ one makes use of the theorem of Ramsey. Let $F = S \setminus E(S)$ and $p = \mathrm{card}(F)$. We prove that S is strongly m-periodic with $m = R(2, 3, p+1) - 1$, where R denotes the function of Ramsey's theorem. Let $s_1, \ldots, s_m$ be a sequence of m elements of S. We define

$$B_0 = \{\{i,j\} \mid 1 \le i < j \le m+1 \text{ and } s_i s_{i+1} \cdots s_{j-1} \in E(S)\},$$

and for any $f \in F$,

$$B_f = \{\{i,j\} \mid 1 \le i < j \le m+1 \text{ and } s_i s_{i+1} \cdots s_{j-1} = f\}.$$

By Ramsey's theorem there exist i_1, i_2, i_3, with $1 \le i_1 < i_2 < i_3 \le m+1$, such that $\{i_1, i_2\}$,$\{i_2, i_3\}$, and $\{i_1, i_3\}$ are all in the same class. This class is certainly B_0 because, otherwise, there will exist $f \in F$ such that

$$f = s_{i_1} \cdots s_{i_3-1} = (s_{i_1} \cdots s_{i_2-1})(s_{i_2} \cdots s_{i_3-1}) = f^2$$

which is a contradiction.

$iii \Rightarrow i$: if a semigroup S is strongly m-periodic, then it satisfies the condition $D(2, m)$ or $C(2, m)$; then, by Theorem 3.7.3, S is finite.

3.8 Permutation and iteration property

In the previous sections we proved that "permutability" and "iteration" are two different "mechanisms" on which some finiteness conditions for semigroups are based. In this section we prove that these two properties can be suitably combined in a more general finiteness condition for finitely generated and periodic semigroups [45].

Theorem 3.8.1. *Let S be a finitely generated and periodic semigroup. S is finite if and only if there exists a pair of integers (n, k) with $k = 2$ or 3, such that any sequence of n elements of S is either permutable or k-iterable on the right.*

Proof. The "only if" part of the theorem is trivial. Let us prove the "if" part. We make the proof for the case $k = 3$ (the case $k = 2$ is included in it). First of all we will prove that all subgroups of S are weakly permutable. Let G be a subgroup of S and $r = 2n^2$. We prove that G is r-weakly-permutable. In fact, any sequence $g_1, \ldots, g_n$ of elements of G is either permutable or 3-iterable on the right, i.e., there exist integers i, j such that $1 \le i \le j \le n$ and

$$g_1 \cdots g_j = g_1 \cdots g_{i-1}(g_i \cdots g_j)^3.$$

Since G is a group by cancellation one has

$$(g_i \cdots g_j)^2 = 1,$$

where 1 is the identity of G. Hence, by Proposition 3.7.2 it follows that G is $2n^2$-weakly permutable. Moreover, G is periodic. Therefore, by Theorems 3.4.1 and 3.4.6 the finitely generated subgroups of S are finite and the semigroup S has a weak $\mathcal{J}$-depth decomposition (see Theorem 3.6.1).

Let us suppose by contradiction that S is infinite; thus K_S is infinite. We denote by H the set of the canonical representatives of the elements of K_S. Since K_S is infinite and closed by factors, then, by Lemma 3.2.2, there exists an irreducible uniformly recurrent word x such that $F(x) \subseteq H$.

We prove now that x is ω-power free. Let $u \in F(x)$. Since S is periodic, there exists an integer $p(u)$ with the property that for any $h \geq p(u)$ there exists $h' < p(u)$ such that

$$\phi(u)^h = \phi(u^h) = \phi(u^{h'}) = \phi(u)^{h'}.$$

Therefore, since x is irreducible, for any $h \geq p(u)$, $u^h \notin F(x)$. Observe now that there exist infinitely many integers i such that

$$\phi(x[i])S^1 \neq \phi(x[i+1])S^1. \tag{3.24}$$

In fact, if this is not the case, then there exists an integer $h > 0$ such that for any $i, j \geq h$

$$\phi(x[i])S^1 = \phi(x[j])S^1,$$

so that

$$\phi(x[i]) \; \mathcal{R} \; \phi(x[i+1]) \quad \text{and} \quad \text{then } \phi(x[i]) \; \mathcal{J} \; \phi(x[i+1]). \tag{3.25}$$

Since $K_S = \cup_{t>0} K_t$, there exists a positive integer t such that $\phi(x[h]) \in K_t$. Since K_t is union of $\mathcal{J}$-classes, by equation (3.25) for any $j \geq h$ one has

$$\phi(x[j]) \in K_t.$$

Since K_t is finite, then there exist two integers i, j, $i < j$, such that $\phi(x[i]) = \phi(x[j])$ and this is absurd since x is irreducible.

Then there exists an integer i such that $\phi(x[i])S^1 \neq \phi(x[i+1])S^1$ and $i > h(n)$, where h is the function in Proposition 2.7.8. By this latter proposition, $x[i]$ has as a suffix the m th term of a bi-ideal sequence $f_{s+1} = f_s g_s f_s$, $s = 1, \ldots, m-1$, with $m \geq n$, $f_1 \in x_{i+1}A^*$, and $g_s \in x_{i+1}A^*$ for $s = 1, \ldots, m-1$. Moreover, the canonical factorization $(w_1, \ldots, w_m)$ of f_m has a derived sequence $(u_1, \ldots, u_n)$ which is either an n-division or an inverse n-division. We remark that since $f_1 \in x_{i+1}A^*$ and $g_s \in x_{i+1}A^*$ for $s = 1, \ldots, m-1$, one has that $(w_1, \ldots, w_m, x_{i+1})$ is an $(m+1)$-sequence.

Let $(u'_1, \ldots, u'_n)$ be the inverse canonical factorization associated with $(u_1, \ldots, u_n)$. From Corollary 2.7.1 either $(u_1, \ldots, u_n)$ or $(u'_1, \ldots, u'_n)$ is an

n-division. By the definition of a derived sequence there exist $n+1$ integers $j_1, j_2, \ldots, j_{n+1}$ such that $1 \le j_1 < j_2 < \cdots < j_{n+1} \le m+1$ and $u_1 = w_{j_1} \cdots w_{j_2-1}, \ldots, u_n = w_{j_n} \cdots w_{j_{n+1}-1}$. We set in the following

$$w_f = w_{j_{n+1}} \cdots w_m; \tag{3.26}$$

moreover $x[i] = \lambda w_1 \cdots w_m$, with $\lambda \in A^*$, so that there exists $u \in A^*$ such that

$$x[i] = uu_1 \cdots u_n w_{j_{n+1}} \cdots w_m = uu_1 \cdots u_n w_f. \tag{3.27}$$

Let us set for all $i \in [1,n]$, $s_i = \phi(u_i)$, and $s'_i = \phi(u'_i)$; we consider separately the two possible subcases.

Case a). $(u_1, \ldots, u_n)$ is an n-division.

By our hypotheses one has that the n-tuple $(s_1, \ldots, s_n)$ is either permutable or 3-iterable on the right. In the first case there exists a nontrivial permutation $\sigma \in S_n$ such that

$$s_1 \cdots s_n = s_{\sigma(1)} \cdots s_{\sigma(n)};$$

moreover, since $(u_1, \ldots, u_n)$ is an n-division one has

$$u_{\sigma(1)} \cdots u_{\sigma(n)} < u_1 \cdots u_n,$$

and this contradicts the fact that $u_1 \cdots u_n$ is irreducible. Now suppose that $(s_1, \ldots, s_n)$ is 3-iterable on the right. Thus there exist integers p, q such that $1 \le p \le q \le n$ and

$$s_1 \cdots s_q = s_1 \cdots s_{p-1}(s_p \cdots s_q)^3;$$

denoting by $\equiv$ the kernel of ϕ, one can write

$$u_1 \cdots u_q \equiv u_1 \cdots u_{p-1}(u_p \cdots u_q)^3 \in u_1 \cdots u_{p-1} u_p \cdots u_q u_p A^*. \tag{3.28}$$

Since $(u_1, \ldots, u_n)$ is a derived sequence of $(w_1, \ldots, w_m)$ and then of the $(m+1)$-sequence $(w_1, \ldots, w_m, x_{i+1})$, by equation (3.26) we can write for a suitable $\xi \in A^*$

$$u_p = u_{q+1} \cdots u_n w_{j_{n+1}} \cdots w_m x_{i+1} \xi = u_{q+1} \cdots u_n w_f x_{i+1} \xi.$$

Thus from equation (3.28) there exists $\zeta \in A^*$ such that

$$u_1 \cdots u_q \equiv u_1 \cdots u_n w_f x_{i+1} \zeta \in u_1 \cdots u_n w_f x_{i+1} A^*. \tag{3.29}$$

One has then from equations (3.27) and (3.29):

$$uu_1 \cdots u_q \equiv uu_1 \cdots u_n w_f x_{i+1} \zeta \in x[i+1] A^*.$$

Hence, $\phi(uu_1 \cdots u_q)S^1 \subseteq \phi(x[i+1])S^1$. Since $\phi(x[i])S^1 \subseteq \phi(uu_1 \cdots u_q)S^1$, one has $\phi(x[i])S^1 \subseteq \phi(x[i+1])S^1$ and then $\phi(x[i])S^1 = \phi(x[i+1])S^1$ which is a contradiction.

Case b). $(u_1, \ldots, u_n)$ is an inverse n-division.

By Corollary 2.7.1, the inverse canonical factorization $(u'_1, \ldots, u'_n)$ is an n-division. Then, as before we may prove that the n-tuple $s'_1, \ldots, s'_n$ cannot be permutable. So it has to be 3-iterable on the right. Therefore, there exist p, q with $1 \leq p \leq q \leq n$, such that

$$s'_1 \cdots s'_q = s'_1 \cdots s'_{p-1}(s'_p \cdots s'_q)^3. \tag{3.30}$$

We can also write, equivalently, the previous equation as

$$u'_1 \cdots u'_q \equiv u'_1 \cdots u'_{p-1}(u'_p \cdots u'_q)^3 = u'_1 \cdots u'_q u'_p \zeta, \tag{3.31}$$

for a suitable $\zeta \in A^*$. Now by the reciprocity law (see Proposition 2.7.1)

$$u'_1 \cdots u'_q = u_{n-q+1} \cdots u_n$$

so that (3.31) becomes:

$$u_{n-q+1} \cdots u_n \equiv u_{n-q+1} \cdots u_n u'_p \zeta. \tag{3.32}$$

Moreover, $u'_1 \cdots u'_p = u_{n-p+1} \cdots u_n$ and

$$u_{n-p+1} = u_{n-p+2} \cdots u_n w_{j_{n+1}} \cdots w_m x_{i+1} \xi = u_{n-p+2} \cdots u_n w_f x_{i+1} \xi,$$

for a suitable $\xi \in A^*$; thus, by using again the reciprocity law, one derives

$$u'_1 \cdots u'_p = u'_1 \cdots u'_{p-1} u'_p = u_{n-p+2} \cdots u_n u'_p = u_{n-p+1} \cdots u_n$$
$$= u_{n-p+2} \cdots u_n w_f x_{i+1} \xi u_{n-p+2} \cdots u_n.$$

Hence, by cancellation, one has

$$u'_p = w_f x_{i+1} \xi u_{n-p+2} \cdots u_n.$$

By substituting this expression of u'_p in equation (3.32) one obtains

$$u_{n-q+1} \cdots u_n \equiv u_{n-q+1} \cdots u_n w_f x_{i+1} \xi u_{n-p+2} \cdots u_n \zeta.$$

From equation (3.27), $x[i] = uu_1 \cdots u_n w_f$, so that it follows that

$$x[i] \equiv x[i+1] \xi u_{n-p+2} \cdots u_n \zeta w_f.$$

This implies that $\phi(x[i])S^1 \subseteq \phi(x[i+1])S^1$ which is a contradiction.

We remark that Theorem 3.8.1 contains Theorem 3.7.1 and the theorem of Restivo-Reutenauer as particular cases. In conclusion, a problem which remains open is whether in Theorem 3.8.1 one can replace the property of iteration on the right with the central iteration.

3.9 Repetitivity

In this section we analyze finiteness conditions for semigroups based on a property satisfied by finite semigroups and called "repetitivity".

A semigroup S is called *repetitive* if any morphism $\phi : A^+ \to S$, where A is a finite alphabet, is a repetitive map. This, trivially, implies that any finitely generated subsemigroup T of S satisfies the following property: any element t of T generated by the product of a sufficiently large number of generators will contain a factor which is an arbitrary large power of an element of T.

Repetitivity does not assure, in general, the finiteness of a finitely generated semigroup. Indeed, we shall see in Sect. 3.9.1 that the additive semigroup of positive integers $\mathbb{N}_+$ is repetitive. In order to obtain finiteness conditions one has to add to repetitivity of the morphisms or of the semigroups, further requirements. As we shall see in Sects. 3.9.2 and 3.9.3, *strong repetitivity* and *uniform repetitivity* will be suitable conditions stronger than repetitivity, which are finiteness conditions for a finitely generated semigroup.

3.9.1 Repetitive morphisms and semigroups

In this section we consider morphisms from a finitely generated semigroup into a semigroup which are k-repetitive maps (see Def. 2.1.1). We can transmit the property of repetitivity from morphisms to semigroups by the following:

Definition 3.9.1. *A semigroup S is* k-repetitive, $k > 1$, *if for each finite alphabet A, each morphism $\phi : A^+ \to S$ is k-repetitive. A semigroup is* repetitive *if it is k-repetitive for each $k > 1$.*

A natural question is whether repetitivity is a finiteness condition for semigroups. The answer is, in general, negative as shown by the following theorem, whose proof is simply derived from van der Waerden's theorem, showing that the semigroup $\mathbb{N}_+$ of positive integers is repetitive.

Theorem 3.9.1. *The additive semigroup $\mathbb{N}_+$ is repetitive.*

Proof. Let A be a finite alphabet and ϕ any morphism from the free semigroup A^+ into $\mathbb{N}_+$. Let us consider a new alphabet $B = \{b_1, b_2, \ldots, b_m\}$, where $m = \max\{\phi(a) \mid a \in A\}$, and a morphism $\psi : A^+ \to B^+$ defined for all $a \in A$ as $\psi(a) = b_l b_{l-1} \cdots b_1$, where $\phi(a) = l$. Let $r \geq 1$ and $w \in A^+$; by van der Waerden's theorem there exists an integer n, depending on card(A) and r, such that if $w \in A^+$ and $|\psi(w)| \geq n$, then $\psi(w)$ contains an arithmetic cadence $\{t_1, t_2, \ldots, t_{r+1}\}$ of order r and rate d. Moreover, let b_p be the letter of B that occurs in such a cadence.

For any $i \in [1, r+1]$ let us denote by u_i the shortest prefix of w such that $|\psi(u_i)| \geq t_i$. Clearly one has that $|\psi(u_i)| = \phi(u_i) = t_i + p - 1$, because $\psi(u_i) = v_i b_p \cdots b_1$ and $|v_i| = t_i - 1$. We can then factorize w as

$$u_1w_1w_2\cdots w_rv,$$

where $v \in A^*$ and w_i is the word such that $u_{i+1} = u_iw_i$ for $i = 1,\ldots,r$. Then one has

$$\phi(w_i) = \phi(u_{i+1}) - \phi(u_i) = t_{i+1} + p - 1 - (t_i + p - 1) = d,$$

which implies $\phi(w_1) = \phi(w_2) = \cdots = \phi(w_r)$.

Although repetitivity is not a finiteness condition for a semigroup, if one considers stronger conditions such as *strong repetitivity* and *uniform repetitivity*, then one obtains, as we shall see in the next sections, new finiteness conditions for finitely generated semigroups.

3.9.2 Strongly repetitive morphisms

Definition 3.9.2. *Let S be a semigroup. We say that a morphism $\phi : A^+ \to S$ is* strongly repetitive *if it satisfies the following condition: for any map $f : \mathbb{N}_+ \to \mathbb{N}_+$ there exists a positive integer M, which depends on f, such that for any $w \in A^+$ if $|w| \geq M$, then w can be factorized as*

$$w = hv_1\cdots v_{f(p)}h',$$

with $p \in \mathbb{N}_+$, $h, h' \in A^*$, $0 < |v_i| \leq p$, $1 \leq i \leq f(p)$, and

$$\phi(v_1) = \phi(v_2) = \cdots = \phi(v_{f(p)}).$$

A strongly repetitive morphism is certainly repetitive, whereas the converse is not generally true. The notion of strongly repetitive morphism can equivalently be given in terms of infinite words as shown by the following:

Lemma 3.9.1. *Let S be a semigroup and $\phi : A^+ \to S$ be a morphism. Then ϕ is strongly repetitive if the following condition is satisfied:*

(C) for any $x \in A^\omega$ there exists a positive integer p such that for any integer $n > 0$, x admits a factorization (that depends on n)

$$x = uw_1w_2\cdots w_nx',$$

with $u \in A^*$, $x' \in A^\omega$, $0 < |w_i| \leq p$, $(i = 1,\ldots,n)$, and

$$\phi(w_1) = \phi(w_2) = \cdots = \phi(w_n).$$

Proof. Let ϕ be a strongly repetitive morphism of A^+ in S and suppose by contradiction that there exists $x \in A^\omega$ with the property that for any $p > 0$ there exists an integer $n > 0$ such that for any factorization

$$x = uw_1w_2\cdots w_nx',$$

with $u \in A^*$, $x' \in A^\omega$, $0 < |w_i| \leq p$ $(i = 1, \ldots, n)$ there exist integers i, j, $1 \leq i < j \leq n$ for which $\phi(w_i) \neq \phi(w_j)$.

We can then introduce the map $f : \mathbb{N}_+ \to \mathbb{N}_+$ defined for any $p > 0$ as $f(p) = n$. Since ϕ is strongly repetitive there exists a positive integer M, depending on the map f, such that any word $w \in A^*$ of length $\geq M$ can be factorized as

$$w = hv_1v_2 \cdots v_{f(p)}h'$$

with $h, h' \in A^*$, $p \in \mathbb{N}_+$, $0 < |v_i| \leq p$, $1 \leq i \leq f(p)$, and

$$\phi(v_1) = \phi(v_2) = \cdots = \phi(v_{f(p)}).$$

This is absurd since any factor of the infinite word x does not satisfy this property.

Conversely, let the condition (C) be satisfied and suppose, by contradiction, that there is a map $f : \mathbb{N}_+ \to \mathbb{N}_+$ such that for any $M > 0$ there exists $w \in A^+, |w| \geq M$ with the property that for any $p > 0$ if

$$w = hv_1v_2 \cdots v_{f(p)}h'$$

with $h, h' \in A^*$ and $0 < |v_i| \leq p, 1 \leq i \leq f(p)$, then there exist integers i, j such that $1 \leq i < j \leq f(p)$ and $\phi(v_i) \neq \phi(v_j)$.

Let H be the set of all words of A^+ satisfying this latter property. Since H is infinite and closed by factors, by Lemma 1.2.2 there exists an infinite word $x \in A^\omega$ such that $F(x) \subseteq H$. This gives rise to a contradiction, since by hypothesis there exists a positive integer p such that (taking $n = f(p)$), x admits a factorization

$$x = uw_1w_2 \cdots w_{f(p)}x'$$

with $u \in A^*$, $x' \in A^\omega$, $0 < |w_i| \leq p$ $(1 \leq i \leq f(p))$, and

$$\phi(w_1) = \phi(w_2) = \cdots = \phi(w_{f(p)}).$$

Hence, $w_1w_2 \cdots w_{f(p)}$ will be a factor of a word of H which is a contradiction.

Definition 3.9.3. *A semigroup S is called* strongly repetitive *if any morphism $\phi : A^+ \to S$, where A is a finite alphabet, is strongly repetitive.*

The following theorem is due to J. Justin [87]. His proof uses the van der Waerden theorem. We report here a different proof due to Brown [16] and based on Brown's lemma (see Corollary 2.3.3).

Theorem 3.9.2. *A finite semigroup is strongly repetitive.*

Proof. Let S be a finite semigroup. The proof is by induction on the integer $p = \text{card}(S)$. If $p = 1$ the result is trivial. Let us then suppose $p \geq 2$ and assume the result true for all finite semigroups whose cardinality is less than p. Then let S be a semigroup whose cardinality is p. We make induction on

$q = \mathrm{card}(A)$. If $q = 1$ the result is trivial, so that we assume the result true up to q and prove it for $q+1$. In the continuation of the proof we may suppose that ϕ is an epimorphism, otherwise the result follows from the inductive hypothesis since $\mathrm{card}(\phi(A^+)) < \mathrm{card}(S)$.

Let $s = a_1 a_2 \cdots \in A^\omega$ be an infinite word on A and let x be a fixed, but arbitrary, letter of A. By Theorem 2.3.1 there exists a uniformly recurrent word $t \in A^\omega$ such that $F(t) \subseteq F(s)$.

Let us first suppose that $x \notin \mathrm{alph}(t)$. Since $F(t) \subseteq F(s)$, one derives that there are arbitrarily long factors of s missing the letter x. Hence, there exists an infinite set H of finite factors of s missing the letter x. The set H is closed by factors. By Lemma 1.2.2 there exists an infinite word $s' \in (A \setminus \{x\})^\omega$ such that $F(s') \subseteq H \subseteq F(s)$. Since $\mathrm{card}(A \setminus \{x\}) = q$, then by using induction in this case the result follows.

Let us then suppose that $x \in \mathrm{alph}(t)$. Let $t = t_1 \cdots t_n \cdots$. Since t is uniformly recurrent, there exists an integer k such that any factor of t of length k contains at least one occurrence of the letter x. Thus, there exists an infinite sequence of integers $i_1, i_2, \ldots$ such that

$$x = t_{i_1} = t_{i_2} = \cdots$$

and $0 < i_{j+1} - i_j \leq k$ for $j = 1, 2, \ldots$. We can always suppose, without loss of generality, that $x = t_1$. Let us now consider the new finite alphabet

$$B = \{t_{i_j} t_{i_j+1} \cdots t_{i_{j+1}-1} \mid j = 1, 2, \ldots\}.$$

We can then consider t as an infinite word on the alphabet B, that is, as an element of B^ω.

Let $\phi(x) = u \in S$. Then $\phi(t_{i_j}) = u$ for $j = 1, 2, \ldots$, so the restriction of ϕ to B^* is a morphism of B^* in uS. Let us first suppose that $\mathrm{card}(uS) < \mathrm{card}(S)$. By induction one derives that there exists an integer m such that for any n, the word t can be factorized as

$$t = \lambda g_1 \cdots g_n t',$$

with $\lambda \in A^*$, $t' \in A^\omega$, $|g_i| \leq mk$, and $\phi(g_1) = \phi(g_2) = \cdots = \phi(g_n)$. Since $F(t) \subseteq F(s)$ the result holds for s. By a similar argument if one considers the alphabet

$$C = \{t_{i_j+1} t_{i_j+2} \cdots t_{i_{j+1}} \mid j = 1, 2, \ldots\},$$

then the restriction of ϕ to C^* is a morphism of C^* in Su, so that if $\mathrm{card}(Su) < \mathrm{card}(S)$ we have finished by the induction hypothesis on $\mathrm{card}(S)$ (one repeats an argument similar to the above one).

Thus let us assume that $\mathrm{card}(uS) = \mathrm{card}(Su) = \mathrm{card}(S)$. Since x was a fixed but arbitrary letter, it follows that $u = \phi(x)$ is a fixed but arbitrary generator of S. This implies that we may suppose that for all $u \in S$ one has $\mathrm{card}(uS) = \mathrm{card}(Su) = \mathrm{card}(S)$. Since S is finite then one has $uS = S = Su$

for all $u \in S$. This implies that S is a right and left simple semigroup and thus a group (see Sect. 3.1).

Let us now regard S as a set. We consider the sequence $s_1 s_2 \cdots \in S^\omega$ with

$$s_i = \phi(a_1 \cdots a_i), \ (i = 1, 2, \ldots).$$

By Brown's lemma there exists an integer k such that for any integer n there are integers $i_1 < i_2 < \cdots$ such that

$$s_{i_1} = s_{i_2} = \cdots,$$

and $i_{j+1} - i_j \leq k$, $1 \leq j \leq n-1$. Let us set

$$g_1 = a_1 \cdots a_{i_1}, \ \cdots, \ g_n = a_{i_{n-1}+1} \cdots a_{i_n}.$$

One has

$$\phi(g_1) = \phi(g_1 g_2) = \cdots = \phi(g_1 g_2 \cdots g_n).$$

This implies, since S is a group,

$$\phi(g_1) = \phi(g_2) = \cdots = \phi(g_n),$$

and that concludes the proof.

A consequence of the above theorem is the following important finiteness condition for semigroups due to Brown [16]:

Theorem 3.9.3. *Let $\phi : S \to T$ be a morphism of semigroups. If T is locally finite and if, for each idempotent $e \in T$, $\phi^{-1}(e)$ is locally finite, then S is locally finite.*

Proof. In order to prove that S is locally finite we have to show that for any finite alphabet A and morphism $\zeta : A^+ \to S$ one has that $\zeta(A^+)$ is finite. Let us consider the morphism $\psi : A^+ \to T$, where $\psi = \zeta\phi$. Since T is locally finite, one has that $T' = \psi(A^+)$ is finite. This implies that there exists a positive integer r such that for every $t \in T'$ one has $t^r \in E(T)$, where $E(T)$ is the set of all idempotents of T. For every idempotent $e \in T'$ and $n \in \mathbb{N}_+$ we denote by $X_{n,e}$ the set

$$X_{n,e} = \{u \in A^+ \mid \zeta(u) \in \phi^{-1}(e) \text{ and } |u| \leq n\}.$$

Since $\phi^{-1}(e)$ is locally finite and $\zeta(X_{n,e}^+)$ is a finitely generated subsemigroup of $\phi^{-1}(e)$, it follows that $\zeta(X_{n,e}^+)$ is finite. Hence, there exists an integer $p(n, e)$ such that if $u \in X_{n,e}^+$ and $|u| > p(n, e)$, then there exists $u' \in A^+$ such that $|u'| < |u|$ and $\zeta(u) = \zeta(u')$. Let us now set

$$f(n) = r \max\{\, p(rn, e) \mid e \in E(T')\}.$$

We shall prove now that any sufficiently long word is reducible with respect to the morphism ζ. Since from Theorem 3.9.2 the morphism $\psi : A^+ \to T'$ is strongly repetitive, then there exists a positive integer M such that if $w \in A^+$ and $|w| \geq M$, then w can be factorized as

$$w = xu_1 \cdots u_{f(s)}y, \tag{3.33}$$

with $s \in \mathbb{N}_+$, $x, y \in A^*$, $1 \leq |u_i| \leq s$, $i = 1, \ldots, f(s)$, and

$$\psi(u_1) = \cdots = \psi(u_{f(s)}) = t \in T'.$$

Now $t^r = e_0 = e_0^2$, so that $f(s) \geq rp(rs, e_0)$. Let us then set $k = f(s)/r \geq p(rs, e_0)$ and rewrite equation (3.33) as

$$w = xv_1 \cdots v_k y,$$

where

$$v_1 = u_1 \cdots u_r, \ldots, v_k = u_{(k-1)r+1} \cdots u_{kr}.$$

Hence, one has $r \leq |v_i| \leq rs$, $i = 1, \ldots, k$, and

$$\psi(v_1) = \cdots = \psi(v_k) = t^r = e_0.$$

Since $v_i \in X_{rs,e_0}$ $(i = 1, \ldots, k)$, one has $|v_1 \cdots v_k| \geq k \geq p(rs, e_0)$. There exists then $z \in A^+$ such that $|z| < |v_1 \cdots v_k|$ and $\zeta(z) = \zeta(v_1 \cdots v_k)$. This implies that $\zeta(xzy) = \zeta(w)$ and $|xzy| < |w|$.

A different proof of the preceding theorem, based on the concept of "factorization forest", is in [151]. An application to the Burnside problem of the previous concepts and results is given by the following [31]:

Proposition 3.9.1. *Let S be a periodic semigroup. S is finite if and only if there exists a finite alphabet A and a strongly repetitive epimorphism ϕ : $A^+ \to S$.*

Proof. The "only if" part is a consequence of Justin's theorem (see Theorem 3.9.2). As regards to the "if" part one uses the periodicity of S and the fact that the epimorphism ϕ is strongly repetitive to show that any sufficiently long word of A^+ is reducible.

Let us set for any $p \in \mathbb{N}_+$, $A^{[1,p]} = A \cup A^2 \cup \cdots \cup A^p$ and $T_p = \phi(A^{[1,p]})$. The set T_p is, obviously, a finite subset of S. For each $t \in T_p$ we denote by ω_t the order of the element t. From the hypothesis of periodicity of S it follows that for all $t \in T_p$, ω_t is finite. Let us set

$$d_p = \max\{\omega_t \mid t \in T_p\} + 1.$$

This yields a map $f : \mathbb{N}_+ \to \mathbb{N}_+$, where for any $p \in \mathbb{N}_+$, $f(p) = d_p$. For any $t \in T_p$ there exists k (which depends on t) such that $k < f(p)$ and

$$t^{f(p)} = t^k.$$

In fact we can write $f(p) = \omega_t + q$ where q is a suitable positive integer depending on p. By the definition of ω_t it follows $t^{\omega_t+q} = t^k$ with $k \leq \omega_t < f(p)$.

Let ϕ denote the strongly repetitive morphism from A^+ onto S. For any $s \in S$, let w_s be the canonical representative of s. We prove now that $S = \phi(A^{[1,M]})$, where M is the integer in the Def. 3.9.2. Indeed, if $|w_s| > M$, then one can factorize w_s as

$$w_s = hv_1 \cdots v_{f(p)}h',$$

with $p \in \mathbb{N}_+$, $h, h' \in A^*$, $0 < |v_i| \leq p$, $1 \leq i \leq f(p)$, and

$$\phi(v_1) = \cdots = \phi(v_{f(p)}) = t \in S.$$

Now $t \in \phi(A^{[1,p]})$ so that $t^{f(p)} = t^k$ with $k < f(p)$. Setting $w' = hv_1 \cdots v_k h'$, we have $\phi(w') = \phi(w_s) = s$ with $|w'| < |w_s|$, which is a contradiction. Hence, $|w_s| \leq M$ and $S = \phi(A^{[1,M]})$, showing that S is finite.

The importance of the notion of strongly repetitive morphism in the theory of locally finite semigroups is shown by the following theorem that was announced in [88] and proved in [89] (an alternative proof, that we report here is in [55]).

Theorem 3.9.4. *Let S be a semigroup. S is locally finite if and only if it is strongly repetitive.*

Proof. The "only if" part of the theorem is a trivial consequence of Theorem 3.9.2. To prove the "if" part it is sufficient to show that a finitely generated and strongly repetitive semigroup S is finite. If S is also periodic, then the result follows from Proposition 3.9.1. If S is not periodic, then there exists at least one element $s \in S$ having an infinite order, i.e., $\langle s \rangle$ is isomorphic to $\mathbb{N}_+$. We shall prove then that $\mathbb{N}_+$ is not strongly repetitive.

Let $c \in A^\omega$ be the infinite word in the two-letter alphabet $A = \{a, b\}$ defined in the proof of Proposition 2.2.2, for any fixed irrational number β, as follows: for all $n > 0$

$$c(n) = \begin{cases} a & \text{if } \{n\beta\} \in [0, 1/2) \\ b & \text{if } \{n\beta\} \in [1/2, 1). \end{cases}$$

The word c can be factorized as

$$c = f_1 a f_2 a \cdots f_j a \cdots,$$

where for any $j \geq 1$, $f_j \in b^*$. From Proposition 2.2.2 it follows that for every $d > 0$ there exists an integer $h(d)$ such that c has not an arithmetical cadence of rate $\leq d$ and of order greater than $h(d)$. This implies that the lengths of the words f_j, $j \geq 1$, have to be limited. Let us then set $L = \sup\{|f_j| \mid j \geq 1\}$ and introduce the prefix code

$$X = \{fa \mid f \in b^* \text{ and } 0 \le |f| \le L\}.$$

Let Y be an alphabet such that $\text{card}(Y) = \text{card}(X)$ and consider a bijection α of Y onto X which can be extended to an isomorphism of Y^+ in X^+. Let us define the morphism $\phi : Y^+ \to \mathbb{N}_+$ from Y^+ in the additive semigroup $\mathbb{N}_+$, by setting

$$\phi(y) = |\alpha(y)|, \text{ for all } y \in Y.$$

Hence, one has for any $w \in Y^+$, $\phi(w) = |\alpha(w)|$. We now prove that the morphism ϕ is not strongly repetitive. Indeed, consider the infinite word c' over Y defined as

$$c' = \alpha^{-1}(f_1 a)\alpha^{-1}(f_2 a)\cdots$$

and the map $g : \mathbb{N}_+ \to \mathbb{N}_+$ defined for any $d > 0$ as $g(d) = h((L+1)d) + 1$. If ϕ is supposed to be strongly repetitive, then there will exist an integer $d > 0$ such that c' contains a factor of the kind

$$v_1 v_2 \cdots v_{g(d)},$$

where $v_i \in Y^+$, $0 < |v_i|_Y \le d$, ($|v_i|_Y$ is the length of v_i with respect to the alphabet Y), $i = 1, \ldots, g(d)$, and

$$\phi(v_1) = \phi(v_2) = \cdots = \phi(v_{g(d)}) = p.$$

This implies that the infinite word c over A has a factor

$$B_1 B_2 \cdots B_{g(d)},$$

where $|B_i| = p \le (L+1)d$, $i \in [1, g(d)]$. Since we can write $B_i = C_i a$, one has

$$B_1 B_2 \cdots B_{g(d)} = C_1 a C_2 a \cdots C_{g(d)} a,$$

so that c contains an arithmetical cadence having a rate equal to $p \le d(L+1)$ and order $g(d) = h((L+1)d) + 1$ which is a contradiction.

3.9.3 Uniformly repetitive semigroups

Definition 3.9.4. *A semigroup S is* uniformly k-repetitive, $k > 1$, *if for each finite alphabet A, each morphism $\phi : A^+ \to S$ is uniformly k-repetitive. A semigroup is* uniformly repetitive *if it is uniformly k-repetitive for each $k > 1$.*

The following theorem, whose proof is based on van der Waerden's theorem, shows that any finite semigroup is uniformly repetitive.

Theorem 3.9.5. *A finite semigroup is uniformly repetitive.*

Proof. Let $\phi : A^+ \to S$ be a morphism from a free semigroup A^+ into a finite semigroup S. We prove the statement by induction on $n = \text{card}(S)$. If $n = 1$, then the statement is trivially true. Let $n > 1$. We prove that for any integer k, any sufficiently long word of A^+ contains a uniform k-power mod ϕ. We recall that a *left (right) zero* of a semigroup S is an element $u \in S$ such that for any $s \in S$, $us = u$ ($su = u$). As one easily verifies, if a semigroup has two different left (right) zeros, then it cannot contain right (left) zeros. Then suppose that the semigroup S contains at most one left zero. Let w be a word such that $|w| \geq lk$. The integer l will be chosen later. Then we can write $w = xw_1w_2 \cdots w_ky$, where $|w_i| = l$, $1 \leq i \leq k$. If $\phi(w_i)$ is a left zero for any i, $1 \leq i \leq k$, then w contains a uniform k-power mod ϕ, since S contains at most one left zero. Suppose that there exists an integer i, $1 \leq i \leq k$, such that $\phi(w_i)$ is not a left zero of S. In such a case for any prefix u of w_i, $\phi(u)$ is not a left zero of S. For any integer j, $1 \leq j \leq l$, call v_j the prefix of length j of w_i. Let p be a positive integer and let $l \geq W(n, p+1)$, where W is the van der Waerden function. From Theorem 2.2.2 there exists an element $s \in S$ and an arithmetic progression $j, j+d, \ldots, j+pd$, such that

$$\phi(v_j) = \phi(v_{j+d}) = \cdots = \phi(v_{j+pd}) = s.$$

Let z_i be the word such that

$$v_{j+id} = v_{j+(i-1)d}\ z_i,$$

$1 \leq i \leq p$. Then one has $s\phi(z_i) = s$. Let $Z = \{z_1, \ldots, z_p\}$. The subsemigroup $\phi(Z^+)$ is properly contained in S, otherwise s is a left zero of S and this is impossible, since s is a left factor of $\phi(w_i)$. Then $\text{card}(\phi(Z^+)) < n$ and by the inductive hypothesis we can choose p such that the word $z_1z_2 \cdots z_p$, as a word on the alphabet Z, contains as a factor a word y which is a uniform k-power mod ϕ. Since the words z_i's have the same length d, the factor y, as a word on the alphabet A, is a uniform k-power mod ϕ.

The preceding theorem has shown that any finite semigroup is uniformly repetitive. The converse of this statement holds for finitely generated semigroups. Indeed, one can prove (see Corollary 3.9.2) that *any finitely generated uniformly repetitive semigroup is finite.* The following proposition is proved in [129].

Proposition 3.9.2. *Let k be an integer greater than* 1. *The following statements are equivalent:*

i. $\mathbb{N}$ *is not uniformly k-repetitive.*

ii. Any finitely generated and uniformly k-repetitive semigroup is finite.

Proof. As $\mathbb{N}$ is finitely generated, then *ii* $\Rightarrow$ *i* is trivially verified. Suppose now that $\mathbb{N}$ is not uniformly k-repetitive. This means that there exists a finite alphabet A, a morphism $\alpha : A^+ \to \mathbb{N}$ and an infinite subset U of A^+ such that each word u of U does not contain as a factor any uniform k-power mod α. By Lemma 1.2.2, there exists an infinite word $t \in A^\omega$, say

$$t = t_1 t_2 \cdots t_i \cdots,$$

such that $F(t) \subseteq F(U)$, so that any factor of t is not a uniform k-power mod α. Let S be a uniformly k-repetitive semigroup generated by a finite set G. Let $\phi : G^+ \to S$ be the canonical epimorphism, from the free semigroup G^+ onto S, defined by $\phi(g) = g$ for each $g \in G$ and suppose, by contradiction, that S is infinite. In this hypothesis, by Proposition 3.2.1, there exists an infinite word $m \in G^\omega$ which is irreducible. Let us factorize m in the following way

$$m = M_1 M_2 \cdots M_i \cdots$$

where $|M_i| = \alpha(t_i)$. Let $h = \max\{\alpha(a) \mid a \in A\}$ and B be a suitable alphabet with a bijection $\beta : B \to G \cup G^2 \cup \cdots \cup G^h$. Consider the morphism $\gamma : B^+ \to S$ defined as follows $\gamma(b) = \phi(\beta(b))$ for each letter b of B; consider also the infinite word x on the alphabet B defined by $x_i = \beta^{-1}(M_i)$ for $i > 0$. As S is uniformly k-repetitive, there exist positive integers r and s such that

$$\gamma(x_{r+1}x_{r+2}\cdots x_{r+s}) = \gamma(x_{r+s+1}x_{r+s+2}\cdots x_{r+2s})$$
$$= \cdots = \gamma(x_{r+(k-1)s+1}x_{r+(k-1)s+2}\cdots x_{r+ks}),$$

i.e.,

$$\phi(M_{r+1}M_{r+2}\cdots M_{r+s}) = \phi(M_{r+s+1}M_{r+s+2}\cdots M_{r+2s})$$
$$= \cdots = \phi(M_{r+(k-1)s+1}M_{r+(k-1)s+2}\cdots M_{r+ks}).$$

As the factors of m are irreducible, this implies that the k words

$$M_{r+1}M_{r+2}\cdots M_{r+s},$$

$$M_{r+s+1}M_{r+s+2}\cdots M_{r+2s},$$

$$\vdots$$

$$M_{r+(k-1)s+1}M_{r+(k-1)s+2}\cdots M_{r+ks},$$

have the same length, as words on the alphabet G, i.e.,

$$\sum_{j=r+1}^{r+s} \alpha(t_j) = \sum_{j=r+s+1}^{r+2s} \alpha(t_j) = \cdots = \sum_{j=r+(k-1)s+1}^{r+ks} \alpha(t_j)$$

or

$$\alpha(t_{r+1}\cdots t_{r+s}) = \alpha(t_{r+s+1}\cdots t_{r+2s}) = \cdots = \alpha(t_{r+(k-1)s+1}\cdots t_{r+ks}).$$

Hence, the infinite word t contains a uniform k-power mod α, and that is a contradiction.

Now, let us introduce the following definition:

Definition 3.9.5. *A word w on A is called an* abelian n-power *if there exist n nonempty words, say $w_1, \ldots, w_n$, such that $w = w_1 \cdots w_n$ and for each $i, j \in \{1, 2, \ldots, n\}$ the word w_i is commutatively equivalent to w_j.*

The following proposition is due to F. M. Dekking [29]:

Proposition 3.9.3. *Let A be a two-letter alphabet. Then there exists an infinite word on A no factor of which is an abelian 4-power.*

Corollary 3.9.1. *The additive semigroup $\mathbb{N}$ is not uniformly 4-repetitive.*

Proof. Let us consider the morphism $\psi : \{a,b\}^+ \to \mathbb{N}$, defined by $\psi(a) = 1$ and $\psi(b) = 2$. An easy computation shows that for any two words u and v, with $|u| = |v|$, if $\psi(u) = \psi(v)$, then u and v are commutatively equivalent. By the previous proposition there exists an infinite word over $\{a,b\}$ no factor of which is an abelian 4-power. Such a word cannot contain a factor of the kind $w_1w_2w_3w_4$, with $|w_1| = |w_2| = |w_3| = |w_4|$ and $\psi(w_1) = \psi(w_2) = \psi(w_3) = \psi(w_4)$, otherwise $w_1w_2w_3w_4$ should be an abelian 4-power.

From Proposition 3.9.2 and Corollary 3.9.1 we have

Proposition 3.9.4. *Let S be a finitely generated semigroup. If S is uniformly k-repetitive for some $k \geq 4$, then it is finite.*

Corollary 3.9.2. *A finitely generated uniformly repetitive semigroup is finite.*

It is an open problem whether $\mathbb{N}$ is uniformly 2-repetitive or uniformly 3-repetitive. This seems to be a difficult problem in combinatorial number theory.

Lemma 3.9.2. *Let S and T be finitely generated semigroups. If T is infinite and $T \times S$ is repetitive, then S is uniformly repetitive.*

Proof. Suppose that $T \times S$ is k-repetitive. We prove that S is k-uniformly repetitive. Let C be a finite alphabet and $\phi : C^+ \to T$ be an epimorphism. Since T is infinite, by Proposition 3.2.1 there exists an infinite word

$$m = m_1 \cdots m_i \cdots$$

on C such that all factors of m are irreducible. Now, let A be any finite alphabet, $\alpha : A^+ \to S$ be a morphism from A^+ into S and

$$t = t_1 \cdots t_i \cdots$$

be an arbitrary infinite word over A. Let $B = C \times A$. Clearly B is finite. Let us consider the morphism ψ from B^+ to $T \times S$ defined by $\psi(c,a) = (\phi(c), \alpha(a))$ for any (c,a) in $C \times A$. Let

$$s = s_1 \cdots s_i \cdots$$

be the infinite word over B, defined by $s_i = (m_i, t_i)$, for $i \geq 1$.

Since $T \times S$ is k-repetitive, there exist k consecutive factors of s, $w_1, \ldots, w_k$, such that $\psi(w_1) = \cdots = \psi(w_k)$. From this easily follows that there exist k consecutive factors $x_1, \ldots, x_k$ of m and k consecutive factors $y_1, \ldots, y_k$ of t such that $\phi(x_1) = \cdots = \phi(x_k)$ and $\alpha(y_1) = \cdots = \alpha(y_k)$. If $|x_i| < |x_j|$ for some $i \neq j$, then we have a contradiction by the irreducibility of the factors of m. This proves that $|x_1| = |x_2| = \cdots = |x_k|$ and so $|y_1| = |y_2| = \cdots = |y_k|$, i.e., α is uniformly k-repetitive. As A and t are arbitrary, this means that S is uniformly k-repetitive.

The following result was proved in [128]; here we have an alternative proof.

Proposition 3.9.5. *Let S and T be finitely generated semigroups. If $T \times S$ is repetitive, then S or T is finite.*

Proof. By Lemma 3.9.2, we have that if T is not finite, then S is uniformly repetitive. By Corollary 3.9.2, we have that S is finite.

4. Finitely Recognizable Semigroups

In this chapter we consider for any semigroup S two important families of parts of S. The first is the family Rec(S) of the *recognizable subsets* of S. A set $X \subseteq S$ is *recognizable* if it is union of classes of a congruence in S of finite index. The second is the family Rat(S) of the *rational subsets* of S defined as the smallest family of parts of S containing the finite parts and closed under the *rational operations.*

When S is a finitely generated free semigroup, then a fundamental theorem due to S. C. Kleene states that $\text{Rec}(S) = \text{Rat}(S)$. Moreover, in this case Rec(S) coincides with the family of parts (or languages) recognizable by a *finite automaton.* As a consequence of a theorem of Myhill and Nerode, proved in Sect. 4.1, a language is recognizable (or regular) if and only if the syntactic semigroup $S(L)$ of L is finite.

In the previous chapter we have considered several finiteness conditions for finitely generated semigroups. As we shall see in more detail in the next chapter, these conditions applied to the syntactic semigroup $S(L)$ of a language L give rise to regularity conditions for the language L.

In this chapter we shall consider conditions for a finitely generated semigroup which are "weaker" in the sense that they do not assure the finiteness of the semigroup S but only that the *finite parts* of S are recognizable. Such a semigroup will be called *finitely recognizable.* Since Rec(S) is closed under Boolean operations one has that a semigroup is finitely recognizable if and only if all the singletons $\{s\}$, with $s \in S$, are recognizable. If $\phi : A^+ \to S$ denotes the canonical epimorphism, then the condition that S is finitely recognizable is equivalent (see Lemma 4.2.1) to the statement that all congruence classes in A^+, modulo the nuclear congruence $\phi\phi^{-1}$, are regular languages. In this way one obtains regularity conditions for a wide class of formal languages.

An important problem, posed by J. Brzozowski in 1979 on the regularity of non-counting (or aperiodic) classes of order $n > 0$, consists precisely in proving that the semigroup $B(r, n, n+1) = A^+/\theta_n$, where θ_n is the congruence generated by the identity $x^n = x^{n+1}$ and $r = \text{card}(A)$, is finitely recognizable. This and other related problems will be considered in Sect. 4.6. A partial positive solution of it will be outlined in the case $n \geq 5$. This result implies that the *word problem* for the free semigroup $B(r, n, n+1)$ in the Burnside variety defined by the equation $x^n = x^{n+1}$ is solvable.

In the following sections we shall present some general results concerning the structure of finitely recognizable semigroups. These semigroups have to be stable and all their subgroups have to be finite. As a consequence of the $\mathcal{J}$-depth decomposition theorem we give more equivalent characterizations of the class of finitely generated and finitely recognizable semigroups which have a $\mathcal{J}$-depth function. This class coincides with the class of finite $\mathcal{J}$-above semigroups. Some applications to the word problem are shown. Finally, a positive solution of a conjecture of Brown, relative to $B(r, 2, 3)$ is given.

4.1 The Myhill–Nerode theorem

In this section we introduce ***recognizable subsets*** of a semigroup; then we prove the classical theorem of Myhill–Nerode which gives a basic characterization of recognizable sets.

Let S be a set and θ be an equivalence relation in S. The index $\operatorname{ind}(\theta)$ of θ is the cardinality of the quotient set S/θ. Let X be a subset of S. We say that θ ***saturates*** X, or that X is ***closed*** with respect to θ, if the following condition is satisfied: for $x, y \in S$

$$(x \in X \text{ and } x\theta y) \Rightarrow y \in X,$$

that is, X is union of equivalence classes of θ.

Definition 4.1.1. *Let S be a semigroup. A subset X of a semigroup S is* recognizable *if it is a union of classes of a congruence in S of finite index.*

In the following the family of recognizable parts of S will be denoted by $\operatorname{Rec}(S)$. Let S be a semigroup and X be a subset of S. The ***syntactic congruence*** of X, denoted by $\equiv_X$, is defined as follows: for any $s, t \in S$

$$s \equiv_X t \Longleftrightarrow \forall u, v \in S^1 (usv \in X \Leftrightarrow utv \in X).$$

For any $s \in S$ let $\operatorname{Cont}_X(s) = \{(u, v) \in S^1 \times S^1 | \; usv \in X\}$; then one has for $s, t \in S$

$$s \equiv_X t \Longleftrightarrow \operatorname{Cont}_X(s) = \operatorname{Cont}_X(t).$$

One can easily verify that $\equiv_X$ is a congruence relation in S. The ***syntactic semigroup*** of X is then defined as $S(X) = S/\equiv_X$. When S is a monoid, the syntactic semigroup $S(X)$ is, in fact, a monoid called the ***syntactic monoid*** of X. The natural epimorphism $\phi : S \to S(X)$ is usually called the ***syntactic morphism.***

For any subset X of S one can consider the right Nerode equivalence $\mathcal{N}_X^r$ of X defined in the following way. For any $s \in S$ set

$$s^{-1}X = \{t \in S \mid st \in X\} \text{ and } Xs^{-1} = \{t \in S \mid ts \in X\}.$$

One has then for $s, t \in S$

$$s \mathcal{N}_X^r t \text{ if and only if } s^{-1}X = t^{-1}X.$$

In a symmetric way the left Nerode equivalence $\mathcal{N}_X^l$ of X is defined as

$$s \mathcal{N}_X^l t \text{ if and only if } Xs^{-1} = Xt^{-1}.$$

The relations $\mathcal{N}_X^r$ and $\mathcal{N}_X^l$ are right and left invariant with respect to the product of S, respectively. We observe that $\equiv_X \subseteq \mathcal{N}_X^r$ and $\equiv_X \subseteq \mathcal{N}_X^l$.

As one can easily verify, the equivalence relations $\equiv_X$, $\mathcal{N}_X^r$ and $\mathcal{N}_X^l$ saturate X.

In the following we denote, when there are not ambiguities, $\mathcal{N}_X^r$ simply by $\mathcal{N}_X$ and call it the *Nerode equivalence* of X.

Proposition 4.1.1. *The syntactic congruence $\equiv_X$ is maximal, with respect to the inclusion, in the set of all the congruences that saturate X.*

Proof. Let θ be a congruence that saturates X. For any $s, t \in S$ one has

$$s \;\theta\; t \Rightarrow \forall u, v \in S^1 \;(usv \;\theta\; utv).$$

Since θ saturates X, from $s \;\theta\; t$ one derives

$$\forall u, v \in S^1 \;(usv \in X \iff utv \in X).$$

Therefore, $s \;\theta\; t \Rightarrow s \equiv_X t$, that is, $\theta \subseteq \equiv_X$.

In a similar way one proves the following:

Proposition 4.1.2. *The Nerode equivalence $\mathcal{N}_X$ is maximal, with respect to the inclusion, in the set of all the right invariant equivalences that saturate X.*

The following important theorem is due to J. Myhill and A. Nerode (see [62]).

Theorem 4.1.1. *Let X be a subset of a semigroup S. The following conditions are equivalent:*

i. $X \in \mathrm{Rec}(S)$.
ii. There exist a finite semigroup T and a morphism $\phi : S \to T$, such that $X = \phi^{-1}(\phi(X))$.
iii. The syntactic semigroup $S(X)$ is finite.
iv. X is saturated by a right invariant equivalence of finite index.
v. The Nerode equivalence $\mathcal{N}_X$ has a finite index.

Proof. *i* ⇒ *ii*: let X be a union of classes of a congruence θ in S of finite index. Let $\phi : S \to S/\theta$ be the natural epimorphism. For any $s \in S$ one has $[s]_\theta = \phi^{-1}(\phi(s))$ so that $X = \phi^{-1}(\phi(X))$.

ii ⇒ *iii*: let T be a finite semigroup and $\phi : S \to T$ be a morphism such that $X = \phi^{-1}(\phi(X))$. The congruence $\phi\phi^{-1}$ has a finite index and saturates X. By Proposition 4.1.1, $\phi\phi^{-1} \subseteq \equiv_X$. Therefore, $\equiv_X$ has a finite index and the syntactic semigroup $S(X)$ is finite.

iii ⇒ *iv*: X saturates $\equiv_X$ which is right invariant. By hypothesis $S(X)$ is finite, therefore, $\equiv_X$ has a finite index.

iv ⇒ *v*: by hypothesis X saturates a right invariant equivalence R of finite index. By Proposition 4.1.2, $R \subseteq \mathcal{N}_X$ and $\mathcal{N}_X$ has a finite index.

v ⇒ *i*: we introduce the relation θ in S defined as: for any $t_1, t_2 \in S$

$$t_1 \,\theta\, t_2 \iff \forall s \in S^1 \ ([st_1]_{\mathcal{N}_X} = [st_2]_{\mathcal{N}_X}).$$

One can easily verify that

$$t_1 \,\theta\, t_2 \iff t_1 \equiv_X t_2.$$

Therefore, θ coincides with $\equiv_X$. Moreover, the relation θ has a finite index. Indeed, let us observe that if $s, s', t \in S$ and $s \,\mathcal{N}_X\, s'$, then since $\mathcal{N}_X$ is right invariant, one has $st \,\mathcal{N}_X\, s't$, or $[st]_{\mathcal{N}_X} = [s't]_{\mathcal{N}_X}$. Thus let $T = \{s_1, s_2, \ldots, s_r\}$ be a finite set of representatives of the classes of $\mathcal{N}_X$, where $r = \mathrm{card}(S/\mathcal{N}_X)$. For any $t \in S$, the class $[t]_\theta$ is then uniquely determined by the sequence

$$([t]_{\mathcal{N}_X}, [s_1t]_{\mathcal{N}_X}, \ldots, [s_rt]_{\mathcal{N}_X}).$$

The number of these sequences is at most r^{r+1}. Therefore, the syntactic congruence $\equiv_X$ has a finite index and $X \in \mathrm{Rec}(S)$, because X saturates $\equiv_X$.

Remark 4.1.1. The statement of the equivalence of the conditions *i*, *ii*, and *iii* is usually referred as *Myhill's theorem* and the equivalence of *i*, *iv*, and *v* as *Nerode's theorem*. Moreover, in a similar way, one can prove that all the conditions of Theorem 4.1.1 are equivalent to the following condition: *vi. the index of the Nerode left equivalence* $\mathcal{N}^l_X$ *is finite.*

Corollary 4.1.1. Rec(S) *is closed under the Boolean operations.*

Proof. Let X be a recognizable subset of S. We first prove that its complement $X^c = S \setminus X$ is recognizable. Indeed, the syntactic congruence of X^c is equal to the syntactic congruence of X, so that $S(X^c) = S(X)$. Hence, by Theorem 4.1.1 the result follows.

Let us now suppose that $X_1, X_2 \in \mathrm{Rec}(S)$. By Theorem 4.1.1 there exist two congruences of finite index θ_1 and θ_2 which saturate X_1 and X_2, respectively. Let $\theta = \theta_1 \cap \theta_2$ and set $m_1 = \mathrm{ind}(\theta_1)$ and $m_2 = \mathrm{ind}(\theta_2)$. One easily verifies that θ is still a congruence and its index is less than or equal to m_1m_2. Since θ saturates X_1 and X_2, then it saturates also their intersection $X_1 \cap X_2$. From Theorem 4.1.1 it follows that $X_1 \cap X_2 \in \mathrm{Rec}(S)$. Since $X_1 \cup X_2 = (X_1^c \cap X_2^c)^c$, one derives that $X_1 \cup X_2 \in \mathrm{Rec}(S)$.

Let us now suppose that the semigroup S is the free monoid A^* over a finite alphabet. We shall prove that $\mathrm{Rec}(A^*)$ coincides with the family of languages definable in terms of ***finite automata.***

Let us recall that a finite (deterministic and complete) ***semi-automaton*** is a triplet

$$(S, A, \lambda),$$

where S is a finite set whose elements are called ***states***, A is a finite alphabet and λ is a map

$$\lambda : S \times A \to S,$$

called ***transition function.*** For any $s \in S$ and $a \in A$ we denote $\lambda(s,a)$ simply by sa. The transition function λ can be extended to $S \times A^*$ by setting, for any $s \in S$, $u \in A^*$, and $a \in A$

$$s\epsilon = s, \quad s(ua) = (su)a.$$

A ***finite automaton*** is defined as a quintuple

$$\mathcal{A} = (S, A, \lambda, s_0, S'),$$

where (S, A, λ) is a finite semi-automaton, s_0 is an element of S, called ***initial state*** and S' is a subset of S called set of the ***terminal*** or ***final states.*** A word $w \in A^*$ is ***accepted*** by $\mathcal{A}$ if $s_0 w \in S'$. We denote by $L(\mathcal{A})$ the set of all the words accepted by $\mathcal{A}$:

$$L(\mathcal{A}) = \{w \in A^* \mid s_0 w \in S'\}.$$

A language $L \subseteq A^*$ is ***acceptable by a finite automaton*** if there exists a finite automaton $\mathcal{A}$ such that $L = L(\mathcal{A})$.

Proposition 4.1.3. *A language $L \subseteq A^*$ is acceptable by a finite automaton if and only if $L \in \mathrm{Rec}(A^*)$.*

Proof. Let $L \subseteq A^*$ be a language acceptable by the finite automaton $\mathcal{A} = (S, A, \lambda, s_0, S')$. We introduce in A^* the equivalence relation θ defined as: for all $u, v \in A^*$

$$u \;\theta\; v \iff s_0 u = s_0 v.$$

The equivalence θ is, trivially, right invariant and saturates L. Indeed, if $u \in L$ and $u \;\theta\; v$ one has $s_0 u = s_0 v$. Since $s_0 u \in S'$, it follows $s_0 v \in S'$ and then $v \in L$. The index of θ is finite because $\mathrm{ind}(\theta) \leq \mathrm{card}(S)$. Thus, by Theorem 4.1.1, $L \in \mathrm{Rec}(A^*)$.

Conversely, suppose that $L \in \mathrm{Rec}(A^*)$. By Theorem 4.1.1 the index of the Nerode equivalence $\mathcal{N}_L$ is finite. We can then consider the finite quotient set $A^*/\mathcal{N}_L$ and define the finite automaton (***Nerode's automaton***)

$$\mathcal{A} = (A^*/\mathcal{N}_L,\ A,\ \lambda,\ [\epsilon]_{\mathcal{N}_L},\ F),$$

where the transition function λ is defined as: for all $u \in A^*$ and $a \in A$

$$[u]_{\mathcal{N}_L} a = [ua]_{\mathcal{N}_L}$$

and $F = \{[u]_{\mathcal{N}_L} \mid u \in L\}$. Note that the definition of λ is not ambiguous since $\mathcal{N}_L$ is right invariant. Moreover, F is a subset of $A^*/\mathcal{N}_L$ since $\mathcal{N}_L$ saturates L. One easily verifies that $L = L(\mathcal{A})$ which concludes the proof.

Proposition 4.1.4. *Finite languages are recognizable.*

Proof. By Corollary 4.1.1, Rec(A^*) is closed under Boolean operations, so that it is sufficient to prove that the language $L = \{w\}$, where $w \in A^*$, is recognizable. This can be done either by constructing a finite automaton accepting L, or by using Theorem 4.1.1, showing that the Nerode equivalence $\mathcal{N}_{\{w\}}$, simply denoted by $\mathcal{N}_w$, has a finite index. Indeed, for $u, v \in A^*$ one has

$$u \,\mathcal{N}_w\, v \iff \forall h \in A^* (uh = w \iff vh = w).$$

Now all the words which are not prefixes of w lie in a same equivalence class. If u and v are prefixes of w, then $u \,\mathcal{N}_w\, v$ if and only if $u = v$. Thus, we have a further number of equivalence classes equal to $|w| + 1$ that is the number of prefixes of w. Hence, ind($\mathcal{N}_w$) $= |w| + 2$ and L is recognizable.

4.2 Finitely recognizable semigroups

Let S be a semigroup and $\mathcal{P}(S)$ be the semigroup of its parts. We recall (see Sect. 3.1) that in $\mathcal{P}(S)$ one can introduce the ***rational operations*** of union ($\cup$), product ($\cdot$), and cross ($^+$) where for any subset X of S, X^+ gives the subsemigroup of S generated by X, i.e.,

$$X^+ = X \cup X^2 \cup \cdots \cup X^n \cup \cdots.$$

The family Rat(S) of the rational subsets of S is defined as *the smallest subset of $\mathcal{P}(S)$ containing the finite parts of S and closed under the rational operations.*

When S is a finitely generated free semigroup, or free monoid, then the following fundamental theorem due to S. C. Kleene holds (see [62]):

Theorem 4.2.1. *(Kleene) Let S be a finitely generated free semigroup (monoid). Then*

$$\text{Rec}(S) = \text{Rat}(S).$$

In the above case, as we have seen in the preceding section, Rec(S) coincides also with the family of parts (or languages) definable in terms of finite automata; these are usually called ***rational*** or ***regular languages.*** In general for a semigroup S the above equality does not hold. The semigroups (or monoids) for which $\mathrm{Rec}(S) = \mathrm{Rat}(S)$ have been called *Kleene's semigroups.* Some closure properties and structural properties of the Kleene semigroups and monoids have been studied in [140, 143]. An interesting proper subclass of the class of Kleene's monoids, called class of ***rational monoids***, was introduced by J. Sakarovitch [143].

A consequence of Kleene's theorem is the following important result due to J. McKnight (see [62])

Theorem 4.2.2. *If a semigroup S is finitely generated, then* $\mathrm{Rec}(S) \subseteq \mathrm{Rat}(S)$.

The inclusion in the preceding theorem is, in general, strict. Moreover, there are examples of finitely generated semigroups having some finite parts which are not recognizable. For this reason we define a semigroup as *finitely recognizable* if its finite parts are recognizable. The class of Kleene's semigroups is a narrow subclass of the finitely recognizable semigroups.

Definition 4.2.1. *A semigroup S is called* finitely recognizable *if all its finite parts are recognizable.*

Since $\mathrm{Rec}(S)$ is closed under Boolean operations, one has that a semigroup is finitely recognizable if and only if all the singletons $\{s\}$, with $s \in S$, are recognizable. If $\phi : A^+ \to S$ denotes the canonical epimorphism, then the condition that S is finitely recognizable is equivalent (see Lemma 4.2.1) to the statement that for any $s \in S$, $\phi^{-1}(s) \in \mathrm{Rec}(A^+)$.

We say that an element s of a semigroup S is *recognizable* if $\{s\} \in \mathrm{Rec}(S)$.

Theorem 4.2.3. *Let S be a semigroup and D a regular $\mathcal{D}$-class of S. If $r \in D$ is recognizable, then the group G_D of D is finite.*

Proof. The $\mathcal{D}$-class D is regular so that it contains at least one idempotent element e (see Theorem 3.1.3). Let $r \in D$ be such that $\{r\} \in \mathrm{Rec}(S)$. By Theorem 3.1.1 there exists a bijection of the $\mathcal{H}$-classes H_e and H_r. More precisely since $e \,\mathcal{D}\, r$, there is $b \in S$ such that $e \,\mathcal{R}\, b$ and $b \,\mathcal{L}\, r$. This implies that there exist elements $h, h', k, k' \in S^1$ such that $eh = b, bh' = e, kb = r$, and $k'r = b$. The mappings

$$x \to kxh \ (x \in H_e) \text{ and } z \to k'zh' \ (z \in H_r) \tag{4.1}$$

are mutually inverse, one-to-one mappings of H_e and H_r upon each other. Since $\{r\} \in \mathrm{Rec}(S)$, the syntactic congruence $\equiv_{\{r\}}$ is of a finite index. If the group G_D of the $\mathcal{D}$-class D is infinite, then so will be H_e. Then there will exist elements $s, t \in H_e$ such that $s \neq t$ and $s \equiv_{\{r\}} t$. Since H_e is a group, let s' be the inverse of s in H_e. One has $ss' = s's = e$ and $ss' = e \equiv_{\{r\}} ts'$. One has $keh = r$ so that

$$r = keh \equiv_{\{r\}} kts'h.$$

This implies that $r = keh = kts'h$. Since $ts' \in H_e$ and the mappings in equation (4.1) are bijections, one derives that $e = ts'$. Therefore, $ts's = es = s = te = t$, which is a contradiction.

Corollary 4.2.1. *If the idempotents of a semigroup S are recognizable, then all subgroups of S are finite.*

Proof. By assumption, any idempotent $e \in S$ is recognizable so that H_e is finite. Hence, all maximal subgroups of S are finite.

In particular one has that in a Kleene's semigroup all subgroups have to be finite. From Theorem 4.2.3 we derive in the case of groups the following corollary due to A.V. Anisimov [4]:

Corollary 4.2.2. *Let G be a group. If there exists an element of G which is recognizable, then G is finite.*

Proof. Let $g \in G$ be such that $\{g\}$ is a recognizable part of G. Since $D_g = G$ is, trivially, a regular $\mathcal{D}$-class and the group of D_g equals G, it follows that G is finite.

Theorem 4.2.4. *Let S be a semigroup. If $s \in S$ is recognizable, then s is stable and $J_s = D_s$.*

Proof. Let $\{s\} \in \mathrm{Rec}(S)$ and suppose that $t \in J_s$ is such that $tS^1 \subseteq sS^1$. It follows that there exist elements $r, u, v \in S^1$ such that $t = sr$ and $s = utv$. One has then for all $n \geq 0, s = u^n s(rv)^n$. Since $\equiv_{\{s\}}$ is of finite index, there exist two integers n and m with $n < m$ and such that $(rv)^n \equiv_{\{s\}} (rv)^m$. Setting $m = n+p$, with $p > 0$, we can write $s = u^n s(rv)^n \equiv_{\{s\}} u^n s(rv)^n(rv)^p$. It follows that $s = srv(rv)^{p-1} = tv(rv)^{p-1}$ and then $sS^1 \subseteq tS^1$. Thus $s \mathcal{R} t$. In a symmetric way one proves that if $S^1 t \subseteq S^1 s$, then $s \mathcal{L} t$.

Now let $t \in J_s$. One has $t = xsy$, $x, y \in S^1$. Hence, $S^1 t S^1 \subseteq S^1 xs S^1 \subseteq S^1 s S^1$ and $S^1 t S^1 \subseteq S^1 sy S^1 \subseteq S^1 s S^1$. Thus $xs, sy \in J_s$. Since $S^1 xs \subseteq S^1 s$ and $syS^1 \subseteq sS^1$ from the stability of s one has $xs \mathcal{L} s$ and $sy \mathcal{R} s$. Hence, $t = xsy \mathcal{R} xs$ which implies $t \in D_s$.

Definition 4.2.2. *A semigroup S is* residually finite *if for any pair s, t of elements of S such that $s \neq t$ there exists a congruence θ in S of finite index such that $(s,t) \notin \theta$.*

Finite semigroups are, trivially, residually finite. Also the free semigroup A^+ on the alphabet A is residually finite. To show this let $u, v \in A^+$, $u \neq v$, and let J be the ideal consisting of all the words of A^+ of length greater than the maximum of the lengths of u and v. The Rees congruence $\rho_J = \{id\} \cup (J \times J)$ is then of a finite index and separates u from v.

Theorem 4.2.5. *Let S be a finitely recognizable semigroup. Then S is residually finite.*

Proof. Let $s \in S$. Since $\{s\} \in \mathrm{Rec}(S)$ the syntactic congruence $\equiv_{\{s\}}$ has a finite index. Let now t be any element of S such that $s \neq t$. One has that $\equiv_{\{s\}}$ separates s from t. In fact, if one supposes that $s \equiv_{\{s\}} t$ it would follow that $s = t$.

Let us remark that the above theorem cannot in general be inverted. Indeed, there exist infinite groups, such as free groups [106, 109], which are residually finite. Moreover, by Corollary 4.2.2, a group is finitely recognizable if and only if it is finite.

An important relationship between the recognizable parts of S and the recognizable parts of A^+ is given by the following (see [47]):

Lemma 4.2.1. *Let S be a semigroup and $\phi : A^+ \to S$ be the canonical epimorphism. For a set $X \subseteq S$ the following two conditions are equivalent:*

i. $X \in \mathrm{Rec}(S)$.
ii. $\phi^{-1}(X) \in \mathrm{Rec}(A^+)$.

Proof. The implication $i \Rightarrow ii$ follows from the fact that recognizability is preserved under inverse morphisms (see [62]). Let us then prove the inverse implication.

Let $Z = \phi^{-1}(X)$ and $\equiv_Z$ be the syntactic congruence of Z. The nuclear congruence $\phi\phi^{-1}$, trivially, saturates the set Z, so that $\phi\phi^{-1} \subseteq \equiv_Z$, since $\equiv_Z$ is the maximal congruence which saturates Z. Thus, denoting by $S(Z) = A^+/\equiv_Z$ the syntactic semigroup of Z, there exists a morphism $\tau : A^+/\phi\phi^{-1} \to S(Z)$, which associates with each congruence class $\phi^{-1}(\phi(x))$, $x \in A^+$, the class $[x]_{\equiv_Z}$. Since $S \cong A^+/\phi\phi^{-1}$, by the isomorphism theorem, it follows that $S(Z) \cong S/\tau\tau^{-1}$. By hypothesis $Z \in \mathrm{Rec}(A^+)$ so that $S(Z)$ is finite and then the congruence $\tau\tau^{-1}$ is of a finite index. Now, as one easily verifies, the congruence $\tau\tau^{-1}$ saturates the set $X = \phi(Z)$ that implies $X \in \mathrm{Rec}(S)$.

4.3 The factor semigroup

Let S be a semigroup and r be an element of S. We denote by $F(r)$, or simply F_r, the set of all factors of r:

$$F_r = \{s \in S \mid r \in S^1 s S^1\}.$$

Let $J_r = S \setminus F_r$; J_r is a two-sided ideal of S. We denote by N_r the Rees quotient semigroup S/J_r of S by J_r. As is well known N_r is isomorphic to the semigroup having as support $F_r \cup \{0\}$ and the product $\circ$ is defined as follows: for $s, t \in F_r \cup \{0\}$, $s \circ t = st$ if $s, t \in F_r$ and $st \in F_r$, $s \circ t = 0$, otherwise. We call N_r also the *factor semigroup* of r.

Lemma 4.3.1. *Let r be an element of a semigroup S. If F_r is finite, then $\{r\} \in \mathrm{Rec}(S)$.*

Proof. We can write $N_r = S/\rho_r$, where ρ_r is the Rees congruence $\rho_r = \{id\} \cup (J_r \times J_r)$. The index of ρ_r is equal to card$(F_r)+1$ so that if F_r is finite, then ρ_r is of finite index. Moreover, ρ_r saturates $\{r\}$ so that $\{r\} \in \text{Rec}(S)$.

Proposition 4.3.1. *Let S be a right stable semigroup having a $\mathcal{J}$-depth function. For any $r \in S$ the factor semigroup N_r satisfies* min$_R$.

Proof. Let $r \in S$ and denote N_r and F_r simply by N and F respectively. Suppose now, by contradiction, that there exists an infinite strictly descending chain of principal right ideals of N:

$$a_1N^1 \supset a_2N^1 \supset \cdots \supset a_nN^1 \supset \cdots, \tag{4.2}$$

with $a_n \in N$ $(n = 1, 2, \ldots)$. For all $n > 0$ one has that $a_n \neq 0$; in fact, otherwise, if $a_n = 0$, for a suitable $n > 0$, then $a_nN^1 = 0 = a_{n+1}N^1$, which is a contradiction. Moreover, from equation (4.2) one derives

$$S^1a_1S^1 \supset S^1a_2S^1 \supset \cdots \supset S^1a_nS^1 \supset \cdots.$$

Indeed, from (4.2) one has that for all $n > 0$, $a_{n+1} = a_nq_n$, with $q_n \in F$, so that for all $n > 0$, $S^1a_nS^1 \supseteq S^1a_{n+1}S^1$. Let us now suppose that there exists an integer n for which $S^1a_nS^1 = S^1a_{n+1}S^1$ so that $a_n\mathcal{J}a_{n+1}$. Since $a_{n+1} = a_nq_n$, with $q_n \in F$, it follows that $a_nS^1 \supseteq a_{n+1}S^1$. Since S is right stable one has $a_nS^1 = a_{n+1}S^1$. Hence, $a_n = a_{n+1}p$, with $p \in F$. Thus $a_nN^1 = a_{n+1}N^1$ which is a contradiction. Since for all $n > 0$ the a_n's are factors of r, one has also that

$$S^1a_nS^1 \supseteq S^1rS^1.$$

Let $d_{\mathcal{J}}(r)$ be the $\mathcal{J}$-depth of r. For $n > d_{\mathcal{J}}(r)$, $S^1a_nS^1 = S^1a_{n+1}S^1$; otherwise there would exist a strictly ascending chain of length greater than $d_{\mathcal{J}}(r)$ and starting with r which is a contradiction. Thus $a_n \ \mathcal{J} \ a_{n+1}$ for $n > d_{\mathcal{J}}(r)$. From this one derives, as above, that $a_nN^1 = a_{n+1}N^1$ which is a contradiction.

Proposition 4.3.2. *Let S be a semigroup. S is finite $\mathcal{J}$-above if and only if for all $r \in S$, F_r is finite.*

Proof. Let us first observe that for any $r \in S$, F_r is a union of $\mathcal{J}$-classes. Indeed, let $u \in F_r$ and $u \ \mathcal{J} \ v$. This implies that $u = hvk$ with $h, k \in S^1$, so that $v \in F_r$. Now suppose that for all $r \in S$, F_r is finite. It follows that the number of $\mathcal{J}$-classes above J_r has to be finite. Thus S is weakly finite $\mathcal{J}$-above. Moreover, since $J_r \subseteq F_r$, one has that J_r is finite. Hence, S is finite $\mathcal{J}$-above. Conversely, suppose that S is finite $\mathcal{J}$-above and r is any element of S. By definition the number of $\mathcal{J}$-classes of the factors of r is finite. Moreover, any $\mathcal{J}$-class in F_r is finite; hence, it follows that F_r is finite.

Let S be a semigroup. If for all $r \in S$, F_r is finite, then from Lemma 4.3.1, S is finitely recognizable. The converse is not, in general, true as shown by the following:

Example 4.3.1. Let A be the alphabet $A = \{a, b, c\}$ and A^+ the free semigroup over A. We consider the ideal $J = A^*cA^*$ and the semigroup S which is the Rees quotient of A^+ by J, $S = A^+/J$. The support of S can be identified by $\{a, b\}^+ \cup \{0\}$, where 0 is the Rees congruence class of J containing all the words of A^+ in which there is at least one occurrence of the letter c. Let $\phi : A^+ \to S$ denote the natural epimorphism where $\phi\phi^{-1}$ is the Rees congruence of J. If $r \in S$ and $r \neq 0$, then $\phi^{-1}(r)$ is a singleton of A^+ and then $\phi^{-1}(r) \in \mathrm{Rec}(A^+)$; moreover, $\phi^{-1}(0) = A^*cA^* \in \mathrm{Rec}(A^+)$. It follows by Lemma 4.2.1 that S is finitely recognizable. However, the set F_0 of the factors of 0 is infinite since any element of $\{a, b\}^+$ is a factor of 0.

One easily verifies that the family $\mathrm{Rec}(S)$ is closed under the product $(\cdot)$ and the operation $(^+)$ so that $\mathrm{Rec}(S) \supseteq \mathrm{Rat}(S)$. Moreover, since S is finitely generated, one has that $\mathrm{Rat}(S) \supseteq \mathrm{Rec}(S)$. Thus S is an example of a Kleene's semigroup which is not finite $\mathcal{J}$-above.

The following theorem gives some characterizations of finitely generated finite $\mathcal{J}$-above semigroups [47].

Theorem 4.3.1. *Let S be a finitely generated semigroup having a finite $\mathcal{J}$-depth function. The following conditions are equivalent:*

i. S has a $\mathcal{J}$-depth decomposition.
ii. S is finite $\mathcal{J}$-above.
iii. S is finitely recognizable.
iv. S is stable and all subgroups are finite.
v. S is right stable and the subgroups of S are locally finite.

Proof. *i* ⇒ *ii*: this implication follows from Proposition 3.6.3.

ii ⇒ *iii*: since S is finite $\mathcal{J}$-above, from Proposition 4.3.2 one has that for all $r \in S$, F_r is finite. Thus from Lemma 4.3.1, S is finitely recognizable.

iii ⇒ *iv*: S is finitely recognizable so that by Corollary 4.2.1 all subgroups of S are finite and, by Theorem 4.2.4, one derives that S is stable.

iv ⇒ *v*: it is trivial.

v ⇒ *i*: by the $\mathcal{J}$-depth decomposition theorem one has that S has a weak $\mathcal{J}$-depth decomposition. Moreover, since S has a $\mathcal{J}$-depth function, from Proposition 3.6.2, S has a $\mathcal{J}$-depth decomposition.

Let us remark that the implication *v* ⇒ *ii* can be also derived by proving that for all $r \in S$, F_r is finite. Indeed, from Proposition 4.3.1 the semigroup N_r satisfies $\min_R$. Moreover, since the subgroups of S are locally finite the same will be true for the subgroups of N_r. Hence, by Theorem 3.6.2, N_r and then F_r will be finite. Thus from Proposition 4.3.2 the semigroup S is finite $\mathcal{J}$-above.

Corollary 4.3.1. *Let S be a finitely generated semigroup. S has a $\mathcal{J}$-depth decomposition if and only if S is finite $\mathcal{J}$-above.*

Proof. The "only if" part is the statement of Proposition 3.6.3. The "if" part derives from Theorem 4.3.1. One has only to observe that a finite $\mathcal{J}$-above semigroup has, trivially, a $\mathcal{J}$-depth function.

A different proof of the fact that a finitely generated finite $\mathcal{J}$-above semigroup has a $\mathcal{J}$-depth decomposition can be derived from a theorem of J. C. Birget and J. Rhodes ([9], Proposition 10.3).

Let A be a finite alphabet of cardinality $r > 0$ and $n, m > 0$, $n \neq m$. We denote by $B(r, n, m)$ the semigroup $A^+/\theta_{n,m}$, where $\theta_{n,m}$ is the congruence in A^+ generated by the relation $\{(x^n, x^m) \mid x \in A^+\}$. It was proved in [46] that for $r > 0$ and $n > 4$ the semigroup $B(r, n, n+1)$ is finite $\mathcal{J}$-above (see Sect. 4.6).

Another important example of finitely generated finite $\mathcal{J}$-above semigroups is given by the free inverse semigroups. We recall that an *inverse semigroup* S is a semigroup such that for any $x \in S$ there exists a unique element $x^{-1} \in S$ such that

$$xx^{-1}x = x \text{ and } x^{-1}xx^{-1} = x^{-1}. \tag{4.3}$$

Let us denote by $FIS(A)$ the free inverse semigroup on the set A, i.e., $FIS(A) = (A \cup A')^+/\theta$, where A' is a copy of A disjoint from A and θ is a suitable congruence relation. The structure and basic properties of $FIS(A)$ were investigated by W. D. Munn [121]; he proved that $FIS(A)$ is finite $\mathcal{J}$-above. From Theorem 4.3.1 it follows that $FIS(A)$ is finitely recognizable.

4.4 Rewriting systems

In this section we give the notion of *semi-Thue system* or *rewriting system* over a free-monoid A^* (free-semigroup A^+) (see [85]). A *semi-Thue* or *rewriting system* is a pair $\mathcal{T} = (A, \pi)$, where A is a finite alphabet and π is a binary relation $\pi \subseteq A^* \times A^*$ ($\pi \subseteq A^+ \times A^+$). We shall simply denote $\mathcal{T}$ by π when there are no ambiguities. Any pair of words $(u, v) \in \pi$ is called a *production* and is also denoted by $u \to v$. We denote by $\Rightarrow_\pi$ the *regular closure* of π, i.e., for any $u, v \in A^*$

$$u \Rightarrow_\pi v \text{ if and only if } \exists (x, y) \in \pi \text{ and } \exists h, k \in A^* : u = hxk \text{ and } v = hyk.$$

If $u \Rightarrow_\pi v$, then v is called a *direct descendant* of u with respect to π. The derivation relation $\Rightarrow_\pi^*$ is defined as the transitive and reflexive closure of $\Rightarrow_\pi$. We observe that the relation $\Rightarrow_\pi^*$ is, trivially, left and right invariant with respect to the product in A^*. If $u \Rightarrow_\pi^* v$, then we say that v is a *descendant* of u. We denote by $L_\pi(u)$ the set of all descendants of u,

$$L_\pi(u) = \{v \mid u \Rightarrow_\pi^* v\}.$$

For any u, $L_\pi(u)$ is also called the *language generated by u* relative to the rewriting system (A, π). Let us denote by $\Leftrightarrow_\pi$ the symmetric closure of $\Rightarrow_\pi$ and by $\Leftrightarrow_\pi^*$ the reflexive and transitive closure of $\Leftrightarrow_\pi$. One has that $\Leftrightarrow_\pi^*$ is a congruence (*Thue's congruence*) which is the smallest congruence of A^* containing π, that is, the congruence generated by π. We shall denote it also by $\langle\pi\rangle$. Two rewriting systems (A_1, π_1) and (A_2, π_2) are *equivalent* if $A_1 = A_2$ and $\langle\pi_1\rangle = \langle\pi_2\rangle$.

The *word problem* for (A, π) consists in deciding whether or not two arbitrary words u and v are congruent mod $\langle\pi\rangle$.

Definition 4.4.1. *The rewriting system* (A, π) *has a* (recursively) solvable, or decidable, word problem *if the relation* $\langle\pi\rangle$ *is a recursive relation (i.e., a recursive subset of* $A^* \times A^*$*).*

In other terms the rewriting system (A, π) has a solvable word problem if there exists an algorithm by means of which one can decide, after a finite number of steps, whether or not any two arbitrarily given words u and v of A^* are equivalent mod $\langle\pi\rangle$, that is, whether or not $(u, v) \in \langle\pi\rangle$. It is well known that if π is a recursive set, then $\langle\pi\rangle$ is a recursively enumerable subset of $A^* \times A^*$, that is, there exists an algorithm which enumerates all the elements of $\langle\pi\rangle$.

When π is a symmetric relation, then (A, π) is called a *Thue system.* In such a case the derivation relation $\Rightarrow_\pi^*$ coincides with $\langle\pi\rangle$, so that it is a congruence of the free monoid A^*. In such a case for any $u \in A^*$,

$$L_\pi(u) = [u]_{\langle\pi\rangle}.$$

If π is a semi-Thue system, then we denote by $T(\pi)$ the Thue system $T(\pi) = \pi \cup \pi^{-1}$; $T(\pi)$ is called the *Thue system induced by* π. One has that $\Rightarrow_{T(\pi)}^* = \Leftrightarrow_{T(\pi)}^* = \Leftrightarrow_\pi^*$.

A class of rewriting systems which is of great interest in relation to the word problem, satisfies the following important property called the *Church–Rosser* property.

Definition 4.4.2. *A semi-Thue system* (A, π) *is* Church–Rosser *if the following condition holds: for any* $u, v \in A^*$, *if* $u \Leftrightarrow_\pi^* v$, *then there exists* $z \in A^*$ *such that* $u \Rightarrow_\pi^* z$ *and* $v \Rightarrow_\pi^* z$.

When a semi-Thue system is Church–Rosser, one has that two words are equivalent if and only if they can derive a common word. Thus the word problem consists in deciding the existence of a common descendant of two given words.

Let (A, π) be a rewriting system. A word $u \in A^*$ is called *irreducible* if there is no word $w \in A^*$ such that $u \Rightarrow_\pi w$. A word v is called a *normal form* for u if $v \in [u]_{\langle\pi\rangle}$ and v is irreducible.

If a rewriting system is Church–Rosser, then in each congruence class mod $\Leftrightarrow_\pi^*$, there is *at most* one normal form. Indeed, if $v_1 \Leftrightarrow_\pi^* v_2$ and v_1, v_2 are normal forms, then from the Church–Rosser property there exists a word $w \in A^*$ such that $v_1 \Rightarrow_\pi^* w$ and $v_2 \Rightarrow_\pi^* w$. From the irreducibility of v_1 and v_2 it follows that $w = v_1 = v_2$.

Definition 4.4.3. *A semi-Thue system (A, π) is* terminating *if there is no infinite derivation*

$$u_1 \Rightarrow_\pi \cdots \Rightarrow_\pi u_n \Rightarrow_\pi \cdots.$$

Let (A, π) be a semi-Thue system. If π is *length-reducing*, i.e., for any $(u, v) \in \pi$, $|u| > |v|$, then it is, trivially, terminating. In such a case (A, π) is called a *length reducing system.*

When (A, π) is terminating, then in each congruence class mod $\Leftrightarrow_\pi^*$, there exists *at least* one normal form. A rewriting system is called *complete* if it is terminating and Church–Rosser. In this case in any congruence class there exists one and only one normal form.

Proposition 4.4.1. *Let $\mathcal{T} = (A, \pi)$ be a complete rewriting system. Suppose that for any word u one can decide whether u is irreducible and, if u is not irreducible, one can effectively find a word u' such that $u \Rightarrow_\pi u'$. Then the word problem is decidable.*

Proof. One can recursively solve the word problem as follows. Let u and v be two words. We denote by $\hat{u}$ and $\hat{v}$ their normal forms. They can be computed in a finite number of steps. Indeed, either u is irreducible or we can compute a descendant u_1 such that $u \Rightarrow_\pi u_1$. Now either u_1 is irreducible or we can compute a descendant u_2 such that $u_1 \Rightarrow_\pi u_2$. In this way we construct a descending chain $u_1 \Rightarrow_\pi u_2 \Rightarrow_\pi \cdots$ which has to terminate since $\mathcal{T}$ is terminating. Hence, there exists an integer n such that u_n is irreducible. Moreover, since $\mathcal{T}$ is Church–Rosser, u_n is the normal form $\hat{u}$ of u. Since $u \Leftrightarrow_\pi^* v$ if and only if $\hat{u} = \hat{v}$, one can effectively decide whether or not $u \Leftrightarrow_\pi^* v$.

4.5 The word problem

Let S be a semigroup. We recall that a *presentation* of S is a semi-Thue system (A, R) where A is an alphabet and R a binary relation in A^+ such that $S \cong A^+/\langle R \rangle$, where $\langle R \rangle$ is the congruence generated by R. If A is finite, then the presentation is called *finitely generated*; if R is finite the presentation is called *finitely related.* If A and R are both finite, then the presentation is called *finite.* If S has a finite presentation, then S is called *finitely presented.*

The presentation (A, R) of S has a *solvable word problem* if the word problem for the semi-Thue system (A, R) is decidable.

The following lemma shows that for finitely generated presentations of the same semigroup S the property of having a solvable word problem is independent from the particular (finitely generated) presentation; in this case one can say simply that S has a *solvable word problem.*

Lemma 4.5.1. *If a given finitely generated presentation of a semigroup has a solvable word problem, then the same is true for every finitely generated presentation.*

Proof. Let (A_1, R_1) and (A_2, R_2) be two finitely generated presentations of the same semigroup $S \cong A_1^+/\langle R_1\rangle \cong A_2^+/\langle R_2\rangle$. We suppose that the word problem for the first presentation is solvable, that is, there exists an algorithm by means of which one can decide whether or not two arbitrary words of A_1^+ represent the same element of S. Let $\phi_1 : A_1^+ \to S$ and $\phi_2 : A_2^+ \to S$ be the two canonical epimorphisms. We choose a map $\psi : A_2 \to A_1^+$ such that for any $a \in A_2$, a, and $\psi(a)$ represent the same element of S, i.e., $\phi_2(a) = \phi_1(\psi(a))$. The map ψ can be extended to a morphism, which we shall still denote by ψ, of A_2^+ in A_1^+. Thus, one has that for any $w \in A_2^+$, $\phi_2(w) = \phi_1(\psi(w))$, that is, w and $\psi(w)$ represent the same element of S. The morphism ψ is effectively computable. Indeed, given the finite list of words $\psi(a)$ with $a \in A_2$, in order to compute $\psi(w)$ one has only to make a finite number of substitutions.

Now let $w_1, w_2 \in A_2^+$ be two arbitrary words. In order to decide whether or not $\phi_2(w_1) = \phi_2(w_2)$ one needs only to compute $\psi(w_1)$ and $\psi(w_2)$ and use the algorithm of the first presentation to decide whether or not $\phi_1(\psi(w_1)) = \phi_1(\psi(w_2))$.

We recall that a variety $\mathcal{V}$ of semigroups is a class of semigroups closed under the formation of subsemigroups, quotients, and (unrestricted) direct products (compare [3]). Hence, one has:

i. If $S \in \mathcal{V}$ and T is a subsemigroup of S, then $T \in \mathcal{V}$.
ii. If $S \in \mathcal{V}$ and θ is a congruence in S, then $S/\theta \in \mathcal{V}$.
iii. If $\{S_\gamma\}$ is any set of semigroups of $\mathcal{V}$, then their direct product $\prod_\gamma S_\gamma \in \mathcal{V}$.

Let X be an alphabet. An ***identity*** over X is a pair (p, q) of words over X usually denoted with an equality $p = q$. One says that a semigroup S satisfies the identity or equation

$$p(x_1, \ldots, x_n) = q(x_1, \ldots, x_n),$$

with $x_i \in X$, $1 \le i \le n$, if for any morphism $\phi : X^+ \to S$ one has $\phi(p) = \phi(q)$, that is,

$$p(\phi(x_1), \ldots, \phi(x_n)) = q(\phi(x_1), \ldots, \phi(x_n)).$$

Let Σ be a set of identities. We denote by $[\Sigma]$ the class of all semigroups satisfying all the identities of the set Σ. A class $\mathcal{C}$ of semigroups is called ***equationally based*** if there exists a set of identities Σ such that $\mathcal{C} = [\Sigma]$.

From a classical theorem of Birkhoff of universal algebra (see [23]) one derives:

Proposition 4.5.1. *A class of semigroups is a variety if and only if it is equationally based.*

If $\mathcal{V}$ is a variety of semigroups, then from the general theory one has that for any cardinal number r there exists in $\mathcal{V}$ a semigroup $F_r(\mathcal{V})$ satisfying the following properties: *i.)* there exists a subset A of $F_r(\mathcal{V})$ having cardinality r which generates $F_r(\mathcal{V})$, *ii.)* any map $f : A \to S$, where S is any semigroup of $\mathcal{V}$, can be extended to a morphism $\phi : F_r(\mathcal{V}) \to S$. The semigroup $F_r(\mathcal{V})$ is also called the *free semigroup with r generators in* $\mathcal{V}$.

Definition 4.5.1. *A variety $\mathcal{V}$ of semigroups is called* finitely based *if it is defined by a finite set of equations.*

The semigroup $B(r, n, m)$ introduced in the preceding section is the free semigroup with r generators in the (finitely based) variety of semigroups satisfying the equation

$$x^n = x^m.$$

Further examples of finitely based varieties are, for instance, the class of commutative semigroups (which satisfy the equation $xy = yx$) or the left zero bands (which satisfy the two equations $x^2 = x$, $xy = x$).

A semigroup $S \in \mathcal{V}$ is *finitely presented in* $\mathcal{V}$ if there exists a finite positive integer r and a relation R in $F_r(\mathcal{V})$ such that R is a finite set and $S \cong F_r(\mathcal{V})/\langle R\rangle$, where $\langle R\rangle$ is the congruence generated by R.

The following important theorem relates the residual finiteness of a semigroup and the word problem. This is a slightly different version of a theorem of T. Evans [63]:

Theorem 4.5.1. *Let S be a finitely presented semigroup in a finitely based variety. If S is residually finite, then the word problem for S is solvable.*

Proof. Let $\mathcal{V}$ be a finitely based variety defined by a finite set of identities (p_i, q_i), $i \in I$, where p_i and q_i are words on a finite alphabet X of cardinality n. Let $S \in \mathcal{V}$ be a finitely presented semigroup in $\mathcal{V}$. One has then $S \cong F_r(\mathcal{V})/\langle R\rangle$, where $\langle R\rangle$ is the congruence generated by the relation $R = \{(r_j, s_j) \mid j \in J\}$, where J is a finite set and r_j and s_j are words on the set of generators of $F_r(\mathcal{V})$.

Let A be a finite alphabet of cardinality $m \geq r$ and denote by $\phi : A^+ \to S$ an epimorphism. One has then $S \cong A^+/\sim$ where $\sim\, = \phi\phi^{-1}$ is the nuclear congruence of ϕ. It is possible to generate $\sim$ as follows. We introduce in A^+ the relations ρ_v, ρ_0, and ρ defined as

$$\rho_v = \{(p_i(u_1, \ldots, u_n), q_i(u_1, \ldots, u_n)) \mid u_h \in A^+,\ h = 1, \ldots, n \text{ and } i \in I\},$$

$$\rho_0 = \{(r_j(a_1, \ldots, a_m), s_j(a_1, \ldots, a_m)) \mid a_h \in A,\ h = 1, \ldots, m \text{ and } j \in J\}$$

and $\rho = \rho_v \cup \rho_0$. Let us denote by $\Rightarrow$ the regular closure of ρ and by $\Leftrightarrow$ the symmetric closure of $\Rightarrow$. One easily derives that the transitive closure $\Leftrightarrow^*$

of $\Leftrightarrow$ is equal to $\sim$. The relation ρ is recursive. In fact, ρ_0 is a finite set of pairs of words of A^+ and ρ_v is an infinite set of pairs of words of A^+ that we can enumerate for instance, according to the value $k = |p_i(u_1, \ldots, u_n)| + |q_i(u_1, \ldots, u_n)|$, i.e., the sum of the lengths. For any pair (u, v) of words of A^+ one can decide whether or not $(u, v) \in \rho$. Indeed, one can certainly decide whether $(u, v) \in \rho_0$ since ρ_0 is a finite set. In order to decide whether or not $(u, v) \in \rho_v$ it is sufficient to compute $|u| + |v| = k$ and to check whether (u, v) belongs to the sublist of ρ_v of weight k. From this it follows that also the relation $\Leftrightarrow$ is recursive and that for all $u \in A^+$ the set $T_u = \{v \in A^+ \mid u \Leftrightarrow^* v\}$ is recursively enumerable, that is, there exists an algorithm $\mathcal{A}$ by means of which one can enumerate all the elements of T_u. The set T_u coincides with the congruence class $[u]$ mod $\sim$ of u. Thus for any pair (u, v) of words of A^+ if $u \sim v$ (this implies that $u = v$ in S), then by means of $\mathcal{A}$ one can generate v after a finite number of steps.

If v is not congruent to u mod $\sim$ (this implies that u and v are not equal in S), then since S is residually finite, there exists a finite semigroup S' and an epimorphism $\psi : S \to S'$, such that $\psi(\phi(u)) \neq \psi(\phi(v))$ (i.e., u and v are different on S'). If we are able to construct by an algorithm $\mathcal{B}$ such a finite semigroup, then we can conclude that u and v are different in S, that is, u and v are not congruent mod $\sim$. We can construct step-by-step all finite semigroups F generated by the elements $g_i = \zeta(a_i)$ $(i = 1, \ldots, m)$, where $\zeta = \phi\psi$, and then check whether or not

$$p_i(\zeta(u_1), \ldots, \zeta(u_n)) = q_i(\zeta(u_1), \ldots, \zeta(u_n)), \ i \in I.$$

This test can be effectively done in view of the finiteness of F. In this way we check the membership of F to the variety. Moreover, one tests whether or not for all $j \in J$,

$$r_j(g_1, \ldots, g_m) = s_j(g_1, \ldots, g_m),$$

(check the presentation). If these two tests are positive, then F in an homomorphic image of S. Now one tests whether or not

$$u(g_1, \ldots, g_m) = v(g_1, \ldots, g_m).$$

If $u(g_1, \ldots, g_m) \neq v(g_1, \ldots, g_m)$, then the procedure stops, otherwise it will continue. Thus there exist two algorithms $\mathcal{A}$ and $\mathcal{B}$: the first stops if $u = v$ in S, whereas the second stops if $u \neq v$ in S. In conclusion, one can decide in a finite number of steps whether or not u and v are congruent mod $\sim$.

We stress that in the preceding theorem the hypothesis that S is finitely presented cannot be weakened to "recursively" presented. In fact there exist finitely generated, recursively presented groups with an unsolvable word problem (see [106], Chap. 4).

Corollary 4.5.1. *Let S be a finitely presented semigroup in a finitely based variety. If S is finitely recognizable, then the word problem for S is solvable.*

Proof. From Theorem 4.2.5, S is residually finite. Thus from Theorem 4.5.1 the result follows.

A recent survey on the main algorithmic problems concerning semigroups and other varieties of classical algebras is in [96].

4.6 On a conjecture of Brzozowski

In 1979 Brzozowski presented at the International Symposium on Formal Language theory in Santa Barbara, California, a paper entitled "Open problems about regular languages" [19] in which six open problems and conjectures are discussed.

Most of these problems such as the "star height" or the "optimality of finite and complete prefix codes" are still open, even though a considerable effort has been made to solve them. The fifth conjecture known as the "regularity of noncounting classes" can be formulated as follows: let A^+ be a finitely generated free semigroup over a finite alphabet A and n be a fixed positive integer. One can introduce in A^+ the congruence relation θ_n generated by the relation $\{(x^n, x^{n+1}) \mid x \in A^+\}$ and the quotient semigroup $M_n = A^+/\theta_n$. For any word $w \in A^+$ the congruence class $[w]_{\theta_n}$ or simply $[w]_n$, of w is a ***noncounting language of order n*** according to R. McNaughton and S. Papert [112].

Conjecture 4.6.1. Let $n > 0$. For any word $w \in A^+$ the congruence class $[w]_n$ is a regular language.

Actually this conjecture was already formulated in 1970 for the case $n = 2$ by Simon [147] in an unpublished manuscript. More recently, the following has been considered:

Conjecture 4.6.2. (***Generalized Brzozowski's conjecture***) Let $n, m > 0$ and $\theta_{n,m}$ be the congruence generated by the relation $\{(x^n, x^{n+m}) \mid x \in A^+\}$. For any word $w \in A^+$ the congruence class $[w]_{\theta_{n,m}}$ is a regular language.

In what follows, when talking about Brzozowski's conjecture, we shall refer to Conjecture 4.6.2, even though the first formulation, as stated in Brzozowski's paper [19], is exactly the statement of Conjecture 4.6.1.

In 1990 [40, 46] the authors proved the Brzozowski conjecture for any $n \geq 5$ and $m = 1$. Actually the stronger result was proved, that for any $n \geq 5$ the semigroup A^+/θ_n is finite $\mathcal{J}$-above and, by Lemma 4.3.1, this enforces the regularity of the congruence classes. Also, the further result that, in the semigroup A^+/θ_n, the ***word problem*** is solvable, for $n \geq 5$ was proved.

Slightly later and in a different way J. McCammond [111] proved the conjecture for $n \geq 6$ and for any $m > 0$. His proof, based on some automata transformations, is very difficult to understand (it is based on forty lemmas proved by simultaneous induction). The importance of the paper stands mainly on his results on Burnside's subgroups, in particular he proved that the maximal subgroups are cyclic of order m.

Subsequently, A. Pereira do Lago [56] improved the result, basing his work on the techniques introduced in [40, 46]. On one hand he proved in a more direct way the conjecture for $n \geq 4$ and for any m, simplifying the original rewriting system, and on the other he underlined the links between such a system and Green's relations on the quotient monoid.

V. Guba proved [74] in 1993 the conjecture for $n \geq 3, m \geq 1$. A more complete version of the work of do Lago containing some further results on the structure of the Burnside semigroup is in [57].

The case $n = 2$ and $m = 1$ is still unsolved and it is not clear how the techniques used up to now will be able to work it out. For $n = 1$ Brzozowski's conjecture is trivially true for the values of the integer m for which the semigroup $A^*/\theta_{1,m}$ is finite. A very recent result of do Lago [58] shows that the conjecture is, in general, false for $n = 1$ and $n = 2$ for a large enough value of m (see Proposition 4.6.1).

4.6.1 Problems and results

We shall now consider the Brzozowski conjecture in a more general context of problems that have been considered in recent contributions.

Let A be a finite alphabet and $n, m > 0$. Let $\pi \subseteq A^+ \times A^+$ be the relation $\pi = \{(x^n, x^{n+m}) \mid x \in A^+\}$. Denote by $B(r, n, n+m)$ the quotient semigroup $B(r, n, n+m) = A^+/\theta_{n,m}$, where $r = \text{card}(A)$. The semigroup $B(r, n, n+m)$ is called the *Burnside semigroup* with r generators defined by the equation $x^n = x^{n+m}$.

As usual we denote by $\Rightarrow_\pi$, or simply by $\Rightarrow$ the regular closure of π, and by $\Leftrightarrow$ the symmetric closure of $\Rightarrow$. One has $\theta_{n,m} = \Leftrightarrow^*$.

One can consider the following problems on the Burnside semigroup.

1. Finiteness. For $r = 1$, $B(1, n, n + m)$ is always finite. For $r > 1$, $B(r, n, n + m)$ is finite for $n = 1$ and $m = 1, 2, 3, 4, 6$. This follows, since $B(r, 1, 1 + m)$ coincides with $S(r, m)$, from the theorem of Green and Rees (see Theorem 3.3.2) and the fact that the Burnside group $G(r, n)$ is finite for $n = 1, 2, 3, 4$, and 6. From this theorem and the results of Adjan, Novikov, Lysionok, and Ivanov (see Chap. 3) one derives that $B(r, 1, 1+m)$ is infinite for any odd integer $m \geq 115$ and, moreover, for any $m \geq 2^{48}$.

For $r > 2$, and $n > 1$, $B(r, n, n+m)$ is infinite. This is an easy consequence of the existence of infinitely many square-free words on an alphabet having at least three letters (see Proposition 2.6.2). However, even if $r = 2$ and $n > 1$, it was proved by Brzozowski et al. [18] that $B(2, n, n + m)$ is infinite.

2. Regularity of congruence classes. By Lemma 4.2.1 one has that the congruence classes are regular if and only if the elements of $B(r, n, n+m)$

are recognizable, and this is equivalent to the requirement that $B(r,n,n+m)$ is finitely recognizable. As we said before, it is known at the present time that for $n \geq 3$ and $m \geq 1$ the elements of $B(r,n,n+m)$ are recognizable.

3. Finiteness $\mathcal{J}$-above. By Proposition 4.3.2 the finiteness $\mathcal{J}$-above of $B(r,n,n+m)$ is also equivalent to the statement that for any $s \in B(r,n,n+m)$ the set F_s of the factors of s is finite. It has been proved that $B(r,n,n+1)$ is finite $\mathcal{J}$-above for any $n \geq 5$ [40, 46]. At the present time we know that $B(r,n,n+m)$ is finite $\mathcal{J}$-above for $n \geq 3$ and $m \geq 1$.

4. The word problem. This problem consists in deciding whether, given two words $u, v \in A^+$, they represent the same element of $B(r,n,n+m)$. As a result of the previous works we know that the word problem is solvable for $n \geq 3$ and $m \geq 1$.

5. Structure. The $\mathcal{H}$-classes of regular $\mathcal{D}$-classes have m elements. For $n \geq 3$ and $m \geq 1$ the group of such a $\mathcal{D}$-class is a cyclic group of order m. Recent results on the structure of maximal groups in $B(r,n,n+m)$, for $n=1$ and $n=2$, are given in Proposition 4.6.1.

6. Extensions and generalizations. Consider the semi-Thue system (A,π), with $\pi = \{(x^n, x^{n+m}) \mid x \in A^+\}$, and denote by $\Rightarrow_\pi^*$ the corresponding derivation relation. For any $w \in A^+$ let $L_\pi(w)$ be the language generated by w,

$$L_\pi(w) = \{v \in A^+ \mid w \Rightarrow_\pi^* v\}.$$

Given $n, m \geq 1$ one can ask whether $L_\pi(w)$ is a regular language. The only known results are in the case $n = m = 1$ (see Chap. 6).

We now present some recent results which link the above problems. These results also give an idea of the approach followed in solving the Brzozowski conjecture, for $n > 4$. We first observe that any finite semigroup is trivially finite $\mathcal{J}$-above; moreover if $B(r,n,n+m)$ is finite, then the equivalence classes are regular languages. Hence, if we have an affirmative answer to Problem 1, then we also have a positive answer to Problems 3 and 2. Unfortunately, as we said, there are only a few cases in which the semigroup $B(r,n,n+m)$ is finite. From Lemma 4.3.1 one has that to find an affirmative answer to Problem 2 it is enough to solve positively Problem 3.

We now observe that in the previous works the decidability of the word problem in the semigroups $B(r,n,n+m)$ has been directly proved, at the present for $n \geq 3$. The proof relies on some properties such as the Church–Rosser property, of suitable rewriting systems that have been shown to be equivalent to the semi-Thue system (A,π) with $\pi = \{(x^n, x^{n+m}) \mid x \in A^+\}$. An important fact underlined in [47] is that: if $B(r,n,n+m)$ is finite $\mathcal{J}$-above, then the word problem in $B(r,n,n+m)$ is solvable. This is a consequence of Proposition 4.3.2, Lemma 4.3.1, and Corollary 4.5.1.

Another question is the following: are Problems 2 and 3 equivalent? A partial answer to this question is given by Theorem 4.3.1. In general finite recognizability does not imply the finiteness $\mathcal{J}$-above (see Example 4.3.1). As we shall see a positive answer to the preceding question will be given for $n \geq 3$ and $m \geq 1$.

We shall give now an outline of the proof of Conjecture 4.6.1 in the case $n > 4$.

Lemma 4.6.1. *The $\mathcal{H}$-classes of $B(r, n, n+1)$ are trivial.*

Proof. Let $S = B(r, n, n+1)$. For any $s \in S$ one has, trivially, that $s^n = s^{n+1}$. Thus any element of S generates a subsemigroup whose period is 1. Suppose now that $s_1 \ \mathcal{H} \ s_2$. Then there exist $h, k \in S$ such that

$$s_1 = hs_2 \text{ and } s_2 = s_1 k.$$

Hence,

$$s_1 = hs_2 = hs_1 k = h^n s_1 k^n = h^n s_1 k^{n+1} = s_1 k = s_2.$$

This shows that the $\mathcal{H}$-classes of S are trivial.

From the preceding lemma one has that the subgroups of $B(r, n, n+1)$ are all trivial. Moreover, $B(r, n, n+1)$ is stable, since by Lemma 3.6.2 periodic semigroups are stable. Therefore, from Theorem 4.3.1, in order to prove that $B(r, n, n+1)$ is finitely recognizable and, therefore, that the noncounting classes are regular, it is enough to show that $B(r, n, n+1)$ has a $\mathcal{J}$-depth function. In fact in [40, 46] the following stronger result is proved.

Theorem 4.6.1. *Let $S = B(r, n, n+1)$, $n \geq 5$. For any $m \in S$ there are only finitely many $\mathcal{J}$-classes of elements $t \in F_m$, where F_m is the set of all factors of m (i.e., S is weakly finite $\mathcal{J}$-above). This implies that S has a $\mathcal{J}$-depth function.*

Outline of the proof. The proof is quite laborious since it requires many technical lemmas. Here we shall only limit ourselves to outline the main ideas and results on which the proof is based.

Fact 1 *(Equivalence Theorem).* One defines a semi-Thue system (A, σ) where the set σ is a recursive subset of $A^+ \times A^+$ which is constructed starting from π according to some rules [46]. The elements of σ are suitable productions of the kind

$$x' x^{n-2} x'' \to x' x^{n-1} x''.$$

with x' suffix and x'' prefix of x. The system (A, σ) is equivalent to (A, π), that is, for any $u, v \in A^+$, $u \Leftrightarrow_\sigma^* v \Leftrightarrow_\pi^* u$.

Fact 2 *(Renormalization Theorem).* Let $[w]_n$ be a noncounting class, where w is its canonical representative. The renormalization theorem states that any word $x \in [w]_n$ can be derived from w in the rewriting system (A, σ), that is,

$$\text{if } x \Leftrightarrow_\pi^* w, \text{ then } w \Rightarrow_\sigma^* x.$$

Let us then consider the rewriting system $\Delta = (A, \Rightarrow_c)$, where $\Rightarrow_c = \Rightarrow_\sigma^{-1}$. One has that Δ is equivalent to (A, σ) and then to (A, π). The rewriting system Δ is Church–Rosser. Indeed, if $u \Leftrightarrow_c^* v \Leftrightarrow_\pi^* u$, then from the renormalization theorem

$$u \Rightarrow_c^* w \text{ and } v \Rightarrow_c^* w,$$

where w is the canonical representative in the class $[u]_n$. Moreover, Δ is terminating because it is length reducing. Hence, Δ is complete. The rewriting system (A, σ) is such that for any $u \in A^+$ one can effectively decide whether or not u is irreducible with respect to the relation $\Rightarrow_c$. Moreover, if u is not irreducible one can effectively construct the set of all direct descendants $\{v \in A^+ \mid u \Rightarrow_c v\}$ [46]. From Proposition 4.4.1 one derives that the word problem for S is solvable.

Fact 3 *(Finiteness of the $\mathcal{J}$-depth).* Let $[w]_n$ be a noncounting class and w its canonical representative. Then for any $x \in F([w]_n)$ there exist $w' \in [w]_n$ and $x' \in F(w')$ such that w' is a direct descendant of w in the rewriting system σ, that is, $w \Rightarrow_\sigma w'$ and, moreover, $\phi(x) \ \mathcal{J} \ \phi(w')$.

Since there are only finitely many direct descendants of w in the rewriting system (A, σ), Fact 3 implies that for any $[w]_n$ there are only finitely many $\mathcal{J}$-classes in $F([w]_n)$. From this it easily follows that $B(r, n, n+1)$ has a $\mathcal{J}$-depth function.

From the preceding theorem and Theorem 4.3.1 one has that S is finite $\mathcal{J}$-above and Problems 2 and 3 are equivalent.

Let us observe that do Lago, in his proof of Conjecture 4.6.2 for $n \geq 4$ and $m > 0$, follows the lines of the proof in [40, 46]. He introduces a simpler rewriting system (A, τ) equivalent to (A, π), whose productions are of the kind

$$x'x^{n-1} \to x'x^{n-1+m}$$

with x' suffix of x. He proved that for $n \geq 4$ and $m \geq 1$ the length-reducing system $(A, \Rightarrow_\tau^{-1})$ is complete. The main difference with respect to the line of the proof in [40, 46] is a direct proof that $S = B(r, n, n+m)$ is finite $\mathcal{J}$-above without using Theorem 3.3 and properties of maximal subgroups of S, which are actually finite (cyclic of order m), as shown by the same author for $n \geq 4$ and by McCammond for $n \geq 6$.

The following interesting proposition has been proved recently [58] by do Lago:

Proposition 4.6.1. *Any maximal group in the free Burnside semigroup $B(r, n, n+m)$ is a free Burnside group satisfying the equation $x^m = 1$.*

If $n \geq 3$ such maximal groups are cyclic groups of order m. However, for $n = 1$ and for $n = 2$ these maximal groups, which are free Burnside groups, may have $2m-1$ generators. Hence, for large enough m they are infinite. From Theorem 4.2.3 it follows that in such cases the Burnside semigroup $B(r,n,n+m)$ is not finitely recognizable and then it is not finite $\mathcal{J}$-above. This implies that for $n=1$ and $n=2$ and large enough m the Conjecture 4.6.2 is false.

Some results relative to Problem 6 will be considered in Chap. 6.

4.7 On a conjecture of Brown

It was proved by Brown [15] that the semigroup $B(r,2,3)$ is the disjoint union of locally finite semigroups. More precisely, let us set $S=B(r,2,3)$ and $E(S)$ be the set of all idempotents of S. Any element $s \in S$ is such that $s^2 \in E(S)$; thus S can be decomposed as a disjoint union of the sets

$$S_e = \{s \in S \mid s^2 = e \in E(S)\}.$$

Theorem 4.7.1. *For each idempotent $e \in E(S)$, S_e is a locally finite subsemigroup.*

A consequence of the preceding theorem is that there exist semigroups, such as $B(r,2,3)$ with $r>1$, which are not locally finite but union of disjoint locally finite subsemigroups. In [17] Brown states that it was asserted by L. N. Shevrin in a private communication that for any $e \in E(S)$, the semigroup S_e is in fact finite. However, as far as he knows no proof of this assertion was published. In the same paper Brown introduces a particular approach to the attempt of proving the finiteness of S_e based on the idea that if s is an element of S and g any generator of S, then not too much "cancellation" can occur in the product gs. However, the results obtained by Brown are not such as to derive the finiteness of S_e. We shall give now a simple proof [50] of the finiteness of the semigroups S_e for any $e \in E(S)$ (see Theorem 4.7.2). This is a consequence of a general lemma (see Lemma 4.7.1) concerning the semigroup $B(r,n,n+m)$.

In the following $\Leftrightarrow^*$ will denote the congruence $\theta_{n,m}$.

Lemma 4.7.1. *Let ϕ be the natural epimorphism $\phi : A^+ \to A^+/\Leftrightarrow^*$ and w, w' be words of A^+ such that $w \Leftrightarrow^* w'$. For any prefix p' (suffix s') of w' there exists a prefix p (a suffix s) of w such that*

$$\phi(p)\ \mathcal{R}\ \phi(p')\quad (\phi(s)\ \mathcal{L}\ \phi(s')).$$

Proof. Suppose first that $w \Leftrightarrow w'$. We have to consider two cases: $w \Rightarrow^{-1} w'$ and $w \Rightarrow w'$. In the first case

$$w = \lambda x^{n+m}\mu,\ w' = \lambda x^n \mu$$

with $\lambda, \mu \in A^*$ and $x \in A^+$. If $p' \in P(\lambda x^n)$, then $p' \in P(w)$ so that the result is trivially achieved by taking $p = p'$. If $p' = \lambda x^n \alpha$, $\alpha \in P(\mu)$, then $p' \Leftrightarrow^* \lambda x^{n+m}\alpha \in P(w)$ and the result is proved. Let us then suppose that $w \Rightarrow w'$, i.e.,

$$w = \lambda x^n \mu \text{ and } w' = \lambda x^{n+m}\mu,$$

with $\lambda, \mu \in A^*$. If $p' \in P(\lambda x^n)$, then $p' \in P(w)$. If $p' = \lambda x^{n+m}\alpha$, $\alpha \in P(\mu)$, then $p' \Leftrightarrow^* \lambda x^n \alpha \in P(w)$. It remains to consider the case

$$p' = \lambda x^n x^k x_1, \quad \text{with } x_1 \in P(x),\ 0 \le k < m.$$

Let $x = x_1 x_2$ and $p = \lambda x^n \in P(w)$. One has

$$p' = px^k x_1 \ \text{ and } \ p = \lambda x^n \Leftrightarrow^* \lambda x^{n+m} = p' x_2 x^r, \text{ where } r + k + 1 = m.$$

Hence, it follows that $\phi(p)\ \mathcal{R}\ \phi(p')$. Suppose now that $w \Leftrightarrow^* w'$, i.e., there exist an integer $n > 0$ and words $w_1, \ldots, w_n$ such that $w = w_1$, $w' = w_n$, and for all $i \in [1, n-1]$, $w_i \Leftrightarrow w_{i+1}$. The proof is by induction on the integer n. For $n = 1$ the result is trivial since $w = w'$. Suppose now that the result is true for all integers $< n$. Since $w_{n-1} \Leftrightarrow w'$, by the preceding result any prefix p' of w' is such that $\phi(p')\ \mathcal{R}\ \phi(p'')$ where $p'' \in P(w_{n-1})$. By induction there exists a prefix p of w such that $\phi(p)\ \mathcal{R}\ \phi(p'')$. Hence, by transitivity, one has $\phi(p)\ \mathcal{R}\ \phi(p')$. A similar proof can be carried out in the case of suffixes of w' to prove that for any suffix s' of w' there exists a suffix s of w such that $\phi(s)\ \mathcal{L}\ \phi(s')$.

Theorem 4.7.2. *Let $S = B(r, n, n+1)$. For each idempotent $e \in E(S)$ the set*

$$S_e = \{s \in S \mid s^n = e \in E(S)\}$$

is finite.

Proof. Let $\phi : A^+ \to S$ be the canonical epimorphism and $e \in E(S)$. We denote by Σ_e the set of the canonical representatives of the elements of S_e. One has that $\phi(\Sigma_e) = S_e$ and ϕ restricted to Σ_e is a bijection. We prove now that the set Σ_e is in fact finite. Let $w = x_e$ be the canonical representative of e. One has that for any $x \in \Sigma_e$, $\phi(x^n) = e = \phi(w)$, i.e., $x^n \Leftrightarrow^* w$. From Lemma 4.7.1 one has that there exist $w_1 \in P(w)$ and $w_2 \in S(w)$ such that

$$\phi(x)\ \mathcal{R}\ \phi(w_1) \text{ and } \phi(x)\ \mathcal{L}\ \phi(w_2).$$

Suppose now by contradiction that Σ_e is infinite. Since the set $P(w)$ is finite then, by the pigeon-hole principle, there will exist an infinite set $\{x_i \mid i \in I\} \subseteq \Sigma_e$ and an element $p \in P(w)$ such that

$$\phi(x_i)\ \mathcal{R}\ \phi(p),\ i \in I.$$

Moreover, again by the pigeon-hole, there will exist an infinite subset $\{y_j \mid j \in J\}$ of $\{x_i \mid i \in I\}$ and $s \in S(w)$ for which

$$\phi(y_j) \mathcal{L} \phi(s), \ j \in J.$$

Thus for $r, t \in J$ and $r \neq t$ one has that $\phi(y_r) \mathcal{H} \phi(y_t)$. By Lemma 4.6.1 the $\mathcal{H}$-classes of S are trivial so that $\phi(y_r) = \phi(y_t)$; this implies $y_r = y_t$ which is a contradiction.

In conclusion, we observe that $S_e \subseteq F_e$, where $F_e = \{s \in S \mid e \in S^1 s S^1\}$ is the set of all factors of e. In the case $n > 2$ the fact that S_e is finite can be derived by the stronger result that for any $s \in S$ the set F_s is finite, since S is finite $\mathcal{J}$-above (see Sect. 4.6).

5. Regularity Conditions

A language L over the finite alphabet A is recognizable or regular if and only if the syntactic monoid $S(L)$ is finite (see Theorem 4.1.1). A property P of languages is said to be ***syntactic*** if whenever two languages L and L' have the same syntactic monoid and L has the property P, then also L' has the property P. In this sense, ***commutativity*** is a syntactic property (L is commutative if and only if its syntactic monoid is commutative), ***periodicity*** is syntactic and also ***rationality*** is a syntactic property. We point out that when a characterization of rationality involves only properties of languages which are syntactic, then this result is more a result on semigroups than on languages. Indeed, a relation $f_1 \equiv_L f_2$, with $f_1, f_2 \in A^*$, can be rewritten as

$$\forall u, v \in A^* \ (uf_1v \in L \Longleftrightarrow uf_2v \in L).$$

Thus the above relation does not depend on the contexts (u, v) which complete the words f_1 and f_2 in L, and it is equivalent to the equality $s_1 = s_2$ in $S(L)$, where $\phi : A^* \to S(L)$ is the syntactic morphism and $\phi(f_1) = s_1$ and $\phi(f_2) = s_2$.

However, as we shall see in the following sections, one can consider conditions formulated by properties which are not syntactic. In such a case it is not possible to pass directly through the syntactic monoid to prove the rationality of the language; the dependence on "contexts" can be overcome by very strong "forcing" arguments, like Ramsey's theorem, which will be used in the proof of some "non-uniform" regularity conditions.

The following lemma, called Nerode's criterion, is a useful tool in dealing with non-uniform conditions.

Lemma 5.0.2. *Let P be a property defined in the class of all the languages over a finite alphabet A and $\mathcal{L}_P$ be the family of the languages satisfying P. If $\mathcal{L}_P$ is finite and for any $a \in A$, $L \in \mathcal{L}_P$ implies $a^{-1}L \in \mathcal{L}_P$, then any language $L \in \mathcal{L}_P$ is regular.*

Proof. Let $L \in \mathcal{L}_P$. By hypothesis, for any $u \in A^*$, $u^{-1}L \in \mathcal{L}_P$. Since $\mathcal{L}_P$ is finite, the sets $u^{-1}L$, $u \in A^*$, are finitely many. By Nerode's theorem (see Theorem 4.1.1) L is regular.

The study of regularity conditions for periodic languages has been called also the ***Burnside problem for languages.*** The "permutation" and the "iteration" conditions for semigroups can be translated into regularity conditions which are syntactic and of interest for the Burnside problem for languages. However, in the following sections we shall analyze some versions of these conditions which cannot be expressed as properties of the syntactic monoid of the language.

We stress that any regularity condition, even though non-syntactic for a language L is, in fact, a finiteness condition for the syntactic monoid $S(L)$. However, not every monoid is the syntactic monoid of a language (see [101], Chap. 6). A monoid M is called *syntactic* if there exists a language L such that $M = S(L)$.

5.1 Uniform conditions

a) Periodicity and permutation

Let L be a language. We say that L is *periodic (permutable)* if the syntactic monoid $S(L)$ is periodic (permutable). The conditions of periodicity and permutability for L can be also expressed as:

Periodicity. For any $u \in A^*$ there exist integers $n, k > 0$ such that for all $x, y \in A^*$ one has

$$xu^n y \in L \Longleftrightarrow xu^{n+k} y \in L. \tag{5.1}$$

Permutability. There exists an integer $n > 1$ such that for any sequence $u_1, \ldots, u_n$ of n words of A^* there exists a nontrivial permutation σ of $\{1, \ldots, n\}$ such that for all $x, y \in A^*$

$$xu_1 \cdots u_n y \in L \Longleftrightarrow xu_{\sigma(1)} \cdots u_{\sigma(n)} y \in L.$$

Let $n > 1$. We say that a language L is *n-permutable*, or satisfies the property $\mathcal{P}_n$, if $S(L)$ is n-permutable. A 2-permutable language L is also called a *commutative language.* One easily verifies that a language L is commutative if and only if L satisfies the following condition:

$$u \in L \text{ and } u \sim v \Longrightarrow v \in L,$$

where $\sim$ denotes the relation of commutative equivalence.

Since the syntactic monoid $S(L)$ of a language $L \subseteq A^*$ is finitely generated, then from the theorem of Restivo and Reutenauer (see Theorem 3.4.1) one has the following regularity condition:

Proposition 5.1.1. *A language L over a finite alphabet is regular if and only if it is periodic and permutable.*

Let us observe that the conditions of periodicity and permutability are formulated in a way which is independent of the contexts (x, y) which complete the words in L (*uniform condition*). Moreover, they are expressed by a double implication $\Longleftrightarrow$.

If one eliminates from the permutation condition the uniformity (even on one side) it is open as to whether one has still a regularity condition for a periodic language. In Sect. 5.3 we shall see a special kind of non-uniform permutation property (called *permutative property*) which is sufficient to assure the regularity of a periodic language.

In Chap. 6 we shall prove that a permutable language is regular under a condition weaker than periodicity. This condition is obtained from the periodicity condition by replacing in equation (5.1) the "double" with a "single" implication.

b) Iteration

A classical result of automata theory, whose proof we report for the sake of completeness, is the following:

Proposition 5.1.2. *A language L is regular if and only if there exists an integer m such that for all $z_1, \ldots, z_m \in A^*$ there exist integers $h, k, 1 \leq h \leq k \leq m$, such that for all $u, v \in A^*$*

$$uz_1 \cdots z_m v \in L \Longleftrightarrow uz_1 \cdots z_{h-1}(z_h \cdots z_k)^n z_{k+1} \cdots z_m v \in L, \tag{5.2}$$

for all $n \geq 0$.

Proof. If L is regular, then from Myhill's theorem (see Theorem 4.1.1) the syntactic congruence of L has a finite index, or, which is the same, the syntactic monoid $S(L)$ is finite. By Proposition 3.7.1, $S(L)$ satisfies the iteration property $C(n, m)$ with $m = \text{card}(S(L)) + 1$ and $n \geq 0$. Let $\phi : A^* \to S(L)$ be the syntactic morphism and $z_1, \ldots, z_m$ be m words of A^*. Denoting $\phi(z_i)$ by s_i, $i = 1, \ldots, m$, then from $C(n, m)$ one has that there exist integers $h, k, 1 \leq h \leq k \leq m$, such that

$$s_1 \cdots s_m = s_1 \cdots s_{h-1}(s_h \cdots s_k)^n s_{k+1} \cdots s_m$$

for all $n \geq 0$. We can rewrite the preceding equation as

$$z_1 \cdots z_m \equiv_L z_1 \cdots z_{h-1}(z_h \cdots z_k)^n z_{k+1} \cdots z_m$$

for all $n \geq 0$ and this obviously implies equation (5.2).

Conversely, if one supposes that equation (5.2) is satisfied, then any word of A^* of length $\geq m$ is reducible because it is syntactically congruent to a shorter word (it is sufficient to make $n = 0$ in (5.2)). This implies from Proposition 3.2.1 that $S(L)$ is finite and then, from Theorem 4.1.1, L is regular.

The preceding theorem is a particular case of more general results which can be derived by using the iteration conditions for the syntactic monoid $S(L)$. In fact, the following propositions, whose proof is a straightforward consequence of Theorem 3.7.1 and Theorem 3.7.3, hold:

Proposition 5.1.3. *A language L is regular if and only if there exist integers m and s such that for any $z_1, \ldots, z_m \in A^*$ there exist integers $h, k, 1 \le h \le k \le m$, such that for all $u, v \in A^*$:*

$$uz_1 \cdots z_m v \in L \Longleftrightarrow uz_1 \cdots z_{h-1}(z_h \cdots z_k)^n z_{k+1} \cdots z_m v \in L, \qquad (5.3)$$

for all $n \ge s$.

Proposition 5.1.4. *A language L is regular if and only if there exists an integer m and $p = 2$ or $p = 3$ such that for any $z_1, \ldots, z_m \in A^*$ there exist integers $h, k, 1 \le h \le k \le m$, such that for all $u, v \in A^*$*

$$uz_1 \cdots z_m v \in L \Longleftrightarrow uz_1 \cdots z_{h-1}(z_h \cdots z_k)^p z_{k+1} \cdots z_m v \in L. \qquad (5.4)$$

By using Theorem 3.7.4 one can restrict the iteration condition to m-sequences of words:

Proposition 5.1.5. *A language L is regular if and only if there exists an integer m and $p = 2$ or $p = 3$ such that for any m-sequence $z_1, \ldots, z_m \in A^*$ there exist integers $h, k, 1 \le h \le k \le m$, such that for all $u, v \in A^*$*

$$uz_1 \cdots z_m v \in L \Longleftrightarrow uz_1 \cdots z_{h-1}(z_h \cdots z_k)^p z_{k+1} \cdots z_m v \in L. \qquad (5.5)$$

The following proposition [102] gives an iteration condition which is uniform but is not syntactic. Indeed, this condition is expressed by a single implication instead of a double implication.

Proposition 5.1.6. *A language L is regular if and only if there exists an integer $m > 0$ such that for any m-sequence $w_1, \ldots, w_m$ there exist integers h, k, $1 \le h \le k \le m$, such that for all $u, v \in A^*$,*

$$uw_1 \cdots w_m v \in L \Longrightarrow uw_1 \cdots (w_h \cdots w_k)^n \cdots w_m v \in L,$$

for all $n \ge 0$.

Proof. The "only if" part of the proposition is trivial. Let us then prove the "if" part. By hypothesis one has that for all $n \ge 0$

$$\mathrm{Cont}_L(w_1 \cdots w_m) \subseteq \mathrm{Cont}_L(w_1 \cdots (w_h \cdots w_k)^n \cdots w_m). \qquad (5.6)$$

We shall first show that

$$\mathrm{Cont}_L(w_1 \cdots (w_h \cdots w_k)^2 \cdots w_m) \subseteq \mathrm{Cont}_L(w_1 \cdots w_m).$$

Indeed, let $u, v \in A^*$ be such that

$$uw_1 \cdots (w_h \cdots w_k)^2 \cdots w_m v \in L. \qquad (5.7)$$

Since $w_1, \ldots, w_m$ is an m-sequence, there exists $\alpha \in A^*$ such that

$$w_1 \cdots (w_h \cdots w_k)^2 \cdots w_m = w_1 \cdots w_m \alpha. \tag{5.8}$$

By canceling on the left on both the sides of the above equation the common prefix $w_1 \cdots w_k$, one has

$$w_h \cdots w_m = w_{k+1} \cdots w_m \alpha$$

and then

$$w_1 \cdots w_m = w_1 \cdots w_{h-1} w_{k+1} \cdots w_m \alpha. \tag{5.9}$$

From equations (5.7) and (5.8) one has $uw_1 \cdots w_m \alpha v \in L$, so that

$$(u, \alpha v) \in \mathrm{Cont}_L(w_1 \cdots w_m).$$

From (5.6) in the case $n = 0$, one gets

$$(u, \alpha v) \in \mathrm{Cont}_L(w_1 \cdots w_{h-1} w_{k+1} \cdots w_m),$$

i.e., by equation (5.9), $uw_1 \cdots w_m v \in L$. Thus one has

$$uw_1 \cdots w_m v \in L \iff uw_1 \cdots (w_h \cdots w_k)^2 \cdots w_m v \in L.$$

By Proposition 5.1.5, the result follows.

5.2 Pumping properties

Let us now introduce some non-uniform iteration conditions for a language that we call *pumping properties* since they are inspired by the "pumping lemma" for regular languages (see [62]).

Definition 5.2.1. *Let L be a language and $x = uvw$. We say that the word v is a* pump *for x, in the context (u, w), relative to L if for any $n \geq 0$*

$$x \in L \iff uv^n w \in L.$$

We say that v is a positive-pump *if the latter condition is satisfied for any $n > 0$.*

We shall not specify the context of a pump v when there are no ambiguities. Let us observe that any pump is also a positive-pump, whereas the converse is not generally true.

Definition 5.2.2. *A language L satisfies the property D_m (property C_m) if for any $x, w_1, w_2, \ldots, w_m, y \in A^*$, there exist i, j, $1 \leq i < j \leq m+1$, such that $w_i \cdots w_{j-1}$ is a pump (positive-pump) for $xw_1w_2 \cdots w_m y$ relative to L. A language L satisfies the* block pumping property (positive block pumping property) *if there exists an integer $m > 0$ such that L satisfies the property D_m (property C_m).*

Definition 5.2.3. *A language L satisfies the* block cancellation property *if there exists an integer $m > 0$ such that for any $x, w_1, w_2, \ldots, w_m, y \in A^*$, there exist i, j, $1 \leq i < j \leq m+1$, such that*

$$xw_1w_2\cdots w_my \in L \Longleftrightarrow xw_1w_2\cdots w_{i-1}w_j\cdots w_my \in L.$$

It is trivial from the definition that if a language satisfies the block pumping property, then it satisfies also the positive block pumping property and the block cancellation property.

In the former definitions the integers i and j depend on the context (x, y) in which the word $w_1w_2\cdots w_m$ is inserted. If they do not depend on the context, then the corresponding properties will be said to be ***uniform***. We observe that ***the uniform block cancellation property and the uniform (positive) block pumping property are syntactic.*** The first property, from Proposition 5.1.2, assures the finiteness of the syntactic monoid $S(L)$ of a language L and then the regularity of L. The second one by Proposition 5.1.3 also implies the regularity of L.

The relevance of the block pumping properties for the Burnside problem for languages is due to the fact that ***if a language L satisfies the property D_m or C_m, then it is periodic.*** Indeed, as shown in the next lemma, for any word w one has that $w^m \equiv_L w^{m+m!}$.

Lemma 5.2.1. *Let L be a language satisfying C_m. Then the syntactic monoid $S(L)$ is periodic.*

Proof. Let $w \in L$. We prove that for any $x, y \in A^*$

$$xw^my \in L \Longleftrightarrow xw^{m+m!}y \in L.$$

Condition C_m, applied to $w_1 = w_2 = \cdots = w_m = w$, shows that there exists an integer $k \leq m$, depending on the context (x, y), such that for any $s > 0$

$$xw^my \in L \Longleftrightarrow xw^{m+sk}y \in L.$$

Since k divides $m!$, there exists an integer $s > 0$ such that $sk = m!$ and the statement is true.

Lemma 5.2.2. *Let L be a language satisfying the property C_m (property D_m), then for any $u \in A^*$, $u^{-1}L$ satisfies the property C_m (property D_m).*

Proof. Let $u \in A^*$. Since L satisfies C_m (satisfies D_m), then for any $x, y, w_1, \ldots, w_m \in A^*$ there exist i, j, with $1 \leq i < j \leq m+1$, such that for any $s > 0$

$$uxw_1\cdots w_my \in L \Longleftrightarrow uxw_1\cdots w_{i-1}(w_i\cdots w_{j-1})^sw_j\cdots w_my \in L;$$

therefore, for any $s > 0$ ($s \geq 0$) one has

$$xw_1\cdots w_my \in u^{-1}L \Longleftrightarrow xw_1\cdots w_{i-1}(w_i\cdots w_{j-1})^sw_j\cdots w_my \in u^{-1}L,$$

and $u^{-1}L$ satisfies C_m (satisfies D_m).

The following theorem was proved in [59] by A. Ehrenfeucht, R. Parikh, and G. Rozenberg.

Theorem 5.2.1. *A language is regular if and only if it satisfies the block pumping property.*

Proof. Let $\mathcal{L}_{D_m}$ be the family of the languages satisfying D_m. Now we prove that $\mathcal{L}_{D_m}$ is finite. For this it is sufficient to prove that there exists an integer N such that for any two languages $L, L' \in \mathcal{L}_{D_m}$ one has

$$L = L' \iff L \cap A^{[N]} = L' \cap A^{[N]}. \tag{5.10}$$

In fact let $N = R(2, m+1, 2) - 1$, where R is the Ramsey function as in Theorem 2.1.1. Suppose by contradiction that there exist two languages $L, L' \subseteq \mathcal{L}_{D_m}$ such that equation (5.10) is not satisfied. Then there exists a word w of minimal length in the symmetric difference $L \triangle L'$ such that $|w| \geq N$. We can write $w = w_1 w_2 \cdots w_N$, with $|w_i| \geq 1$, for $i = 1, \ldots, N$. Let $X = \{1, \ldots, N+1\}$. Consider a partition of $P_2(X)$ in two subsets P_1, P_2 where $\{i, j\} \in P_1$ if and only if

$$w_1 w_2 \cdots w_N \in L \iff w_1 w_2 \cdots w_{i-1} w_j \cdots w_N \in L,$$

with $1 \leq i < j \leq N+1$, and $P_2 = P_2(X) \setminus P_1$. By Ramsey's theorem there exists a subset $Y = \{j_1, j_2, \ldots, j_{m+1}\}$ of X, such that $1 \leq j_1 < j_2 < \cdots < j_{m+1} \leq N+1$ and

$$\text{either } P_2(Y) \subseteq P_1 \text{ or } P_2(Y) \subseteq P_2.$$

Since L satisfies D_m, one has $P_2(Y) \subseteq P_1$. Let $(u_1, \ldots, u_m)$ be the derived sequence of $(w_1, \ldots, w_N)$ corresponding to the increasing sequence $j_1 < j_2 < \cdots < j_{m+1}$ (see Def. 2.1.3). By construction $w = x u_1 \cdots u_m y$ and, since $P_2(Y) \subseteq P_1$, for any $i, j \in \{1, \ldots, m\}$ with $i < j$, one has

$$x u_1 u_2 \cdots u_m y \in L \iff x u_1 u_2 \cdots u_{i-1} u_j \cdots u_m y \in L. \tag{5.11}$$

Since L' satisfies D_m, there exist $i, j \in \{1, \ldots, m\}$ such that

$$x u_1 u_2 \cdots u_m y \in L' \iff x u_1 u_2 \cdots u_{i-1} u_j \cdots u_m y \in L'. \tag{5.12}$$

Let $w' = x u_1 u_2 \cdots u_{i-1} u_j \cdots u_m y$. Since w is of minimal length in $L \triangle L'$ and $|w'| < |w|$, one has $w' \in L \Leftrightarrow w' \in L'$. Then, by equations (5.11) and (5.12), one derives $w \in L \Leftrightarrow w \in L'$, which is a contradiction. Thus equation (5.10) is proved and this implies that in $\mathcal{L}_{D_m}$ there are only finitely many languages.

From Lemma 5.2.2 for any $u \in A^*$, $L \in \mathcal{L}_{D_m}$ implies that $u^{-1}L \in \mathcal{L}_{D_m}$. Hence, by Lemma 5.0.2, any language in $\mathcal{L}_{D_m}$ is regular.

We observe that the proof of the previous theorem uses only the block cancellation property which is weaker than the block pumping property. In fact we have proved the following stronger result:

Theorem 5.2.2. *If a language satisfies the block cancellation property, then it is regular.*

A weaker block cancellation property can be introduced by replacing in Def. 5.2.3 the double implication $\Longleftrightarrow$ with the single implication $\Longrightarrow$. More precisely a language L satisfies the *weak block cancellation property* if there exists an integer m such that for any word $w \in L$ such that $w = xw_1 \cdots w_m y$ with $x, y, w_1, \ldots, w_m \in A^*$ there exist i, j such that $1 \leq i < j \leq m+1$ and $xw_1 \cdots w_{i-1} w_j \cdots w_m y \in L$. If a language satisfies this property for m, then it satisfies the property for any $n \geq m$.

Proposition 5.2.1. *Let L be a language. If L and its complement L^c satisfy the weak block cancellation property, then they are regular.*

Proof. By Theorem 5.2.2 it is sufficient to show that L satisfies the block cancellation property. Let n be an integer such that L and L^c both satisfy the weak block cancellation property for n. Let $w = xw_1 \cdots w_n y$ be any word of A^*. We define i, j with $1 \leq i < j \leq n+1$ as follows.

Since L satisfies the weak block cancellation property, if $w \in L$, then let i, j be such that $1 \leq i < j \leq n+1$ and $xw_1 \cdots w_{i-1} w_j \cdots w_n y \in L$. In a similar way since L^c satisfies the weak block cancellation property, if $w \in L^c$, then let i, j be such that $1 \leq i < j \leq n+1$ and $xw_1 \cdots w_{i-1} w_j \cdots w_n y \in L^c$. Thus $w \in L$ implies that $xw_1 \cdots w_{i-1} w_j \cdots w_n y \in L$; moreover, $w \notin L$ or, equivalently, $w \in L^c$, implies that $xw_1 \cdots w_{i-1} w_j \cdots w_n y \in L^c$, i.e., $xw_1 \cdots w_{i-1} w_j \cdots w_n y \notin L$. Hence, for any $w = xw_1 \cdots w_n y \in A^*$ there exist i, j with $1 \leq i < j \leq n+1$ and

$$w \in L \Longleftrightarrow xw_1 \cdots w_{i-1} w_j \cdots w_n y \in L.$$

Thus L satisfies the block cancellation property.

The above proposition was used in [138] to prove that two complementary languages which are *supports of rational power series* are regular.

A problem left open in [59] is whether the positive block pumping implies regularity. We shall give now a positive answer to this question [159] (see Theorem 5.2.3).

Lemma 5.2.3. *Let m be a positive integer and $\mathcal{L}_{C_m}$ be the family of languages satisfying C_m. Then for any $k > 0$ there exists an integer M (depending on k) such that for any $(x_1, y_1, L_1), (x_2, y_2, L_2), \ldots, (x_k, y_k, L_k) \in A^* \times A^* \times \mathcal{L}_{C_m}$, and for any $w_1, w_2, \ldots, w_M \in A^*$ there exist integers i, j, $1 \leq i < j \leq M+1$, such that $w_i \cdots w_{j-1}$ is a positive pump for $x_s w_1 w_2 \cdots w_M y_s$ relative to L_s for $s \in \{1, 2, \ldots, k\}$.*

Proof. The proof is by induction on k. If $k = 1$, then the statement is true for $M = m$, since L_1 satisfies C_m. Let $k > 1$. By the induction hypothesis there exists an integer M' that satisfies the statement for $k - 1$. Then let us set $M = R(2, M' + 1, 2) - 1$, where R is the function of Ramsey's theorem. Let $w_1, w_2, \ldots, w_M \in A^*$ and $X = \{1, 2, \ldots, M + 1\}$. Let us consider the bipartition (Y_1, Y_2) of $P_2(X)$, where Y_1 is the set

$$\{\{i, j\} \mid w_i \cdots w_{j-1} \text{ is a positive pump for } x_k w_1 \cdots w_M y_k \text{ relative to } L_k\}$$

and $Y_2 = P_2(X) \setminus Y_1$.

By Ramsey's theorem there exists $Y \in P_{M'+1}(X)$ such that

$$\text{either } P_2(Y) \subseteq Y_1 \text{ or } P_2(Y) \subseteq Y_2.$$

Let $Y = \{j_1, j_2, \ldots, j_{M'+1}\}$, with $1 \leq j_1 < j_2 < \cdots < j_{M'+1} \leq M + 1$ and $(v_1, \ldots, v_{M'})$ be the corresponding derived sequence of $(w_1, \ldots, w_M)$. Since $M' \geq m$, there exist integers s, t, with $1 \leq s < t \leq M' + 1$ such that $v_s \cdots v_{t-1}$ is a positive pump for $x_k w_1 \cdots w_M y_k$ relative to L_k. Since $v_s \cdots v_{t-1} = w_{j_s} \cdots w_{j_t - 1}$, one has that

$$\{j_s, j_t\} \in Y_1 \cap P_2(Y) \neq \emptyset,$$

hence, $P_2(Y) \subseteq Y_1$. This means that for any two integers i, j, with $1 \leq i < j \leq M' + 1$, $v_i \cdots v_{j-1}$ is a positive pump for $x_k w_1 \cdots w_M y_k$ relative to L_k. By the induction hypothesis, we can choose i, j, with $1 \leq i < j \leq M' + 1$, such that for any $s \in \{1, \ldots, k-1\}$, $v_i \cdots v_{j-1}$ is a positive pump for $x_s w_1 \cdots w_M y_s$ relative to L_s.

Lemma 5.2.4. *The languages satisfying C_m are finitely many.*

Proof. Suppose that there exists an integer $k > 0$ such that for any $L_1, L_2 \subseteq A^*$, satisfying C_m, one has

$$L_1 \cap A^{[k]} = L_2 \cap A^{[k]} \iff L_1 = L_2,$$

then the statement is trivially true. Let us suppose by contradiction that there exist infinitely many languages of A^* satisfying C_m; then such an integer k does not exist. Hence, for any $k > 0$ there exist at least a word $w_k \in A^*$ and two languages $L_{1,k}$, $L_{2,k}$, satisfying C_m, such that

$$L_{1,k} \cap A^{[k]} = L_{2,k} \cap A^{[k]}$$

and

$$w_k \in L_{1,k}, \; w_k \notin L_{2,k}.$$

Moreover, we may suppose that w_k has minimal length, that is, for any $l < |w_k|$ one has

$$L_{1,k} \cap A^{[l]} = L_{2,k} \cap A^{[l]}.$$

Moreover, $|w_k| \geq k$.

Let us consider now the language $L = \{w_1, w_2, \ldots, w_k, \ldots\}$. Since L is infinite, by Theorem 2.3.1, there exists a uniformly recurrent word $\alpha \in A^\omega$ such that $F(\alpha) \subseteq F(L)$. Let $K : \mathbb{N} \to \mathbb{N}$ be the recurrency function of α and $D : \mathbb{N} \to \mathbb{N}$ be the function associated with the infinite word α as defined in Proposition 2.7.7. Let M be a positive integer satisfying Lemma 5.2.3 for $k = 4$ and set $N = (M+1) \cdot K(D(M))$. Let u be a factor of α such that $|u| = N$. Since $F(\alpha) \subseteq F(L)$, there exist $w \in L$, $x, y \in A^*$ such that $w = xuy$. Moreover, there exist two languages $L_1, L_2 \subseteq A^*$, satisfying C_m, such that for any $l < |w|$

$$L_1 \cap A^{[l]} = L_2 \cap A^{[l]}$$

and

$$w \in L_1, \quad w \notin L_2.$$

Since $|u| = N$, we can write $u = u_1 u_2 \cdots u_{M+1}$, with $|u_i| \geq K(D(M))$, with $i = 1, \ldots, M+1$; therefore, there exists a word $v \in F(\alpha)$, with $|v| = D(M)$, such that

$$u_i = \lambda_i v \mu_i, \ \lambda_i, \mu_i \in A^*, \ 1 \leq i \leq M+1.$$

Hence, as $w = xuy$, one has

$$w = \lambda v_1 v_2 \cdots v_{M+1} \mu,$$

with $\lambda, \mu \in A^*$, $v_i \in vA^*$, $|v| = D(M)$, and $v_1 v_2 \cdots v_{M+1} \in F(\alpha)$. By Lemma 5.2.3 there exist integers i, j, with $1 \leq i < j \leq M+1$, such that $z = v_i \cdots v_{j-1}$ is a positive pump for $w = \lambda v_1 v_2 \cdots v_{M+1} \mu$ relative to L_1 and L_2. Therefore, setting $x' = \lambda v_1 v_2 \cdots v_{i-1}$, $y' = v_j \cdots v_{M+1} \mu$, one has $w = x'zy'$, $y' \in vA^*$, $zv \in F(\alpha)$, and for any $s > 0$

$$w = x'zy' \in L_t \Longleftrightarrow x'z^s y' \in L_t, \tag{5.13}$$

for $t = 1, 2$.

By Lemma 5.2.1, the syntactic monoids $S(L_1)$ and $S(L_2)$ are periodic. This implies that for any $w \in A^*$ there exist positive integers n, k, n', k' such that $w^n \equiv_{L_1} w^{n+k}$ and $w^{n'} \equiv_{L_2} w^{n'+k'}$. We can obviously suppose that $n = n'$ and $k = k'$, otherwise n and n' could be replaced by $\max(n, n')$ and k, k' by $\operatorname{lcm}(k, k')$ and the former relations still would hold. Then there exist $n, k > 0$ such that

$$z^n \equiv_{L_t} z^{n+k} \tag{5.14}$$

for $t = 1, 2$. From equation (5.13), for $s = k$, we have

$$x'zy' \in L_t \Longleftrightarrow x'z^k y' \in L_t \tag{5.15}$$

for $t = 1, 2$. Now let z' be the maximal common prefix of z^n and y'. Since v is a common prefix of z^n and y' we have that $|z'| \geq |v| \geq D(M)$. Let $y'' \in A^*$ such that $y' = z'y''$. We will prove now that there exists $\lambda \in A^*$ such that

$$x'z'y'' \in L_t \iff x'z^n\lambda y'' \in L_t \tag{5.16}$$

and

$$x'z^kz'y'' \in L_t \iff x'z^kz^n\lambda y'' \in L_t \tag{5.17}$$

for $t = 1, 2$.

The proof is by induction on $q = |z^n| - |z'|$. If $q = 0$, then $z' = z^n$ and equations (5.16) and (5.17) hold for $\lambda = \epsilon$. Let $q > 0$ and z' be a proper prefix of z^n. Let z'' be the suffix of z' such that $|z''| = D(M)$ and let $a \in A$ be such that $z'a$ is a prefix of z^n. Since $|v| = D(M) = |z''|$, $z \in vA^*$, and $z''a \in F(z^n)$, one can easily see that $z''a \in F(zv) \subseteq F(\alpha)$.

Then by Proposition 2.7.7 we can write

$$z'' = zf_M,$$

where $\lambda \in A^*$ and f_M is the M th term of a bi-ideal sequence $f_{i+1} = f_ig_if_i$, $i \in \{1, \ldots, M-1\}$ such that $f_1 \in aA^*$ and $g_i \in aA^*$ for $i \in \{1, \ldots, M-1\}$. Let $(w_1, w_2, \ldots, w_M)$ be the canonical factorization of f_M. Then one has

$$z'' = zw_1w_2\cdots w_M,$$

with $w_i \in aA^*$, for $i \in \{1, \ldots, M\}$, and $w_i \in w_{i+1}\cdots w_MaA^*$, for $i \in \{1, \ldots, M-1\}$ (see (2.8)). Since z'' is a suffix of z', one has

$$z' = \zeta'w_1w_2\cdots w_M,$$

for a suitable $\zeta' \in A^*$.

We apply Lemma 5.2.3 to the four triplets $(x'\zeta', y'', L_1)$, $(x'\zeta', y'', L_2)$, $(x'z^k\zeta', y'', L_1)$, and $(x'z^k\zeta', y'', L_2)$. Therefore, there exist integers i, j with $1 \le i < j \le M+1$ such that

$$x'\zeta'w_1\cdots w_My'' \in L_t \iff$$

$$x'\zeta'w_1\cdots w_{i-1}(w_i\cdots w_{j-1})^2w_j\cdots w_My'' \in L_t \tag{5.18}$$

and

$$x'z^k\zeta'w_1\cdots w_My'' \in L_t \iff$$

$$x'z^k\zeta'w_1\cdots w_{i-1}(w_i\cdots w_{j-1})^2w_j\cdots w_My'' \in L_t \tag{5.19}$$

for $t = 1, 2$.

Now, since $w_i \in w_j\cdots w_MaA^*$ for $i < j$, for a suitable $\xi \in A^*$, $w_i\cdots w_M = w_j\cdots w_Ma\xi$ and, substituting on the right side of equations (5.18) and (5.19), we have

$$x'z'y'' \in L_t \iff x'z'a\xi y'' \in L_t \tag{5.20}$$

and

$$x'z^kz'y'' \in L_t \iff x'z^kz'a\xi y'' \in L_t \tag{5.21}$$

for $t = 1, 2$.

Therefore, since $z'a$ is a prefix of z^n and $|z^n| - |z'a| = q - 1$, by the induction hypothesis there exists $\xi' \in A^*$ such that

$$x'z'a\xi y'' \in L_t \iff x'z^n\xi'\xi y'' \in L_t \tag{5.22}$$

and

$$x'z^kz'a\xi y'' \in L_t \iff x'z^kz^n\xi'\xi y'' \in L_t \tag{5.23}$$

for $t = 1, 2$. So if we set $\lambda = \xi'\xi$, comparing (5.20) with (5.22) and (5.21) with (5.23), we obtain equations (5.16) and (5.17), respectively. Moreover, from equations (5.14), (5.15), and (5.17), considering that $z'y'' = y'$, one derives

$$x'zy' \in L_t \Leftrightarrow x'z^ky' = x'z^kz'y'' \in L_t \Leftrightarrow x'z^{k+n}\lambda y'' \in L_t \Leftrightarrow x'z^n\lambda y'' \in L_t$$

for $t = 1, 2$. Therefore, from (5.16), it follows that

$$x'zy' \in L_t \iff x'y' \in L_t \tag{5.24}$$

for $t = 1, 2$. We remember that $L_1 \cap A^{[l]} = L_2 \cap A^{[l]}$, for $l < |w|$, and, therefore,

$$x'y' \in L_1 \iff x'y' \in L_2,$$

hence, by equation (5.24)

$$x'zy' \in L_1 \iff x'zy' \in L_2.$$

Since $x'zy' = w$, we have

$$w \in L_1 \iff w \in L_2,$$

that is a contradiction, because $w \in L_1$ and $w \notin L_2$.

Theorem 5.2.3. *A language is regular if and only if it satisfies the positive block pumping property.*

Proof. If L satisfies the positive block pumping property, then there exists an integer m such that L satisfies C_m. From Lemmas 5.0.2, 5.2.2, and 5.2.4 it follows that L is regular. Conversely, if L is regular, then it satisfies the uniform block pumping property (see Proposition 5.1.2) and then C_m for a suitable $m > 0$.

Remark 5.2.1. If in Def. 5.2.2 the words $w_1, \ldots, w_m$ are *letters instead of blocks*, one defines the usual pumping property of automata theory. This condition is very far from implying the regularity. In fact it has been proved [59] that there exist uncountably many non-regular languages which satisfy this property.

Remark 5.2.2. One can consider some extensions of block pumping properties where the "pump" starts only from a given integer $s > 1$. If a language L satisfies this kind of property, then one easily derives that L is a periodic language. Moreover, in the uniform case one can prove, by Proposition 5.1.3 that L is a regular language. An open problem is to see what occurs in the non-uniform case.

Remark 5.2.3. One can easily prove that in Theorems 5.2.1 and 5.2.3 the block and the positive block pumping property respectively, can be restricted to m-sequences.

5.3 Permutative property

In this section we present some non-uniform conditions for a language that correspond to the permutation properties for semigroups.

Definition 5.3.1. *Let m and k be fixed integers such that $m \geq k > 1$. A language L is* (m,k)-permutative *if there exists a permutation $\sigma \in S_k \setminus \{id\}$ such that whenever a word w is written as $w = ux_1 \cdots x_m v$ there exists a derived sequence $(y_1, \ldots, y_k)$ of $(x_1, \ldots, x_m)$ such that $x_1 \cdots x_m = u'y_1 \cdots y_k v'$, $u', v' \in A^*$, and*

$$w = uu'y_1 \cdots y_k v'v \in L \Longleftrightarrow uu'y_{\sigma(1)} \cdots y_{\sigma(k)} v'v \in L.$$

The language L is called permutative *if there exist integers m and k, $m \geq k$, such that L is (m,k)-permutative.*

When $k = 2$, that is, σ is the permutation (12), one obtains the *transposition property* of [136]. A language L has this property if, for some m, every time one distinguishes m successive factors in some word w, then there exist two consecutive blocks of these factors which, when transposed, give a word w' such that $w \in L \Longleftrightarrow w' \in L$. Note that the transposition property, when $m = 2$, is equivalent to the commutativity of L.

One can easily prove that a rational language is always permutative. Furthermore, there are languages which do not have this property: for example, the set of palindromes; consider a palindrome of the form

$$aba^2b \cdots a^{m-1}ba^mba^mba^{m-1} \cdots ba^2ba$$

and set $x_1 = ab, x_2 = a^2b, \ldots, x_m = a^mb$. No permutation of the x's other than the identity keeps this word a palindrome.

The following characterization of rationality was proved by Restivo and Reutenauer in [139].

Theorem 5.3.1. *A language is rational if and only if it is periodic and permutative.*

Proof. The proof can be articulated in the following points.

(*i*) Let k be a fixed integer ≥ 2 and let $\sigma \in \mathcal{S}_k \setminus \{id\}$. For $m \geq k$ we set $n(m) = R(k+1, m+1, 2) - 1$ where R is the function of Ramsey's theorem. We denote by $\mathcal{L}_{m,p}$ the family of languages over the alphabet A such that

- L is (m,k)-permutative,
- For any word x of length shorter than $n(m)$ one has

$$\forall u, v \in A^* \ (ux^p v \in L \Longleftrightarrow ux^{p+p!}v \in L). \tag{5.25}$$

(*ii*) Let L be a periodic language which is (m,k)-permutative. We show that $L \in \mathcal{L}_{m,p}$ for some p with $p + p! \geq 2n(m)$. Indeed, let W be the set of the words of length shorter than $n(m)$. For each x in W, as L is periodic, there exist i_x, j_x with $j_x \geq 1$ such that

$$\forall u, v \in A^* (ux^{i_x} v \in L \Longleftrightarrow ux^{i_x + j_x} v \in L). \tag{5.26}$$

It follows that for any $k \geq 0$

$$\forall u, v \in A^* (ux^{i_x} v \in L \Longleftrightarrow ux^{i_x + kj_x} v \in L).$$

Let

$$i = \max\{i_x \mid x \in W\}, \ j = \max\{j_x \mid x \in W\}.$$

Choose p such that $p \geq i$, $p \geq j$, and $p + p! \geq 2n(m)$. Then, if $|x| \leq n(m)$, that is, $x \in W$, since j_x divides $p!$, there exists a positive integer k such that $kj_x = p!$. Thus one has, for all $u, v \in A^*$,

$$ux^p v = ux^{p-i_x} x^{i_x} v \in L \Longleftrightarrow ux^{p-i_x} x^{i_x + p!} v = ux^{p+p!} v \in L.$$

Hence, L is in $\mathcal{L}_{m,p}$ with $p + p! \geq 2n(m)$.

(*iii*) By (*ii*), we know that any periodic language which is (m,k)-permutative is in some $\mathcal{L}_{m,p}$ with $p + p! \geq 2n(m)$. It will thus suffice to show that any such $\mathcal{L}_{m,p}$ is finite. Indeed, if L is in $\mathcal{L}_{m,p}$, then so is $a^{-1}L$ with $a \in A$, (as it may be easily verified), so one may apply Nerode's criterion (see Lemma 5.0.2) to conclude that any language in $\mathcal{L}_{m,p}$ is rational.

(*iv*) Let $p + p! \geq 2n(m)$, $\mathcal{L} = \mathcal{L}_{m,p}$, and $n = n(m)$. Furthermore, let $r = \text{card}(A)$ and $N = N(r, p+p!, n)$ be the integer of Shirshov's theorem. Let $L, L' \in \mathcal{L}$. We show that

$$L = L' \Leftrightarrow L \cap A^{[N-1]} = L' \cap A^{[N-1]}.$$

(i.e., if L and L' "agree" on all the words of length $< N$, then they coincide). This will ensure that $\mathcal{L}$ is finite and will conclude the proof.

(*v*) Let $<_a$ be the alphabetic order in A^*. Since $<_a$ is a well order, one can make induction with respect to it.

(*vi*) We show by induction that, for any word w, $w \in L \Longleftrightarrow w \in L'$. We already know that if $|w| < N$, then $w \in L \Longleftrightarrow w \in L'$. Let $|w| \geq N$. Suppose w contains as a factor the $(p + p!)$ th power of some nonempty word x, i.e.,

$w = ux^{p+p!}v$, with $|x| \le n-1$. As L and L' are both in $\mathcal{L} = \mathcal{L}_{m,p}$ we have, by equation (5.25),

$$(ux^{p+p!}v \in L \Longleftrightarrow ux^p v \in L) \text{ and } (ux^{p+p!}v \in L' \Longleftrightarrow ux^p v \in L').$$

Now, $ux^p v$ is smaller than w (with respect to the length), thus, by the induction hypothesis, we have

$$ux^p v \in L \Longleftrightarrow ux^p v \in L'.$$

Taking these equivalences together we obtain

$$w \in L \Longleftrightarrow w \in L'.$$

(vii) Suppose now that w contains no $(p+p!)$ th power of a nonempty word of length at most $n-1$. Then, by Corollary 2.4.3, w has a factor which is n-divided:

$$w = ux_1 \cdots x_n v, \tag{5.27}$$

where $(x_1, \ldots, x_n)$ is an n-division of $x_1 \cdots x_n$. Let $X = \{1, 2, \ldots, n+1\}$. Let $P_{k+1}(X)$ be the set of all subsets of X of cardinality $k+1$. Define a subset I of the set $P_{k+1}(X)$ as follows: $\{i_1, i_2, \ldots, i_{k+1}\}$ with $1 \le i_1 < i_2 < \cdots < i_{k+1} \le n+1$, is in I if the corresponding derived sequence $(y_1, \ldots, y_k)$ of $(x_1, \ldots, x_n)$ is such that $x_1 \cdots x_n = u' y_1 \cdots y_k v'$ and

$$uu' y_{\sigma(1)} \cdots y_{\sigma(k)} v' v \in L.$$

Let us remark that, since $(x_1, \ldots, x_n)$ is an n-division of $x_1 \cdots x_n$, by equation (5.27), one has

$$uu' y_{\sigma(1)} \cdots y_{\sigma(k)} v' v <_a w;$$

hence, by induction on the order $<_a$, I remains unchanged if in its definition L is replaced by L'.

Let $J = P_{k+1}(X) \setminus I$. Then, by Ramsey's theorem, there exists $Y \subset X$, with $\text{card}(Y) = m+1$, and such that either $P_{k+1}(Y) \subseteq I$ or $P_{k+1}(Y) \subseteq J$. The word w may be rewritten as

$$w = u'' z_1 \cdot \cdot z_m v'',$$

where $(z_1, \ldots, z_m)$ is the derived sequence of $(x_1, \ldots, x_n)$ corresponding to the set $Y = \{j_1, \ldots, j_{m+1}\}$ with $1 \le j_1 < \cdots < j_{m+1} \le n+1$. For any $i_1, \ldots, i_{k+1} \in Y$, with $i_1 < i_2 < \cdots < i_{k+1}$, the corresponding derived sequence $(t_1, \ldots, t_k)$ of $(x_1, \ldots, x_n)$ is by Lemma 2.1.1 a derived sequence of the sequence $(z_1, \ldots, z_m)$ so that we can write $w = u'' u''' t_1 \cdots t_k v''' v''$. The condition $P_{k+1}(Y) \subseteq I$ implies that

$$u'' u''' t_{\sigma(1)} t_{\sigma(2)} \cdots t_{\sigma(k)} v''' v'' \in L, \tag{5.28}$$

whereas the condition $P_{k+1}(Y) \subseteq J$ implies that

$$u''u'''t_{\sigma(1)}t_{\sigma(2)}\cdots t_{\sigma(k)}v'''v'' \notin L. \tag{5.29}$$

The language L is (m,k)-permutative; hence, if $w \in L$, then there are some $i_1,\ldots,i_{k+1} \in Y$ such that equation (5.28) holds; in this case, $P_{k+1}(Y) \subseteq I$. Hence, if $w \notin L$, because of the permutative property, there are some $i_1,\ldots,i_{k+1}$ such that equation (5.29) holds; in this case, $P_{k+1}(Y) \subseteq J$. Thus, $w \in L \Longleftrightarrow P_{k+1}(Y) \subseteq I$. Because of the above remark, the same holds with L' in place of L. Hence,

$$w \in L \Longleftrightarrow P_{k+1}(Y) \subseteq I \Longleftrightarrow w \in L'.$$

This concludes the proof.

Similar to the permutative property is the ***non-uniform permutation property***: a language L has this property if, for some m (depending only on L), for any words $u, x_1, \ldots, x_m, v$ there exists a permutation $\alpha \in \mathcal{S}_m \setminus \{id\}$, such that

$$ux_1\cdots x_m v \in L \Longleftrightarrow ux_{\alpha(1)}\cdots x_{\alpha(m)}v \in L.$$

Problem 5.3.1. Does Theorem 5.3.1 still hold with the permutation property instead of the permutative property?

Remark 5.3.1. If, in the permutation property, the permutation α depends on $x_1,\ldots,x_m$ only (and not on the contexts (u,v)), then the property becomes syntactic and the answer to the problem is positive; indeed, the syntactic monoid of the language L has the permutation property and, thus by Proposition 5.1.1, L is regular.

6. Well Quasi-orders and Regularity

In this chapter we give some regularity conditions that can be expressed in terms of ***well quasi-orders***. There exist various characterizations of this concept which was often rediscovered by different authors (see [100]); nowadays there exists a large literature on this subject. Well quasi-orders are important in mathematics and in many areas of theoretical computer science as well. Basic theorems such as Higman's theorem (see Theorem 6.2.1) and Simon's theorem (see Theorem 6.2.2) give a deep insight into combinatorics on words and language theory.

As shown by Ehrenfeucht et al. [60], regular languages can be characterized as the closed parts of monotone well quasi-orders in the set of all the words over a finite alphabet (generalized Myhill's theorem); moreover, monotone quasi-orders can be naturally associated with the derivation relations of suitable rewriting (or semi-Thue) systems, so that an interesting problem is to determine under which conditions a semi-Thue system, belonging to a given class, is such that its derivation relation is a well quasi-order. Ehrenfeucht et al., for instance, characterized [60] the ***unitary rewriting systems*** whose derivation relations are well quasi-orders (see Theorem 6.4.1). A further example is given by ***copying systems***. As an application of the generalized Myhill's theorem we shall prove in the last section that languages on a binary alphabet generated by copying systems are regular.

In the following we consider two further classes of rewriting systems. The first class consists of rewriting systems whose productions are of the kind $u^m \to u^{m+k}$, $m > 0$, $k > 0$, and $u_1u_2\cdots u_n \to u_{\sigma(1)}u_{\sigma(2)}\cdots u_{\sigma(n)}$, $u_{\sigma(1)}u_{\sigma(2)}\cdots u_{\sigma(n)} \to u_1u_2\cdots u_n$, for any sequence of words $u_1,\ldots,u_n \in A^+$, $n > 1$, and $u \in A^+$, where σ is a nontrivial permutation of $\{1,\ldots,n\}$; the permutation σ and the pair (m,k) depend on the words $u_1,\ldots,u_n$, and u, respectively.

The second class consists of rewriting systems whose productions are of the kind $a^m \to a^{m+k}$, $m > 0$, $k > 0$, and $u_1u_2\cdots u_n \to u_{\sigma(1)}u_{\sigma(2)}\cdots u_{\sigma(n)}$, for any sequence of words $u_1,\ldots,u_n \in A^+$, $n > 1$, and $a \in A$, where σ is a fixed nontrivial permutation of $\{1,\ldots,n\}$. We prove that the derivation relations induced by the preceding systems are well quasi-orders. Two main corollaries are the following regularity conditions (see Theorems 6.5.2

and 6.6.2). i) *A permutable and quasi-periodic language is regular* and ii) *An almost-commutative language which is quasi-periodic on the letters is regular.*

6.1 Well quasi-orders

Let S be a set and $\leq$ a quasi-order in S. As usual we denote by $\sim$ the equivalence relation $\sim \; = \; \leq \cap \leq^{-1}$. Two elements $s, t \in S$ are *comparable* if $s \leq t$ or $t \leq s$. Two elements are said to be *incomparable* if they are not comparable.

A part X of S is *upper-closed*, or simply *closed*, with respect to $\leq$, if the following condition is satisfied:

$$\text{if } x \in X \text{ and } x \leq y, \text{ then } y \in X.$$

We shall denote by $\mathrm{cl}(X)$ the *closure* of X,

$$\mathrm{cl}(X) = \{s \in S \mid \exists\, x \in X \text{ such that } x \leq s\},$$

so that X is closed if and only if $X = \mathrm{cl}(X)$. For any $X \subseteq S$ one has $X \subseteq \mathrm{cl}(X)$. Moreover, if $Y \subseteq X$, then $\mathrm{cl}(Y) \subseteq \mathrm{cl}(X)$. A closed set X is called *finitely generated* if there exists a finite subset F of X such that $\mathrm{cl}(F) = X$.

A quasi-order in S is called a *well quasi-order* (wqo) if every nonempty subset X of S has at least one minimal element in X but no more than a finite number of (non-equivalent) minimal elements. It is clear that a well ordered set is well quasi-ordered. There exist several conditions which characterize the concept of well quasi-order and that can be assumed as equivalent definitions (see [80]).

Theorem 6.1.1. *Let S be a set quasi-ordered by $\leq$. The following conditions are equivalent:*

i. *$\leq$ is a well quasi-order.*
ii. *the* ascending chain condition *holds for the closed subsets of S.*
iii. *every infinite sequence of elements of S has an infinite ascending subsequence.*
iv. *if $s_1, s_2, \ldots, s_n, \ldots$ is an infinite sequence of elements of S, then there exist integers i, j such that $1 \leq i < j$ and $s_i \leq s_j$.*
v. *there exists neither an infinite strictly descending sequence in S (i.e., $\leq$ is well-founded), nor an infinity of mutually incomparable elements of S.*
vi. *S has the* finite basis property, *i.e., every closed subset of S is finitely generated.*

Proof. *vi* $\Rightarrow$ *ii*: let $(I_n)_{n\geq 0}$ be a sequence of closed subsets such that

$$I_0 \subseteq I_1 \subseteq \cdots \subseteq I_n \subseteq I_{n+1} \subseteq \cdots.$$

Then their union $I = \cup_{n \geq 0} I_n$ is still a closed subset of S, so that, by hypothesis, I is generated by a finite subset F. Thus there exists an integer $n \geq 0$ for which $F \subseteq I_n$, so that $I = \text{cl}(F) \subseteq I_n \subseteq I$. This implies $I_m = I$ for all $m \geq n$.

ii ⇒ *iii*: let $\sigma = s_1, s_2, \ldots, s_n, \ldots$ be an infinite sequence of elements of S. We set $\Sigma = \{s_1, \ldots, s_n, \ldots\}$ and consider the following ascending chain, with respect to inclusion, of closed subsets:

$$\text{cl}(\{s_1\}) \subseteq \text{cl}(\{s_1, s_2\}) \subseteq \cdots \subseteq \text{cl}(\{s_1, \ldots, s_m\}) \subseteq \cdots.$$

By hypothesis this chain has to saturate, that is, there exists an integer n such that

$$\text{cl}(\{s_1, \ldots, s_n\}) = \text{cl}(\{s_1, \ldots, s_m\})$$

for all $m \geq n$. This implies that $\Sigma \subseteq \text{cl}(\Sigma) = \text{cl}(\{s_1, \ldots, s_n\})$. Hence, for each $s \in \Sigma$ there exists an element $t \in \{s_1, \ldots, s_n\}$ such that $t \leq s$. From the pigeon-hole principle there must exist an element $t_1 \in \{s_1, \ldots, s_n\}$ such that t_1 is in the relation $\leq$ with infinitely many elements s_{i_h}, $h > 0$, of the sequence σ. The procedure begins again with the sequence $(s_{i_h})_{h>0}$ and t_1 is the first element of an infinite ascending subsequence of σ.

iii ⇒ *iv*: trivial.

iv ⇒ *v*: trivial.

v ⇒ *i*: let X be a nonempty subset of S. If $x \in X$, then either x is minimal in X with respect to $\leq$ or there exists $x_1 \in X$ such that $x_1 < x$. By repeating the same argument one can construct a strictly descending chain $x > x_1 > x_2 > \cdots > x_n > \cdots$. By hypothesis this chain has to terminate, so that there exists an integer n such that x_n is minimal in X. Thus the set $\mathcal{M}$ of the minimal elements with respect to $\leq$ in X is not empty. Moreover, by *v*. the quotient M of $\mathcal{M}$ by the equivalence $\sim$ is finite.

i ⇒ *vi*: let X be a closed subset of S and $\mathcal{M}$ be the set of the minimal elements with respect to $\leq$ in X. We denote by $\min(X)$ the finite subset of X obtained from the set $\mathcal{M}$ by taking one element from each equivalence class, mod $\sim$. By *i* the set $\min(X)$ is nonempty and finite. Moreover, $\text{cl}(\min(X)) \subseteq X$. Conversely, one has also the inverse inclusion $X \subseteq \text{cl}(\min(X))$. Indeed, by *i* any strictly descending chain of elements of X has to terminate in one element which is minimal. Hence, X is finitely generated by $\min(X)$.

If ρ and σ are two relations on the sets S and T respectively, then the direct product $\rho \otimes \sigma$ is the relation on $S \times T$ defined as

$$(a, b)\ \rho \otimes \sigma\ (c, d) \text{ if and only if } a\ \rho\ c \text{ and } b\ \sigma\ d.$$

Proposition 6.1.1. *Every subset of a wqo set is wqo. If S and T are well quasi-ordered by the relations $\leq_1$ and $\leq_2$, respectively, then $S \times T$ is well quasi-ordered by $\leq_1 \otimes \leq_2$. If $\leq_1$ is a wqo in S, then any other quasi-order $\leq_2$ on S such that $\leq_1 \subseteq \leq_2$ is a wqo.*

Proof. Let S be a set well quasi-ordered by $\leq$. It is trivial from the definition that any subset of S is well quasi-ordered by $\leq$.

Suppose now that the sets S and T are well quasi-ordered by the relations $\leq_1$ and $\leq_2$, respectively. The relation $\leq = \leq_1 \otimes \leq_2$ is a quasi-order in $S \times T$ which is a well quasi-order. Indeed, let

$$(s_1, t_1), (s_2, t_2), \ldots, (s_n, t_n), \ldots$$

be an infinite sequence of elements of $S \times T$. Since S is well quasi-ordered by $\leq_1$, then by condition *iii* of Theorem 6.1.1, the sequence $s_1, \ldots, s_n, \ldots$ has an infinite ascending subsequence

$$s_{j_1} \leq_1 s_{j_2} \leq_1 \cdots \leq_1 s_{j_n} \leq_1 \cdots,$$

with $1 \leq j_1 < j_2 < \cdots < j_n < \cdots$. Since T is well quasi-ordered by $\leq_2$, the infinite sequence $t_{j_1}, t_{j_2}, \ldots, t_{j_n}, \ldots$ of elements of T has an infinite ascending subsequence

$$t_{k_1} \leq_2 t_{k_2} \leq_2 \cdots \leq_2 t_{k_n} \leq_2 \cdots,$$

with $k_i \in \{j_1, j_2, \ldots\}$, $i = 1, 2, \ldots$, and $k_1 < k_2 < \cdots < k_n < \cdots$. Thus the sequence (s_{k_1}, t_{k_1}), $(s_{k_2}, t_{k_2}) \ldots$, $(s_{k_n}, t_{k_n}), \ldots$ is such that

$$(s_{k_1}, t_{k_1}) \leq (s_{k_2}, t_{k_2}) \leq \cdots \leq (s_{k_n}, t_{k_n}) \leq \cdots.$$

This proves by condition *iii* of Theorem 6.1.1, that $\leq$ is a wqo of $S \times T$.

Let us now suppose that $\leq_1$ is a wqo in S and $\leq_2$ is any other quasi-order such that $\leq_1 \subseteq \leq_2$. Let us prove that $\leq_2$ is a wqo. Indeed, let $s_1, \ldots, s_n, \ldots$ be an infinite sequence of elements of S. By condition *iv* of Theorem 6.1.1 there exist integers i, j such that $1 \leq i < j$ and $s_i \leq_1 s_j$. This implies $s_i \leq_2 s_j$ and, therefore, $\leq_2$ is a wqo.

Lemma 6.1.1. *Let $\leq$ be a wqo on S. The following conditions are equivalent:*

i. $\sim = \leq \cap \leq^{-1}$ *is an equivalence of finite index.*
ii. $\leq^{-1}$ *is a wqo.*
iii. $\leq^{-1}$ *is well-founded.*

Proof. *i* $\Rightarrow$ *ii*: an equivalence of finite index is trivially a wqo. Since $\sim \subseteq \leq^{-1}$ it follows from Proposition 6.1.1 that $\leq^{-1}$ is a wqo. The implication *ii* $\Rightarrow$ *iii* is a consequence of Theorem 6.1.1 (see condition *v*). *iii* $\Rightarrow$ *i*: suppose, by contradiction, that $\sim$ has not a finite index. There will exist elements $s_1, s_2, \ldots, s_n, \ldots$ such that for all positive integers $i, j, i \neq j$, s_i is not equivalent to s_j, mod $\sim$. Since $\leq$ is a wqo there exists an infinite subsequence $s_{r_1}, s_{r_2}, \ldots, s_{r_n}, \ldots$ such that $r_1 < r_2 < \cdots < r_n < \cdots$ and

$$s_{r_1} < s_{r_2} < \cdots < s_{r_n} < \cdots$$

which contradicts the hypothesis that $\leq^{-1}$ is well-founded.

Let $\leq$ be a quasi-order on the set S. We can consider the set Σ of all infinite sequences of elements of S, $y : \mathbb{N} \to S$,

$$y = y_0, y_1, \ldots, y_n, \ldots$$

where $y_i = y(i)$ for all $i \geq 0$. The quasi-order on S naturally induces a quasi-order $\leq$ on Σ as follows. Let $y, z \in \Sigma$. If for all $i \geq 0$, $y_i \sim z_i$, then we set $y \sim z$, otherwise we define $i = \min\{j \in \mathbb{N} \mid y_j \not\sim z_j\}$ and set $y < z$ if $y_i < z_i$. One easily verifies that the relation $\leq$ in Σ defined by $y \leq z$ if and only if $y \sim z$ or $y < z$, is a quasi-order in Σ.

Let $\leq$ be a quasi-order on S and $x \in \Sigma$. We call x *bad* if for all i, j, $0 \leq i < j$, one has $x_i \not\leq x_j$. Many proofs on well quasi-orders, as well as the proof of Higman's theorem (see Theorem 6.2.1), are based on the following proposition essentially due to Nash-Williams (see [100]):

Proposition 6.1.2. *Let $\leq'$ be a well-founded quasi-order on S. Let $\leq$ be a quasi-order on S which is not a well quasi-order. Then there exists a bad sequence (with respect to $\leq$) which is minimal with respect to the order $\leq'$.*

Proof. The order $\leq$ is not a wqo so that the negation of the statement *iv* in Theorem 6.1.1 has to be true. Thus there exists at least one infinite sequence $y : \mathbb{N} \to S$, which is bad. Let Σ_0 be the set of all such sequences. We construct then a sequence $x : \mathbb{N} \to S$ inductively, as follows: let x_0 be a minimal element with respect to $\leq'$ in the set $\{y_0 \mid y \in \Sigma_0\}$. For all $i \geq 0$ one defines $\Sigma_{i+1} = \{y \in \Sigma_i \mid y_i = x_i\}$ and x_{i+1} is any minimal element with respect to $\leq'$, in the set $\{y_{i+1} \mid y \in \Sigma_{i+1}\}$. Let us prove that $x \in \Sigma_0$, that is, for all i, j, $i < j$, one has $x_i \not\leq x_j$. Suppose in fact, by contradiction, that there exist i, j, $0 \leq i < j$, for which $x_i \leq x_j$. This is absurd since $x_0, x_1, \ldots, x_j$ are the first $j + 1$ terms of a sequence of Σ_0. By construction the sequence x is minimal in Σ_0 with respect to the order $\leq'$.

If one takes $\leq \, = \, \leq'$ one obtains the following:

Corollary 6.1.1. *Let $\leq$ be a well-founded quasi-order on S which is not a well quasi-order. Then there exists a minimal bad sequence with respect to $\leq$.*

6.2 Higman's theorem

In this section we suppose that the set S quasi-ordered by $\leq$ is a semigroup. In this case a natural notion is that of *monotone* quasi-order that is, a quasi-order which is invariant under right and left multiplication. A special type of monotone quasi-order is a *divisibility* order which is a monotone quasi-order satisfying the property that any factor f (or divisor) of any element $s \in S$ is smaller or equal to s.

The importance of the division orderings is due to a basic theorem of G. H. Higman (see Theorem 6.2.1) showing that a division ordering which is a well quasi-order in a generating set of S is also a well quasi-order in S. A consequence of Higman's theorem is that a finitely generated free semigroup is well quasi-ordered by the subsequence ordering.

Definition 6.2.1. *A quasi-order $\leq$ in a semigroup S is* monotone on the right (on the left) *if for all $x_1, x_2, y \in S$, $x_1 \leq x_2$ implies $x_1 y \leq x_2 y$ ($y x_1 \leq y x_2$).*

A quasi-order is *monotone* if it is monotone on the right and on the left. In this case for all $x_1, x_2, x_3, x_4 \in S$ if $x_1 \leq x_2$ and $x_3 \leq x_4$, then $x_1 x_3 \leq x_2 x_4$. One has, in particular, that a monotone equivalence is a congruence in S.

Lemma 6.2.1. *Let S be a semigroup and $\leq$ be a monotone quasi-order in S. If T_1 and T_2 are two subsets of S and $\leq$ is a wqo on T_1 and T_2, then $\leq$ is a wqo on $T_1 T_2$.*

Proof. By Proposition 6.1.1, $\leq \otimes \leq$ is a wqo on $T_1 \times T_2$. Moreover, since $\leq$ is monotone, $(s_1, s_2) \leq \otimes \leq (t_1, t_2)$ implies $s_1 s_2 \leq t_1 t_2$. Therefore, no bad sequence of elements of $T_1 T_2$ exists, since it generates a bad sequence of elements of $T_1 \times T_2$ with respect to $\leq \otimes \leq$. This implies that $\leq$ is a wqo on $T_1 T_2$.

Definition 6.2.2. *A quasi-order $\leq$ in S is a* divisibility, *or* division, order *if it is monotone and, moreover, for all $s \in S$ and $x, y \in S^1$, $s \leq xsy$.*

The ordering by divisibility in abstract algebras was studied by Higman [80] who proved a very general theorem that in the case of semigroups becomes:

Theorem 6.2.1. *Let S be a semigroup quasi-ordered by a divisibility order $\leq$. If there exists a generating set of S well quasi-ordered by $\leq$, then S will also be so.*

Proof. Let $\leq$ be a divisibility order in S which is a wqo in the generating set A of S. We prove that $\leq$ is a wqo in S. With each $s \in S$ we can associate the integer $|s|$, that we call the *length* of s defined as

$$|s| = \min\{k \mid s \in A^k\}.$$

Let us consider in S the quasi-order $\leq'$ defined as: for $u, v \in S$, $u \leq' v \iff |u| \leq |v|$. This order is, obviously, well-founded.

Suppose, now, that $\leq$ is not a wqo. Then by Proposition 6.1.2 there exists a bad sequence

$$\beta = x_0, x_1, \ldots, x_n, \ldots$$

with respect to $\leq$ which is minimal with respect to $\leq'$. Since A is well quasi-ordered by $\leq$, in β there can occur at most a finite number of elements of A.

We can then always suppose that all the elements of β have a length greater than 1 so that we can write

$$x_i = a_i y_i, \; i \geq 0, \text{ and } 0 < |y_i| < |x_i|,$$

with $a_i \in A$, $i \geq 0$. Let us consider the sequence $a_0, a_1, \ldots, a_n, \ldots$. Since A is well quasi-ordered by $\leq$, from Theorem 6.1.1 there exists an infinite ascending subsequence

$$a_{i_1} \leq a_{i_2} \leq \cdots \leq a_{i_n} \cdots,$$

with $0 \leq i_1 < i_2 < \cdots < i_n < \cdots$. Let us consider the sequence

$$\gamma = x_0, x_1, \ldots, x_{i_1 - 1}, y_{i_1}, y_{i_2}, \ldots,$$

One has that $\gamma <' \beta$ since $y_{i_1} <' x_{i_1}$. However, γ is still a bad sequence. Indeed, if $x_s \leq y_{i_k}$, with $s \in \{0, \ldots, i_1 - 1\}$ and $k \geq 1$, then, since $y_{i_k} \leq a_{i_k} y_{i_k} = x_{i_k}$, one obtains $x_s \leq x_{i_k}$ which is a contradiction. If $y_{i_s} \leq y_{i_k}$ with $1 \leq s < k$, then since $\leq$ is monotone and $a_{i_s} \leq a_{i_k}$, one would have $x_{i_s} = a_{i_s} y_{i_s} \leq a_{i_k} y_{i_k} = x_{i_k}$ which is again absurd. Thus one reaches a contradiction with the minimality of β.

The theorem of Higman has several applications in combinatorics on words as well as in language theory (see, for instance, [104, 148]). A more general version of this theorem, in the case of finitely generated free monoids, will be given in Sect. 6.4.

Now we prove a lemma that will be useful later.

Lemma 6.2.2. *Let S be a monoid, with identity* 1, *and $\leq$ be a monotone quasi-order on S. Let $T \subseteq S$ be such that $1 \leq t$ for any $t \in T$ and $\leq$ be a wqo on T. Then $\leq$ is a wqo on T^*.*

Proof. Let us prove that $\leq$ is a divisibility order on T^*. Indeed, let $x, y, s \in T^*$. Since $1 \leq t$, for any $t \in T$, by the monotonicity of $\leq$, one has $1 \leq x, y$. Hence, $s \leq xs, sy$ and then $s \leq xs \leq xsy$. Since $\leq$ is a wqo on T, by Theorem 6.2.1 $\leq$ is a wqo on T^+ and then on T^*.

A remarkable consequence of Theorem 6.2.1 is the following. Let $S = A^+$ be the free semigroup generated by the alphabet A quasi-ordered by a relation $\leq$. This order can be extended to A^+ as follows. Let u and v be words; we set $u \leq v$ if some subsequence of v majorizes u letter-by-letter, i.e.,

$$u = a_1 \cdots a_n, \; a_i \in A, \; i = 1, \ldots, n$$

and

$$v \in A^* b_1 A^* b_2 A^* \cdots A^* b_n A^*, \; b_i \in A, \; i = 1, \ldots, n,$$

with

$$a_i \leq b_i, \; i = 1, \dots, n.$$

This is, trivially, a divisibility order so that by Theorem 6.2.1 if A is wqo by $\leq$ (as, for instance, when A is finite), then S is itself wqo.

A special divisibility ordering in A^+, called *subsequence ordering* and denoted by $|$, is defined as: for $u, v \in A^+$

$$u \mid v \Longleftrightarrow u = a_1 \cdots a_n \text{ and } v \in A^* a_1 A^* \cdots A^* a_n A^*, \quad a_i \in A, \; i = 1, \dots, n.$$

Lemma 6.2.3. *The subsequence ordering is the smallest divisibility ordering in A^+.*

Proof. We have to prove that $| \;\subseteq\; \leq$ where $\leq$ is any divisibility ordering in A^+. Let $u, v \in A^+$ and suppose that $u \mid v$. One has $u = a_1 \cdots a_n$ and $v = \lambda_1 a_1 \lambda_2 a_2 \lambda_3 \cdots \lambda_n a_n \lambda_{n+1}$ with $\lambda_i \in A^*$, $i = 1, \dots, n+1$. Since $\leq$ is a divisibility ordering, one has

$$a_1 \leq \lambda_1 a_1, \; a_2 \leq \lambda_2 a_2, \dots, \; a_{n-1} \leq \lambda_{n-1} a_{n-1}, \; a_n \leq \lambda_n a_n \lambda_{n+1},$$

so that since $\leq$ is monotone, one derives

$$u = a_1 \cdots a_n \leq \lambda_1 a_1 \lambda_2 a_2 \lambda_3 \cdots \lambda_n a_n \lambda_{n+1} = v.$$

A consequence of the fact that the subsequence ordering $|$ is a wqo is that *any infinite language L on a finite alphabet A is such that there exist at least two words $u, v \in L$ such that $u \neq v$ and u is a subsequence of v.*

For any $u \in A^+$ we define

$$\text{Sub}(u) = \{v \in A^+ \mid v \mid u\}.$$

For any $n > 0$ one can introduce the relation $\sim_n$ in A^+ defined as: for all $u, v \in A^+$

$$u \sim_n v \Longleftrightarrow \text{Sub}(u) \cap A^{[n]} = \text{Sub}(v) \cap A^{[n]}.$$

In other terms u and v are equivalent mod $\sim_n$ if they have the same set of all subsequences of length $\leq n$. For instance, in the case $A = \{a, b\}$ and $n = 3$ one has $ababab \sim_3 bababa$.

One can easily verify that $\sim_n$ is a congruence in A^+ of a finite index. Indeed, the index is upper-bounded by 2^M, where M is the number of all words of length $\leq n$. The following important theorem, due to Simon [148], whose proof we omit, holds:

Theorem 6.2.2. *Let $n > 0$ and $f, g \in A^+$. If $f \sim_n g$, then there exists $h \in A^+$ such that*

$$f \mid h, \; g \mid h \text{ and } f \sim_n h \sim_n g.$$

In other words, any two elements in an equivalence class of $\sim_n$ have a common upper bound (relative to the subsequence ordering) in that equivalence class. For instance, if $A = \{a, b\}$ and $n = 4$, then $f = a^2b^4a^4b^2 \sim_4 a^3b^3a^3b^3 = g$ and $h = a^3b^4a^4b^3$.

6.3 The generalized Myhill theorem

A characterization of recognizable subsets of a semigroup can be obtained in terms of well quasi-orders. In fact, the following theorem, usually called the generalized Myhill theorem, holds [60].

Theorem 6.3.1. *A subset X of a semigroup S is recognizable if and only if X is closed with respect to a monotone well quasi-order in S.*

A more general version of Theorem 6.3.1 will be proved below (as Theorem 6.3.2). The link between congruences and monotone well quasi-orders is given by the following:

Proposition 6.3.1. *An equivalence in a semigroup S is a congruence of a finite index if and only if it is a monotone wqo.*

Proof. A congruence in S of finite index is, trivially, a monotone wqo. Conversely, let θ be a monotone wqo in S which is an equivalence, that is, a wqo which is a congruence. Since $\theta^{-1} = \theta$, by Lemma 6.1.1, θ has a finite index.

Let $X \subseteq S$; we introduce the following relation $\leq_X$ in S defined as: for $s, t \in S$,

$$s \leq_X t \text{ if and only if } \forall h, k \in S^1 (hsk \in X \Rightarrow htk \in X).$$

One also verifies that $\leq_X$ is a monotone quasi-order and X is closed with respect to $\leq_X$. Moreover, the equivalence relation $\leq_X \cap (\leq_X)^{-1}$ coincides with the syntactic congruence $\equiv_X$ of X and $\leq_X$ is maximal (with respect to inclusion) in the set of all monotone quasi-orders with respect to which X is closed. The relation $\leq_X$ is called the *Myhill quasi-order* relation relative to X.

Proposition 6.3.2. *A subset X of a semigroup S is recognizable if and only if $\leq_X$ is a well quasi-order.*

Proof. ($\Leftarrow$) It is trivial by the fact that X is closed with respect to the monotone well quasi-order $\leq_X$, so that by the generalized Myhill theorem the result follows.

($\Rightarrow$) Let $X \in \mathrm{Rec}(S)$. From Theorem 6.3.1 the set X is $\leq$-closed, where $\leq$ is a monotone wqo. Since $\leq \subseteq \leq_X$, from Proposition 6.1.1 it follows that $\leq_X$ is a wqo.

Recognizable parts of a semigroup can also be described (see Theorem 4.1.1) by equivalence relations monotone on the right or on the left, called *right* and *left congruences*, respectively. From Theorem 4.1.1 *a part X of a semigroup S is recognizable if and only if it is closed with respect to a right (or left) congruence of a finite index* (Nerode's theorem). In terms of Nerode's equivalences the property of being recognizable becomes: X is recognizable if and only if $\mathcal{N}_X^r$ or $\mathcal{N}_X^l$ is of finite index.

For any subset X of S we can then introduce the quasi-order $\leq_X^r$ that we call the *(right) Nerode quasi-order relative to* X, defined as: for $s, t \in S$

$$s \leq_X^r t \text{ if and only if } s^{-1}X \subseteq t^{-1}X.$$

One easily verifies the following facts:

1. $\leq_X^r$ is a quasi-order monotone on the right;
2. X is closed with respect to $\leq_X^r$;
3. the equivalence $\leq_X^r \cap (\leq_X^r)^{-1}$ is the Nerode equivalence $\mathcal{N}_X^r$;
4. $\leq_X^r$ is maximal in the set of all quasi-orders monotone on the right with respect to which X is a closed part;
5. $\leq_X \subseteq \leq_X^r$.

Similarly, the quasi-order $\leq_X^l$ is defined as: for $s, t \in S$

$$s \leq_X^l t \text{ if and only if } Xs^{-1} \subseteq Xt^{-1}.$$

The quasi order $\leq_X^l$ satisfies properties which are symmetric to those of $\leq_X^r$. Since $\leq_X \subseteq \leq_X^r$ and $\leq_X \subseteq \leq_X^l$, one has that if X is recognizable, then $\leq_X^r$, as well as $\leq_X^l$, is a well quasi-order.

One can ask the question whether, in analogy with Nerode's theorem, the fact that $\leq_X^r$ is a well quasi-order assures that X is a recognizable part of S. The answer is negative. In fact, when $S = A^*$ is a finitely generated free monoid, one can prove that there exists a very large class of languages L over A which are not regular and such that $\leq_X^r$ is a wqo. For instance, when $A = \{a, b\}$ the language $L = \{a^n b^m \mid n \geq m \geq 0\}$ is not regular and is such that $\leq_L^r$ is a wqo [52].

A partial generalization of Myhill's theorem and of Theorem 6.3.1 as well, is given by the following [52]:

Theorem 6.3.2. *Let X be a subset of S. The following conditions are equivalent*

i. *X is recognizable.*

ii. *X is a closed part of a wqo $\leq_1$ monotone on the right and of a wqo $\leq_2$ monotone on the left.*

iii. *The quasi-orders $\leq_X^r$ and $\leq_X^l$ are wqo.*

Proof. The implication $i \Rightarrow ii$ is trivial since if X is recognizable, then it is a closed part of a congruence of finite index. The implication $ii \Rightarrow iii$ follows from the fact that $\leq_1 \subseteq \leq_X^r$ and $\leq_2 \subseteq \leq_X^l$. Let us then prove the implication $iii \Rightarrow i$. If X is not recognizable, then there exists an infinite sequence

$$x_1, x_2, \ldots, x_n, \ldots$$

of elements of S such that for all positive integers i, j, $i \neq j$, one has that x_i is not equivalent to x_j, mod $\mathcal{N}_X^r$. Since $\leq_X^r$ is a wqo, there exists an infinite subsequence $y_1, y_2, \ldots, y_n, \ldots$ of the above sequence which is strictly increasing with respect to $\leq_X^r$:

$$y_1 <_X^r y_2 <_X^r \cdots <_X^r y_n <_X^r \cdots .$$

This implies that for all $i \geq 1$, $y_i^{-1}X \subset y_{i+1}^{-1}X$. Now, as one easily verifies, any set $y_i^{-1}X$ is a closed part of $\leq_X^l$ so that we have an infinite strictly increasing chain of closed sets of $\leq_X^l$. By Theorem 6.1.1 this contradicts the fact that $\leq_X^l$ is a wqo.

We now give an application of Theorem 6.3.2. Let S be a semigroup. We say that S satisfies the maximal condition M_R (maximal condition M_L) on right (on left) ideals if any chain of right (left) ideals, ordered by inclusion, terminates. We recall that S is quasi-ordered by the relations $\leq_{\mathcal{R}}$ and $\leq_{\mathcal{L}}$ introduced in Chap. 3. Moreover, one easily derives from the definition, that $\leq_{\mathcal{R}}$ and $\leq_{\mathcal{L}}$ are monotone on the left and on the right, respectively.

Lemma 6.3.1. *In a semigroup S the following two conditions are equivalent:*

i. S satisfies the condition M_R (condition M_L).
ii. the quasi-order $\leq_{\mathcal{R}}^{-1}$ (quasi-order $\leq_{\mathcal{L}}^{-1}$) is a wqo.

Proof. Let us first prove that a subset R of S is a closed set of $\leq_{\mathcal{R}}^{-1}$ if and only if R is a right ideal of S. Indeed, let R be a closed set of $\leq_{\mathcal{R}}^{-1}$, $r \in R$ and $s \in S$. Since $rsS^1 \subseteq rS^1$, one has $r \leq_{\mathcal{R}}^{-1} rs$, so that $rs \in R$ and R is a right ideal. Conversely, let R be a right ideal of S and $r \in R$ be such that $r \leq_{\mathcal{R}}^{-1} s$. One has then $sS^1 \subseteq rS^1$, so that $s = rv$ with $v \in S^1$, which implies $s \in R$. Hence, R is a closed set of $\leq_{\mathcal{R}}^{-1}$. Let us now suppose that S satisfies M_R. If $R_1 \subseteq R_2 \subseteq \cdots \subseteq R_n \subseteq \cdots$ is an ascending chain of closed sets of $\leq_{\mathcal{R}}^{-1}$, then by M_R the chain has to terminate so that by Theorem 6.1.1 $\leq_{\mathcal{R}}^{-1}$ is a wqo. Conversely, suppose that $\leq_{\mathcal{R}}^{-1}$ is a wqo and $R_1 \subseteq R_2 \subseteq \cdots \subseteq R_n \subseteq \cdots$ be an ascending chain of right ideals. Since the sets R_i, $i > 0$, are closed with respect to $\leq_{\mathcal{R}}^{-1}$, it follows by Theorem 6.1.1 that the chain has to terminate, so that the condition M_R has to be satisfied. A symmetric argument shows that S satisfies M_L if and only if $\leq_{\mathcal{L}}^{-1}$ is a wqo.

Corollary 6.3.1. *Let S be a semigroup satisfying M_R and M_L. If J is a two-sided ideal of S, then $J \in \mathrm{Rec}(S)$.*

Proof. By the preceding lemma $\leq_{\mathcal{R}}^{-1}$ and $\leq_{\mathcal{L}}^{-1}$ are wqo. Let J be a two-sided ideal of S. As we have seen in the proof of Lemma 6.3.1, J is a closed set of $\leq_{\mathcal{R}}^{-1}$ and $\leq_{\mathcal{L}}^{-1}$. Since $\leq_{\mathcal{R}}^{-1}$ and $\leq_{\mathcal{L}}^{-1}$ are monotone on the left and on the right, respectively, then by Theorem 6.3.2 the result follows.

6.4 Quasi-orders and rewriting systems

Now we consider the case when S is the free monoid generated by a finite alphabet A. Let (A, π) be a rewriting system and $\Rightarrow_\pi^*$ its derivation relation. One easily verifies that $\Rightarrow_\pi^*$ is a monotone quasi-order. From Theorem 6.1.1 and Proposition 6.3.2 we derive the following:

Proposition 6.4.1. *A language over a finite alphabet A is regular if and only if it is $\Rightarrow_\pi^*$-closed where (A, π) is a rewriting system such that $\Rightarrow_\pi^*$ is a well quasi-order.*

Proof. ($\Leftarrow$). Obvious from Theorem 6.3.1 since $\Rightarrow_\pi^*$ is monotone.

($\Rightarrow$). Let $L \subseteq A^*$ be a regular language. By Kleene's theorem L is recognizable. Thus by Proposition 6.3.2, the Myhill quasi-order $\leq_L$ is a wqo. Let us consider the rewriting system (A, π), with $\pi = \leq_L$. Since $\leq_L$ is monotone one has $\Rightarrow_\pi = \leq_L$. Moreover, since $\leq_L$ is reflexive and transitive, one has $\Rightarrow_\pi^* = \leq_L$. Since L is $\leq_L$-closed the result follows.

An interesting problem is to determine under which conditions a semi-Thue system (A, π) belonging to a given class is such that its derivation relation $\Rightarrow_\pi^*$ is a wqo. Let us, for instance, consider the class of *unitary* semi-Thue systems. A semi-Thue system (A, π) is called unitary when π is a finite set of productions of the kind

$$\epsilon \to q, \; q \in I \subseteq A^+.$$

Such a system is then determined by the finite set $I \subseteq A^+$. Let us simply denote by $\Rightarrow_I^*$ the derivation relation in these systems. If $I = A$, then $\Rightarrow_A^*$ is equal to the subsequence ordering $|$ which is a divisibility ordering. Thus, $\Rightarrow_A^*$ is a wqo by the Higman theorem.

Definition 6.4.1. *Let $I \subseteq A^*$. We say that I is* subword unavoidable *if there exists a positive integer k_0, such that any word $u \in A^*$, $|u| > k_0$, can be written as $u = xwy$, where $w \in I$ and $x, y \in A^*$. If k_0 is the minimal integer satisfying this property, then it is called the* subword avoidance bound *of I.*

In [60] Ehrenfeucht et al. gave the following interesting characterization of the finite sets I such that $\Rightarrow_I^*$ is a wqo.

Theorem 6.4.1. *The derivation relation $\Rightarrow_I^*$ of the unitary semi-Thue system associated with the finite set $I \subseteq A^+$ is a wqo if and only if I is subword unavoidable.*

Before proving this theorem we need some preliminary definitions and lemmas. First of all we define inductively the sequence $(I_i)_{i \geq 0}$ of subsets of A^*, as follows:

$$I_0 = I^*,$$

$$\vdots$$

$$I_{n+1} = \Big(\bigcup_{a_1,\ldots,a_k \in A,\ a_1 \cdots a_k \in I} I_n a_1 I_n a_2 \cdots a_{k-1} I_n a_k I_n \Big)^*,$$

$$\vdots$$

Lemma 6.4.1. *For any $n \geq 0$ the derivation relation $\Rightarrow_I^*$ is a wqo on I_n.*

Proof. By induction on n. Since I is finite, by Lemma 6.2.2, one derives that $\Rightarrow_I^*$ is a wqo on $I_0 = I^*$. Suppose, by the inductive hypothesis, that $\Rightarrow_I^*$ is a wqo on I_n. By Lemma 6.2.1, $\Rightarrow_I^*$ is a wqo on $I_n a_1 I_n a_2 \cdots a_{k-1} I_n a_k I_n$, for any $a_1 a_2 \cdots a_k \in I$, with $a_1, \ldots, a_k \in A$. Since a finite union of well quasi-ordered subsets is a wqo subset, one derives that $\Rightarrow_I^*$ is a wqo on

$$\bigcup_{a_1,\ldots,a_k \in A,\ a_1 \cdots a_k \in I} I_n a_1 I_n a_2 \cdots a_{k-1} I_n a_k I_n.$$

Therefore, by Lemma 6.2.2, $\Rightarrow_I^*$ is a wqo on I_{n+1}.

Lemma 6.4.2. *For any $n \geq 0$,*

i. if $uv \in I_n$ and $w \in I$, then $uwv \in I_{n+1}$,
ii. if $uv \in I_n$, $|u| \leq n$, and $w \in I$, then $uwv \in I_n$.

Proof. i: it follows directly from the definition.

ii: by induction on n. If $n = 0$, then $u = \epsilon$ and the statement follows, since $I_0 = I^*$. Assume now that the statement holds for some $n \geq 0$. Let $uv \in I_{n+1}$, $|u| \leq n+1$, and $w \in I$, then $uv = w_1 a_1 w_2 a_2 \cdots a_k w_{k+1}$, where $w_j \in I_n$, $j = 1, \ldots, k+1$, and $a_1 a_2 \cdots a_k \in I^*$, with $a_1, \ldots, a_k \in A$. Then there exists an integer j, $1 \leq j \leq k+1$, such that

$$u = w_1 a_1 w_2 a_2 \cdots a_{j-1} w_j', \; v = w_j'' a_{j+1} \cdots a_k w_{k+1},$$

with $w_j' w_j'' = w_j$. Hence,

$$uwv = w_1 a_1 w_2 a_2 \cdots w_j' w w_j'' \cdots a_k w_{k+1}.$$

By *i*, $w_j' w w_j'' \in I_{n+1}$. If $j = 1$, then $uwv \in I_{n+1}$, since $a_1 w_2 a_2 \cdots a_k w_{k+1} \in I_{n+1}$ and I_{n+1} is closed under concatenation. If $j \geq 1$, then $|w_j'| \leq n$ and, by the induction hypothesis, one has $w_j' w w_j'' \in I_n$. This implies $uwv \in I_n a_1 I_n a_2 \cdots I_n a_k I_n$, that is, $uwv \in I_{n+1}$.

Let us set, for any $n \geq 0$,

$$R(I_n) = \bigcup_{a_1, a_2, \ldots, a_k \in A,\ k \leq n} I_n a_1 I_n a_2 \cdots a_{k-1} I_n a_k I_n.$$

Lemma 6.4.3. *Let I be a finite subword unavoidable subset of A^* and k_0 the subword avoidance bound of I. Then $A^* = R(I_{k_0})$.*

Proof. Suppose, by contradiction, that $A^* \setminus R(I_{k_0})$ is not empty. Choose a word $x \in A^* \setminus R(I_{k_0})$ of minimal length. We have $|x| > k_0$. Then in the first $k_0 + 1$ letters we find, as a factor, a word of I. Thus $x = uwv$, with $w \in I$ and $|u| \leq k_0$. Since x has minimal length in $A^* \setminus R(I_{k_0})$, $uv \in R(I_{k_0})$. Thus we can write

$$uv = w_1 a_1 w_2 a_2 \cdots w_k a_k w_{k+1},$$

where $a_j \in A$, for $1 \leq j \leq k$, and $w_i \in I_{k_0}$, for $1 \leq i \leq k+1$. Then for some i, $1 \leq i \leq k+1$, one has

$$u = w_1 a_1 \cdots w_i', \; v = w_i'' a_i \cdots w_k a_k w_{k+1} \text{ and } w_i = w_i' w_i''.$$

Since $|u| \leq k_0$, one has $|w_i'| \leq k_0$. By Lemma 6.4.2, $w_i' w w_i'' \in I_{k_0}$. Thus $x \in R(I_{k_0})$, which is a contradiction.

Proof of Theorem 6.4.1. Let k_0 be the subword avoidance bound of I. By Lemma 6.4.3

$$A^* = \bigcup_{a_1, a_2, \ldots, a_k \in A, \; k \leq k_0} I_{k_0} a_1 I_{k_0} a_2 \cdots a_{k-1} I_{k_0} a_k I_{k_0}.$$

By Lemma 6.4.1, $\Rightarrow_I^*$ is a wqo on I_{k_0}. Therefore, from Lemma 6.2.1, $\Rightarrow_I^*$ is a wqo on A^*.

Theorem 6.4.1 is then a nontrivial extension of Higman's theorem in the case of a finitely generated free semigroup. A different extension was given by D. Haussler [78].

In the following sections we will give further classes of semi-Thue systems such that their derivation relation is a wqo.

6.5 A regularity condition for permutable languages

We introduce a class of rewriting systems [48] π whose productions are defined as follows. Let n be an integer > 1. For any sequence of words $u_1, \ldots, u_n \in A^+$, and $u \in A^+$

$$u_1 u_2 \cdots u_n \to u_{\sigma(1)} u_{\sigma(2)} \cdots u_{\sigma(n)}, \quad u_{\sigma(1)} u_{\sigma(2)} \cdots u_{\sigma(n)} \to u_1 u_2 \cdots u_n \quad (6.1)$$

$$u^m \to u^{m+k}, \quad m > 0, k > 0, \quad (6.2)$$

where $\sigma \in S_n \setminus \{id\}$ and the pair (m, k) depends on the words $u_1, \ldots, u_n$, and u, respectively. Let us observe that each rewriting system π of the preceding class depends on the integer $n > 1$ and on the two maps $f : (A^+)^n \to S_n \setminus \{id\}$, $g : A^+ \to \mathbb{N}_+ \times \mathbb{N}_+$. Hence, a particular rewriting system of the class should be denoted by $\pi_{n,f,g}$. However, we drop the subscripts when no confusion arises.

The following lemma is a slight generalization of a famous lemma of Dickson (compare [104]).

Lemma 6.5.1. *Let $q > 0$ and m_i, k_i be integers such that $m_i \geq 0$ and $k_i > 0$ $(i = 1, \ldots, q)$. Let us consider in $\mathbb{N}^q$ the relation $\Rightarrow$ defined as: let $(r_1, \ldots, r_q)$, $(s_1, \ldots, s_q) \in \mathbb{N}^q$, $(r_1, \ldots, r_q) \Rightarrow (s_1, \ldots, s_q)$ if and only if there exists $i \in [1, q]$ such that $r_i \geq m_i$, $s_i = r_i + k_i$, and $s_j = r_j$ for $j \neq i$. The reflexive and transitive closure $\Rightarrow^*$ of $\Rightarrow$ is a well quasi-order of $\mathbb{N}^q$.*

Proof. We give first the proof when $q = 1$. Let $m \geq 0$ and $k > 0$. We introduce in $\mathbb{N}$ the relation $\Rightarrow$ defined as: for $x, y \in \mathbb{N}$, $x \Rightarrow y$ if and only if $x \geq m$ and $y = x + k$. Let $\Rightarrow^*$ be the reflexive and transitive closure of $\Rightarrow$. In $\mathbb{N}$ there are only finitely many minimal elements with respect to $\Rightarrow^*$, namely all $n \in \mathbb{N}$ such that $n < m + k$. Let us denote by M the set of all these minimal elements. We prove that $\Rightarrow^*$ is a well quasi-order. In fact, this order is, trivially, well-founded. Let X be any infinite subset of $\mathbb{N}$. For any $x \in X$ there exists a minimal element $t \in M$ such that $t \Rightarrow^* x$. Since M is finite, there exists an infinite subset Y of X and a unique minimal element $t \in M$ such that $t \Rightarrow^* y$ for all $y \in Y$. If y_1 and y_2 are two distinct elements of Y one has $y_1 = t + r_1 k$ and $y_2 = t + r_2 k$, with $r_1, r_2 \geq 0$. If $r_1 < r_2$, then one has $y_1 \Rightarrow^* y_2$. The remaining part of the proof, when $q > 1$, is an obvious consequence of the fact that the direct product of finitely many wqo sets is a wqo set (see Proposition 6.1.1).

Let us observe that from the preceding proposition in the particular case $m_i = 0$ and $k_i = 1$ $(i = 1, \ldots, q)$ one obtains Dickson's lemma.

Theorem 6.5.1. *The derivation relation $\Rightarrow_\pi^*$ is a well quasi-order.*

Proof. The proof is by contradiction. Suppose that $\Rightarrow_\pi^*$, which we simply denote by $\leq$, is not a wqo. We consider the alphabetic order $\leq_a$ in A^+; this order is trivially well-founded. Hence, by Proposition 6.1.2, there exists a bad sequence $x : \mathbb{N} \to A^+$ which is minimal with respect to the order $\leq_a$. Let us now prove that for all $i \geq 0$, the word x_i does not contain n-divided factors. In fact suppose, by contradiction, that

$$x_i = x u_1 u_2 \cdots u_n y, \quad x, y \in A^*$$

and that for all permutations $\tau \in S_n$, $\tau \neq id$, one has

$$x u_{\tau(1)} u_{\tau(2)} \cdots u_{\tau(n)} y <_a x u_1 u_2 \cdots u_n y. \tag{6.3}$$

From (6.1) $u_1 u_2 \cdots u_n \to u_{\sigma(1)} u_{\sigma(2)} \cdots u_{\sigma(n)}$ and $u_{\sigma(1)} u_{\sigma(2)} \cdots u_{\sigma(n)} \to u_1 u_2 \cdots u_n$, where $\sigma \in S_n$ is a suitable nontrivial permutation depending on the words $u_1, \ldots, u_n$.

Let us set $y_i = xu_{\sigma(1)}u_{\sigma(2)}\cdots u_{\sigma(n)}y$; one has from equations (6.1) and (6.3)

$$y_i <_a x_i, \quad x_i \leq y_i \text{ and } y_i \leq x_i.$$

Let us now prove that for all j, $j < i$, one has $x_j \not\leq y_i$ and for all $j > i$, $y_i \not\leq x_j$. In fact, in the first case if $x_j \leq y_i$, since $y_i \leq x_i$, then we reach the contradiction $x_j \leq x_i$. Similarly, in the second case, if $y_i \leq x_j$, since $x_i \leq y_i$, then we reach the contradiction $x_i \leq x_j$. It follows that the sequence z defined as $z_i = y_i$ and $z_h = x_h$, for $0 \leq h \neq i$, is a bad sequence. Since $y_i <_a x_i$, one has $z <_a x$ and this contradicts the minimality of x.

Let us now consider the language X formed by the elements of the sequence x,

$$X = \bigcup_{i \in \mathbb{N}} \{x_i\}.$$

As we have seen before, $F(X)$ does not contain n-divided words. From Theorem 2.5.1, X is a bounded language, that is, there exist words $w_1, \ldots, w_q$, such that $X \subseteq w_1^* \cdots w_q^*$. Hence, for any word $x_i \in X$ there exist non-negative integers $r_1, \ldots, r_q$ such that $x_i = w_1^{r_1} \cdots w_q^{r_q}$. Thus any element of X is uniquely determined by the q-tuple $(r_1, \ldots, r_q) \in \mathbb{N}^q$. Let (m_i, k_i) $(i = 1, \ldots, q)$ be the set of pairs of integers such that $w_i^{m_i} \to w_i^{m_i+k_i}$ according to equation (6.2). We can then consider in $\mathbb{N}^q$ the relation $\Rightarrow$ defined as in Lemma 6.5.1. Let us observe that

$$\text{if } (r_1, \ldots, r_q) \Rightarrow (s_1, \ldots, s_q), \text{ then } w_1^{r_1} \cdots w_q^{r_q} \Rightarrow_\pi w_1^{s_1} \cdots w_q^{s_q}.$$

By Lemma 6.5.1 the reflexive and transitive closure $\Rightarrow^*$ of $\Rightarrow$ is a well quasi-order. Let the set M be defined as

$$M = \{(r_1, \ldots, r_q) \in \mathbb{N}^q \mid w_1^{r_1} \cdots w_q^{r_q} \in X\}.$$

By the statement v of Theorem 6.1.1 there will exist two different q-tuples $(r_1, \ldots, r_q)$, $(s_1, \ldots, s_q) \in M$ such that $(r_1, \ldots, r_q) \Rightarrow^* (s_1, \ldots, s_q)$. Let $x_i = w_1^{r_1} \cdots w_q^{r_q}$ and $x_j = w_1^{s_1} \cdots w_q^{s_q}$; then one would obtain $x_i \leq x_j$ which is a contradiction.

From Theorem 6.5.1 one obtains a further proof [48] of the theorem of Restivo and Reutenauer (see Theorem 3.4.1). Indeed, let S be a finitely generated semigroup and $\phi : A^+ \to S$ be the canonical epimorphism. If S is periodic and permutable, then one derives that there exists a rewriting system π, whose productions are defined by equations (6.1) and (6.2), such that $\Rightarrow_\pi^* \subseteq \phi\phi^{-1}$. By Theorem 6.5.1, $\Rightarrow_\pi^*$ is a wqo; this trivially implies, by Proposition 6.3.1, that the congruence $\phi\phi^{-1}$ is of a finite index. Hence, S is finite.

We shall now prove that if L is a permutable language, then L is regular under a condition which is weaker than that of periodicity. This condition, which is not syntactic, is obtained by replacing in equation (5.1) the *double* implication with a *single* implication. More precisely let us give the following definition:

Definition 6.5.1. *A language L is called* quasi-periodic *if for all $u \in A^*$ there exist integers $n, k > 0$ such that for all $x, y \in A^*$ one has*

$$xu^n y \in L \Longrightarrow xu^{n+k} y \in L.$$

L is called co-quasi-periodic *if the complement L^c of L is quasi-periodic.*

One can easily prove that a language is periodic if and only if it is both quasi-periodic and co-quasi-periodic. We give now an example of a non-periodic language which is quasi-periodic.

Example 6.5.1. Let $A = \{a, b, c\}$ be a three-letter alphabet and $(u_i)_{i \geq 1}$ be an infinite sequence of mutually distinct cube-free words on the alphabet $\{b, c\}$. For instance, one can take as u_i the prefix of length i of the Thue–Morse word on the alphabet $\{b, c\}$ (see Sect. 2.6). Let L be the language

$$L = \bigcup_{n \geq 1} u_n a^n a^* u_n.$$

We prove that L is a quasi-periodic non-periodic language. First let $w \in A^* \setminus a^*$; one easily verifies that if $i \geq 3$, then for all $x, y \in A^*$ one has that $xw^i y \notin L$ that implies $w^3 \equiv_L w^4$. Moreover, one can prove that for all $x, y \in A^*$, $xay \in L \Rightarrow xa^2 y \in L$. From this it follows that L is quasi-periodic. However, the language L is not periodic. In fact suppose that there exist n and k such that $a^n \equiv_L a^{n+k}$. One would derive $u_{n+k} a^{n+k} u_{n+k} \equiv_L u_{n+k} a^n u_{n+k}$ that implies $u_{n+k} a^n u_{n+k} \in L$ which is a contradiction.

Theorem 6.5.2. *A language L is regular if and only if L is permutable and quasi-periodic or co-quasi-periodic.*

Proof. The "only if" part of the theorem is trivial by the fact that if L is regular, then the syntactic monoid $S(L)$ is finite and then periodic and permutable. Let us then prove the "if" part. First of all we observe that we can limit ourselves to prove the result when L is quasi-periodic. In fact suppose that we have already proved the theorem when the language is quasi-periodic and suppose that L is co-quasi-periodic. One has then that $S(L) = S(L^c)$ is permutable and L^c is quasi-periodic; hence, L^c, as well as L, is regular. Let us then suppose that L is permutable and quasi-periodic. This implies that there exists $n > 1$ such that for any sequence of words $u_1, \ldots, u_n$ of A^+ there exists a permutation $\sigma \in S_n$, $\sigma \neq id$, such that

$$u_1 u_2 \cdots u_n \equiv_L u_{\sigma(1)} u_{\sigma(2)} \cdots u_{\sigma(n)}; \tag{6.4}$$

moreover, for any word $u \in A^*$ there exist integers $m > 0$ and $k > 0$ such that for all $x, y \in A^*$

$$xu^my \in L \Longrightarrow xu^{m+k}y \in L. \tag{6.5}$$

We can then introduce a rewriting system π whose productions are defined as: for any sequence of words $u_1, \ldots, u_n \in A^+$, and $u \in A^+$

$$u_1u_2\cdots u_n \to u_{\sigma(1)}u_{\sigma(2)}\cdots u_{\sigma(n)}, \quad u_{\sigma(1)}u_{\sigma(2)}\cdots u_{\sigma(n)} \to u_1u_2\cdots u_n \tag{6.6}$$

$$u^m \to u^{m+k}, \quad m > 0, k > 0, \tag{6.7}$$

where the permutations σ and the integers m and k are those in the equations (6.4) and (6.5), respectively. From equations (6.4) and (6.5) one has that L is $\Rightarrow_\pi^*$-closed, so that from Theorem 6.5.1 and Proposition 6.4.1 the result follows.

6.6 Almost-commutative languages

We introduce a class of rewriting systems π' as follows. Let m, k be positive integers, $n > 1$, and $\sigma \in \mathcal{S}_n \setminus \{id\}$ be a fixed nontrivial permutation. The productions of π' are then defined as: for any letter $a \in A$ and words $w_1, \ldots, w_n \in A^+$

$$a^m \to a^{m+k}, \quad w_1w_2\cdots w_n \to w_{\sigma(1)}w_{\sigma(2)}\cdots w_{\sigma(n)}. \tag{6.8}$$

Any particular rewriting system depends on the integers m, k and on the fixed permutation $\sigma \in \mathcal{S}_n \setminus \{id\}$, so that it should be denoted by $\pi'_{m,k;\sigma}$; however we drop the subscripts when there is no confusion. Let $\Rightarrow_{\pi'}$ be the regular closure of π' and $\Rightarrow_{\pi'}^*$ be the reflexive and transitive closure of $\Rightarrow_{\pi'}$. In the following we shall prove (see Theorem 6.6.1) that the derivation relation $\Rightarrow_{\pi'}^*$ is a well quasi-order [48]. We now need some preliminaries.

Definition 6.6.1. *A semigroup S is* n-almost commutative, $n > 1$, *if there exists a permutation $\sigma \in \mathcal{S}_n \setminus \{id\}$ such that for any sequence $s_1, \ldots, s_n$ of elements of S one has*

$$s_1s_2\cdots s_n = s_{\sigma(1)}s_{\sigma(2)}\cdots s_{\sigma(n)}.$$

S is said to be almost-commutative *if it is n-almost commutative for a suitable $n > 1$.*

We denote by $\mathcal{Q}_n$ (by $\mathcal{Q}$) the class of n-almost-commutative (almost-commutative) semigroups. We remark that for each $n > 1$, the property $\mathcal{Q}_n$ implies $\mathcal{P}_n$ ($\mathcal{P}_n$ is the class of n-permutable semigroups); for $n = 2$, $\mathcal{Q}_2$ and $\mathcal{P}_2$ are equivalent to the commutative property.

Definition 6.6.2. *A language L is* n-almost-commutative, $n > 1$, (almost-commutative) *if its syntactic monoid $S(L)$ is n-almost-commutative (almost-commutative).*

Example 6.6.1. Let $S = \{a, b, 0\}$ be a three-element semigroup, where 0 is a zero element and the multiplication table is given by: $a^2 = ba = 0$ and $ab = a$, $b^2 = b$. The semigroup S is not commutative since $ab \neq ba$. However, S is 3-almost-commutative. Indeed, for any $x, y, z \in S$ one has $xyz = xzy$. Moreover, one easily verifies that S is the syntactic monoid of the language $L = ab^*aA^* \cup bA^*$, so that L is a 3-almost-commutative, non-commutative, language.

The following lemma is due to P. Perkins [125]:

Lemma 6.6.1. *A semigroup S is almost-commutative if and only if there exist integers n and j such that $1 \leq j < j+1 \leq n$ and for any sequence $s_1, \ldots, s_n$ of elements of S one has*

$$(s_1 \cdots s_{j-1})s_j s_{j+1}(s_{j+2} \cdots s_n) = (s_1 \cdots s_{j-1})s_{j+1} s_j (s_{j+2} \cdots s_n).$$

It is also possible to prove that if $S \in \mathcal{Q}_m$, then S satisfies the above condition with $n \geq 2m + 2$.

Theorem 6.6.1. *The derivation relation $\Rightarrow^*_{\pi'}$ is a well quasi-order.*

Proof. Let us first observe that

$$w_1 w_2 \cdots w_n \Rightarrow_{\pi'} w_{\sigma(1)} w_{\sigma(2)} \cdots w_{\sigma(n)} \Rightarrow^*_{\pi'} w_1 w_2 \cdots w_n.$$

In fact from equation (6.8) one obtains

$$w_{\sigma(1)} w_{\sigma(2)} \cdots w_{\sigma(n)} \Rightarrow^*_{\pi'} w_{\sigma^i(1)} w_{\sigma^i(2)} \cdots w_{\sigma^i(n)},$$

for all $i > 0$. Thus, by taking i equal to the order of the permutation σ, we get the result. Let $\sim_{\pi'} = \Rightarrow^*_{\pi'} \cap (\Rightarrow^*_{\pi'})^{-1}$ be the congruence associated with the monotone quasi-order $\Rightarrow^*_{\pi'}$; one has then that for all words $w_1, \ldots, w_n \in A^+$,

$$w_1 w_2 \cdots w_n \sim_{\pi'} w_{\sigma(1)} w_{\sigma(2)} \cdots w_{\sigma(n)}.$$

It follows that the quotient semigroup $S = A^+/\sim_{\pi'}$ will be n-almost-commutative. By Lemma 6.6.1 there exist integers j and n' such that $1 \leq j < j+1 \leq n'$ and for any sequence $u_1, \ldots, u_{n'}$ of words of A^+ one has

$$(u_1 \cdots u_{j-1})u_j u_{j+1}(u_{j+2} \cdots u_{n'}) \sim_{\pi'} (u_1 \cdots u_{j-1})u_{j+1} u_j (u_{j+2} \cdots u_{n'}).$$

Let $A = \{a_1, \ldots, a_h\}$; we totally order its letters by $a_1 <_{lex} a_2 <_{lex} \cdots <_{lex} a_h$ and extend this order to the alphabetic order of A^+. We prove that the quasi-order $\Rightarrow^*_{\pi'}$, that we simply denote by $\leq$, is a wqo. The proof is by contradiction. Suppose that $\leq$ is not a wqo. By Proposition 6.1.2 there exists an infinite bad sequence $x : \mathbb{N} \rightarrow A^+$, i.e., such that for all integers i, j, $0 \leq i < j$, $x_i \not\leq x_j$, which is minimal with respect to $\leq_a$ in the set of all bad sequences. We prove now that if $|x_i| \geq 2n'$, then

$$x_i = \lambda a_1^{i_1} a_2^{i_2} \cdots a_h^{i_h} \mu,$$

with $\lambda, \mu \in A^*$, $|\lambda| = |\mu| = n'$, and $i_r \geq 0$ $(r = 1, \ldots, h)$. Suppose in fact, by contradiction, that there exist letters a and b such that $a <_{lex} b$ and $x_i = \lambda ubav\mu$, with $u, v \in A^*$; then we can factorize x_i as

$$x_i = \lambda ubav\mu = (u_1 \cdots u_{j-1}) u_j u_{j+1} (u_{j+2} \cdots u_{n'}),$$

with $u_i \in A^+$, $i = 1, \ldots, n'$, by taking

$$\lambda u = u_1 \cdots u_{j-1},\ v\mu = u_{j+2} \cdots u_{n'},\ b = u_j,\ a = u_{j+1}.$$

Note that the factorization is possible since $|\lambda| = |\mu| = n'$. Hence, $x_i \Rightarrow_{\pi'}^* y_i = \lambda uabv\mu <_a x_i$. One can easily verify that, for all $j < i$, one has $x_j \not\leq y_i$ and, for all $j > i$, one has $y_i \not\leq x_j$. Let us then consider the sequence z defined as $z_i = y_i$ and $z_h = x_h$, for $0 \leq h \neq i$. The sequence z is then a bad sequence; however, since $y_i <_a x_i$, one has $z <_a x$ and this contradicts the minimality of x.

Let us now consider the language X formed by the elements of the sequence x, i.e., $X = \bigcup_{i \in N} \{x_i\}$. We proved that if $|x_i| \geq 2n'$, then $x_i = \lambda a_1^{i_1} a_2^{i_2} \cdots a_h^{i_h} \mu$, with $|\lambda| = |\mu| = n'$ and $i_r \geq 0$ $(r = 1, \ldots, h)$. Thus any element of X of length $\geq 2n'$ is uniquely determined by the $(2+h)$-tuple $(\lambda, \mu; i_1, \ldots, i_h) \in A^{n'} \times A^{n'} \times \mathbb{N}^h$. We can then introduce in this set the relation $\Rightarrow$ defined as: let $(\lambda, \mu; i_1, \ldots, i_h), (\lambda', \mu'; j_1, \ldots, j_h) \in A^{n'} \times A^{n'} \times \mathbb{N}^h$ then

$$(\lambda, \mu; i_1, \ldots, i_h) \Rightarrow (\lambda', \mu'; j_1, \ldots, j_h)$$

if and only if $\lambda = \lambda'$, $\mu = \mu'$, and there exists $r \in [1, h]$ such that $i_r \geq m$, $j_r = i_r + k$, and $j_s = i_s$ for $s \neq r$. Let $\Rightarrow^*$ be the reflexive and transitive closure of $\Rightarrow$. Let us observe that if $(\lambda, \mu; i_1, \ldots, i_h) \Rightarrow^* (\lambda', \mu'; j_1, \ldots, j_h)$, then

$$\lambda a_1^{i_1} a_2^{i_2} \cdots a_h^{i_h} \mu \Rightarrow_{\pi'}^* \lambda' a_1^{j_1} a_2^{j_2} \cdots a_h^{j_h} \mu' = \lambda a_1^{j_1} a_2^{j_2} \cdots a_h^{j_h} \mu.$$

Let

$$M = \{(\lambda, \mu; i_1, \ldots, i_h) \in A^{n'} \times A^{n'} \times \mathbb{N}^h \mid \lambda a_1^{i_1} a_2^{i_2} \cdots a_h^{i_h} \mu \in X\}.$$

Since M is infinite and $|\lambda| = |\mu| = n'$, there will be an infinite subset of elements having the same pair (λ_0, μ_0). By Lemma 6.5.1, it follows that there will exist two $(2+h)$-tuples $(\lambda_0, \mu_0; i_1, \ldots, i_h)$, $(\lambda_0, \mu_0; j_1, \ldots, j_h) \in M$ such that $(\lambda_0, \mu_0; i_1, \ldots, i_h) \Rightarrow^* (\lambda_0, \mu_0; j_1, \ldots, j_h)$. This implies

$$\lambda_0 a_1^{i_1} a_2^{i_2} \cdots a_h^{i_h} \mu_0 \Rightarrow_{\pi'}^* \lambda_0 a_1^{j_1} a_2^{j_2} \cdots a_h^{j_h} \mu_0,$$

which is a contradiction.

Definition 6.6.3. *A language L is* quasi-periodic on the letters *if for any $a \in A$ there exist integers $m > 0$ and $k \geq 1$ such that $a^m \leq_L a^{m+k}$. L is* co-quasi-periodic on the letters *if L^c is quasi-periodic on the letters.*

A corollary of Theorem 6.6.1 is the following:

Theorem 6.6.2. *Let L be a language which is quasi-periodic or co-quasi-periodic on the letters. If L is almost-commutative, then L is regular.*

Proof. Let L (or L^c) be almost-commutative. Thus, there exist an integer $n > 1$ and a permutation $\sigma \in S_n \setminus \{id\}$ such that for any sequence $u_1, \ldots, u_n$ of words of A^+ one has $u_1 u_2 \cdots u_n \equiv_L u_{\sigma(1)} u_{\sigma(2)} \cdots u_{\sigma(n)}$. If L is quasi-periodic on the letters, then one has that for any $a \in A$ there exists a pair of integers (m_a, k_a) such that $a^{m_a} \leq_L a^{m_a + k_a}$. Let us take $m = \max\{m_a \mid a \in A\}$ and $k = \text{lcm}\{k_a \mid a \in A\}$; it follows that for any $a \in A$, $a^m \leq_L a^{m+k}$. Let us then consider the rewriting system π' defined as: for any letter $a \in A$ and words $u_1, \ldots, u_n \in A^+$, $a^m \to a^{m+k}$ and $u_1 u_2 \cdots u_n \to u_{\sigma(1)} u_{\sigma(2)} \cdots u_{\sigma(n)}$. The language L is $\Rightarrow^*_{\pi'}$-closed so that, by Theorem 6.6.1 and Proposition 6.4.1, the result follows. Let us now suppose that L is co-quasi-periodic on the letters. One has that L^c is quasi-periodic on the letters and that $S(L) = S(L^c)$ is almost-commutative. Hence, L^c and then L are regular.

Let $k \geq 0$. We say that a language L over the alphabet A satisfies the *iteration property* $(IP)_k$ if the following condition is satisfied:

$(IP)_k$. There exists a positive integer $N_{L,k}$ such that each $w \in L$ of length at least $N_{L,k}$ admits a factorization $w = w_1 u w_3 v w_2$ satisfying the following conditions:

i) $|uv| \geq 1$,
ii) $|w_1 u v w_2| \leq N_{L,k}$,
iii) $|w_1|, |w_2| \geq k$,
iv) $w_1 u^n w_3 v^n w_2 \in L$, for all $n > 0$.

We remark that the condition *ii)* expresses the fact that the words u and v of the iterative pair (u, v) are localized "near" the ends of the word w.

The property $(IP)_0$ will be simply denoted by IP and called *iteration property*. A language L satisfies the ***strong iteration property*** SIP if for all $k \geq 0$, L satisfies $(IP)_k$. We say that a class of languages satisfies the iteration property IP (SIP) if any language in the class satisfies IP (SIP).

A class $\mathcal{L}$ of languages is closed with respect to the *quotient operations* if

$$L \in \mathcal{L} \Rightarrow \forall u \in A^* \ (u^{-1}L \in \mathcal{L} \text{ and } Lu^{-1} \in \mathcal{L}).$$

Classes of languages closed with respect to quotient operations are, for instance, rational cones. We recall that a *rational cone*, or simply *cone*, is any family of languages closed under rational transductions (see [7]).

Lemma 6.6.2. *Let $\mathcal{L}$ be a class of languages closed with respect to quotient operations. If $\mathcal{L}$ satisfies IP, then $\mathcal{L}$ satisfies SIP.*

Proof. Let $L \in \mathcal{L}$ and $k > 0$. Since $\mathcal{L}$ is closed with respect to the quotient operations, then for any $x \in A^*$ one has that $x^{-1}L$ and Lx^{-1} belong to $\mathcal{L}$. Let $P_k(L)$ ($S_k(L)$) denote the set of all prefixes (suffixes) of length k of the words of L. For any pair $(p, s) \in P_k(L) \times S_k(L)$ the language $L_{p,s} = (p^{-1}L)s^{-1} \in \mathcal{L}$ satisfies the iteration property IP with an integer $N = N_{p,s}$. Let us then define

$$N_{L,k} = \max\{N_{p,s} \mid (p, s) \in P_k(L) \times S_k(L)\} + 2k.$$

Let $w \in L$ be such that $|w| \geq N_{L,k}$. We can write w as $w = pw's$, with $|p| = |s| = k$ and $w' \in L_{p,s}$ and $|w'| \geq N_{p,s}$. By IP, w' can be factorized as: $w' = xuzvy$, with $|uv| > 1$, $|xuvy| \leq N_{p,s}$, and $xu^nzv^ny \in L_{p,s}$, for all $n > 0$. One has then $pxu^nzv^nys \in L$, for all $n > 0$ and $|pxuvys| \leq N_{p,s} + 2k \leq N_{L,k}$. Since $|px|, |ys| \geq k$ the result follows.

The following theorem holds [48].

Theorem 6.6.3. *Let L be an almost-commutative language satisfying the iteration property SIP. Then L is quasi-periodic on the letters.*

Proof. Let L be an almost-commutative language. We prove that the letters of A are uniformly quasi-periodic, that is, there exist positive integers m and k such that for all $a \in A$

$$a^m \leq_L a^{m+k}, \tag{6.9}$$

where $\leq_L$ denotes the Myhill quasi-order relative to L. The syntactic monoid $S(L)$ is almost-commutative, so that, by Lemma 6.6.1, there exist integers n and j such that $1 \leq j < j + 1 \leq n$ and for any sequence $u_1, \ldots, u_n$ of words of A^+ one has

$$(u_1 \cdots u_{j-1})u_j u_{j+1}(u_{j+2} \cdots u_n) \equiv_L (u_1 \cdots u_{j-1})u_{j+1}u_j(u_{j+2} \cdots u_n). \tag{6.10}$$

By hypothesis L satisfies SIP. Let us consider the integer $N_{L,n}$ and take an even integer $m \geq 2N_{L,n}$. Let then $a \in A$ and $x, y \in A^*$ be such that $w = xa^my \in L$. By SIP xa^my admits a factorization

$$xa^my = w_1uw_3vw_2$$

satisfying the following conditions:

$$i)\ |uv| \geq 1,\ ii)\ |w_1uvw_2| \leq N_{L,n}, \tag{6.11}$$

$$iii)\ |w_1|, |w_2| \geq n \text{ and } iv)\ w_1u^rw_3v^rw_2 \in L, \text{ for all } r > 0. \tag{6.12}$$

Let us first assume that $|x| < n$. One has $w_1 = x\lambda$, $\lambda \in A^*$, and $a^my = \lambda uw_3vw_2$. Since $|\lambda u| \leq |w_1u| < |w_1uvw_2| \leq N_{L,n} \leq m/2$, one derives that $u = a^i$ for a suitable integer $i \geq 0$. In a similar way if also $|y| < n$, then one has $w_2 = \mu y$, $\mu \in A^*$, $a^m = \lambda uw_3v\mu$, and $v = a^j$ for a suitable integer $j \geq 0$. Moreover, $1 \leq i + j < |w_1uvw_2| \leq N_{L,n}$. Hence, $i + j$ divides $N_{L,n}!$. Let m' be such that $(i + j)m' = N_{L,n}!$. One has then in view of iv)

$$w_1 a^{i(m'+1)} w_3 a^{j(m'+1)} w_2 = x\lambda a^i w_3 a^j \mu a^{(i+j)m'} y = x a^{m+k} y \in L,$$

with $k = N_{L,n}!$. Thus if $|x|$ and $|y|$ are less than n, then the result is proved. We assume now that $|x|$ or $|y|$ are greater than or equal to n. Let us assume $|x| \geq n$. We can write $x = x'\xi$, $\xi \in A^*$, with $|x'| = n$. We shall prove now that

$$w = xa^m y \in L \Longrightarrow x'a^{m/2}\xi a^{m/2} y \in L.$$

If $\xi = \epsilon$, then the result is trivial. Let us then suppose that $\xi \in A^+$. We can factorize w in n consecutive nonempty factors as

$$w = x'\xi a^{m/2} a^{m/2} y = (u_1 \cdots u_{j-1}) u_j u_{j+1} (u_{j+2} \cdots u_n),$$

where

$$x' = u_1 \cdots u_{j-1},\ \xi = u_j,\ a^{m/2} = u_{j+1},\ a^{m/2} y = u_{j+2} \cdots u_n.$$

Let us observe that such a factorization is always possible (whatever j is) since $j \leq n-1$ and $|x'| = n$. Hence, by equation (6.10), it follows that $x'a^{m/2}\xi a^{m/2} y \in L$. Let us assume $|y| \geq n$. We can write $y = \eta y'$ with $\eta \in A^*$ and $|y'| = n$. By a symmetric argument one proves that $w' = x'a^{m/2}\xi\eta a^{m/2} y' \in L$. Now $|w| = |w'| > m \geq 2N_{L,n}$. By *SIP*, w' admits a factorization $x'a^{m/2}\xi\eta a^{m/2} y' = w_1 u w_3 v w_2$ satisfying conditions (6.11) and (6.12). In view of *iii*) we can write $w_1 = x'\lambda$, $\lambda \in A^*$, and $w_2 = \mu y'$, $\mu \in A^*$, so that it follows $a^{m/2}\xi\eta a^{m/2} = \lambda u w_3 v \mu$. Moreover, from *ii*) one easily derives $|\lambda u|, |v\mu| < m/2$. This implies $\lambda, \mu \in \{a\}^*$, $u = a^i$, and $v = a^j$ for suitable integers i and j such that $1 \leq i+j < |w_1 u v w_2| \leq N_{L,n}$. Hence, $i+j$ divides $N_{L,n}!$. Let m' be such that $(i+j)m' = N_{L,n}!$. One has then in view of *iv*) $u' = w_1 a^{i(m'+1)} w_3 a^{j(m'+1)} w_2 \in L$. In u' the number of occurrences of the letter a is increased by $N_{L,n}!$ with respect to the number of occurrences of a in w'. Let us now set $\alpha = a^{m'+1}$; we can write

$$u' = w_1 \alpha^i w_3 \alpha^j w_2 = x'\alpha^i a^{m/2-i} \xi\eta \alpha^j a^{m/2-j} y' \in L.$$

Since L is almost commutative, then by making the same transformations as before but in the reversed order, setting $k = N_{L,n}!$, we obtain

$$w'' = x'\xi\alpha^i a^{m/2-i} \alpha^j a^{m/2-j} \eta y' = x a^{(i+j)m'} a^m y = x a^{m+k} y \in L.$$

Finally, we have to consider the case when $|x| \geq n$ and $|y| < n$. In this case one easily derives that $w' = x'a^{m/2}\xi a^{m/2} y \in L$, with $|x'| = n$. Since $|w| = |w'| > m \geq 2N_{L,n}$, then by *SIP*, w' admits a factorization $w' = x'a^{m/2}\xi a^{m/2} y = w_1 u w_3 v w_2$ satisfying conditions (6.11) and (6.12). It follows that $u = a^i$ and $v = a^j$ for suitable integers i and j such that $1 \leq i+j < N_{L,n}$. Let m' be such that $(i+j)m' = N_{L,n}!$. One has then in view of *iv*) $u' = w_1 a^{i(m'+1)} w_3 a^{j(m'+1)} w_2 = w_1 \alpha^i w_3 \alpha^j w_2 \in L$, where $\alpha = a^{m'+1}$. The remaining part of the proof is similar to the preceding one.

Proposition 6.6.1. *Every almost-commutative language in a class of languages closed with respect to the quotient operations and satisfying the iteration property IP is regular.*

Proof. Let L be an almost commutative language in a class $\mathcal{L}$ of languages closed with respect to the quotient operations and satisfying the iteration property IP. By Lemma 6.6.2, L satisfies the strong iteration property SIP, so that by Theorem 6.6.3, L is quasi-periodic on the letters. Hence, from Theorem 6.6.2 the result follows.

Since a rational cone is a family of languages closed with respect to the quotient operations the following holds:

Corollary 6.6.1. *Every almost-commutative language in a rational cone which satisfies the iteration property IP is regular.*

We remark that *one-counter languages* as well as *linear languages* belong to rational cones (see [7]) which satisfy the property IP, so that from Corollary 6.6.1 we derive the following:

Corollary 6.6.2. *Almost-commutative one-counter languages and almost commutative linear languages are regular.*

In particular one has:

Corollary 6.6.3. *Commutative one-counter languages and commutative linear languages are regular.*

The following theorem proved in [103], with a different technique, holds:

Theorem 6.6.4. *A commutative language satisfying IP is regular.*

Proof. Let L be a commutative language satisfying IP. We prove that L satisfies SIP. Suppose that L satisfies IP and let N_L be the corresponding positive integer. For any positive integer k we set $N_{L,k} = 2k + 2N_L$. Let $w \in L$ be a word of length at least $N_{L,k}$. We can write $w = x\lambda\mu y$, with $|x|, |y| \geq k$ and $|\lambda|, |\mu| \geq N_L$. Since L is commutative, then $w' = \lambda xy\mu \in L$. Moreover, as $|w'| \geq N_L$, by IP we can write $w' = w'_1 u w'_3 v w'_2$, with $|uv| > 1$, $|w'_1 uvw'_2| \leq N_L$, and $w'_1 u^n w'_3 v^n w'_2 \in L$, for all $n > 0$. Since $|w'_1 u|, |vw'_2| \leq N_L$, there exist $\alpha, \beta \in A^*$ such that

$$\lambda = w'_1 u\alpha, \ \mu = \beta v w'_2, \ \text{ and } w'_3 = \alpha xy\beta.$$

Thus we have $w'_1 u^n \alpha xy\beta v^n w'_2 \in L$, for all $n > 0$, and, by commutativity, $xw'_1 u^n \alpha\beta v^n w'_2 y \in L$, for all $n > 0$. Therefore, setting $w_1 = xw'_1$, $w_2 = w'_2 y$, and $w_3 = \alpha\beta$, one has $w = w_1 u w_3 v w_2$, $|w_1|, |w_2| \geq k$, $|w_1 uvw_2| \leq N_{L,k}$, and $w_1 u^n w_3 v^n w_2 \in L$, for all $n > 0$. From Theorem 6.6.3, L is quasi-periodic on the letters, so that by Theorem 6.6.2 the result follows.

We consider now the family Qrt of *quasi-rational* languages which is defined as the closure by substitution of the family Lin of linear languages. More precisely set $\mathrm{Lin}_1 = \mathrm{Lin}$ and, for any $k > 1$

$$(\mathrm{Lin})_k = \mathrm{Lin} \odot (\mathrm{Lin})_{k-1},$$

where $\odot$ denotes the substitution operator (see [7]). The family Qrt is then defined as

$$\mathrm{Qrt} = \bigcup_{k>0} (\mathrm{Lin})_k.$$

The following noteworthy theorem due to J. Kortelainen [99] holds:

Theorem 6.6.5. *Any commutative quasi-rational language is regular.*

By the above theorem and the lemma of Perkins we can prove the following more general proposition in which "commutative" is replaced by "almost-commutative":

Theorem 6.6.6. *Any almost-commutative quasi-rational language is regular.*

Proof. We observe first that Qrt is a cone. Indeed, Lin is a cone; moreover, it is well known that if $\mathcal{L}$ and $\mathcal{M}$ are two cones, then $\mathcal{L} \odot \mathcal{M}$ will be so. Thus for any $k > 0$, $(\mathrm{Lin})_k$ is a cone; hence, it follows that Qrt is a cone. If L is an almost-commutative language in Qrt, then, by Lemma 6.6.1, there exist integers n and j such that $1 \le j < j+1 \le n$ and for any sequence $u_1, \ldots, u_n$ of words of A^+ one has

$$(u_1 \cdots u_{j-1}) u_j u_{j+1} (u_{j+2} \cdots u_n) \equiv_L (u_1 \cdots u_{j-1}) u_{j+1} u_j (u_{j+2} \cdots u_n). \quad (6.13)$$

Let $\alpha, \beta \in A^n$; we prove that the language $L_{\alpha,\beta} = (\alpha^{-1} L)\beta^{-1} \in \mathrm{Qrt}$ is commutative, that is, for any $x, y \in A^+$ one has that $xy \equiv_{L_{\alpha,\beta}} yx$. Indeed, let $z, h \in A^*$ and suppose that $zxyh \in L_{\alpha,\beta}$; one has then $\alpha zxyh\beta \in L$. From equation (6.13) since $|\alpha|, |\beta| = n$, one has that $\alpha zyxh\beta \in L$; hence, $zyxh \in L_{\alpha,\beta}$. It follows then that $xy \equiv_{L_{\alpha,\beta}} yx$. By Theorem 6.6.5 one has that $L_{\alpha,\beta}$ is regular. The language L is given, up to a finite set, by

$$\bigcup_{\alpha,\beta \in A^n} \alpha L_{\alpha,\beta} \beta;$$

hence, L is regular since it is a finite union of regular languages.

One can ask the question whether in Theorem 6.6.2 one can substitute "almost-commutative" languages with "permutable" languages. In general the answer to this question is negative; however, for 3-permutable languages the following holds [102]:

Theorem 6.6.7. *Let π be any rewriting system defined by the productions: for any letter $a \in A$ and words $w_1, w_2, w_3 \in A^+$, $a^m \to a^{m+k}$, $w_1w_2w_3 \to w_{\sigma(1)}w_{\sigma(2)}w_{\sigma(3)}$, $w_{\sigma(1)}w_{\sigma(2)}w_{\sigma(3)} \to w_1w_2w_3$, where $\sigma \in S_3 \setminus \{id\}$ depends on the triplet w_1, w_2, w_3 and the pair (m,k) depends on the letter a. The derivation relation $\Rightarrow_\pi^*$ is a well quasi-order.*

Proof. We denote by $\leq$ the derivation relation $\Rightarrow_\pi^*$ and by $\sim$ the equivalence relation $\leq \cap \leq^{-1}$. By hypothesis, for any $a \in A$, $a^m \leq a^{m+k}$, where the pair (m,k) depends on a. Since A is finite, we can suppose, without loss of generality, that the pair (m,k) does not depend on the letter a. Suppose, by contradiction, that $\leq$ is not a wqo; then there exists an infinite bad sequence

$$f_1, f_2, \ldots, f_n, \ldots,$$

which is minimal with respect to the alphabetic ordering. We prove that no word f_i of the sequence contains a 3-divided factor. In fact suppose, by contradiction, that for some $i \geq 1$, $f_i = xw_1w_2w_3y$, $x, y \in A^*$, $w_1, w_2, w_3 \in A^+$, and $w_1w_2w_3$ is 3-divided. By hypothesis there exists a nontrivial permutation $\sigma \in S_3$ such that $w_1w_2w_3 \sim w_{\sigma(1)}w_{\sigma(2)}w_{\sigma(3)}$. Set $g_i = xw_{\sigma(1)}w_{\sigma(2)}w_{\sigma(3)}y$. One has $g_i \sim f_i$. Moreover, since $w_1w_2w_3$ is 3-divided, g_i strictly precedes f_i in the alphabetic ordering. The sequence

$$f_1, f_2, \ldots, f_{i-1}, g_i, f_{i+1}, \ldots, f_n, \ldots,$$

is a bad sequence that contradicts the minimality of the bad sequence $(f_n)_{n\geq 1}$. Therefore, the words f_i's do not contain 3-divided factors. The sequence $(f_n)_{n\geq 1}$ obviously contains arbitrarily long words. Thus, by Proposition 2.4.1, there are infinitely many words in the sequence of the kind $xa^{m+k}a^{m+k}y$, with $x, y \in A^*$ and $a \in A$. Any such word f can be factorized as $f = (xa^{m+k})(a^{m+k})(y)$. Thus $f \sim g$, where g is obtained from f by a suitable nontrivial permutation of S_3. In any case one obtains a word g such that g has a prefix or a suffix of the kind a^{m+k}, with $a \in A$. In this way one can construct a sequence $(g_n)_{n\geq 1}$, where, for any $n \geq 1$, $g_n \sim f_n$ and $|f_n| = |g_n|$; this implies that such a new sequence is bad too. Moreover, infinitely many words of the sequence $(g_n)_{n\geq 1}$ have a prefix (or a suffix) of the kind a^{m+k}, with $a \in A$. Since A is finite, there exists a letter $a \in A$ such that a^{m+k} occurs infinitely many times as a prefix (or as a suffix) of the words of $(g_n)_{n\geq 1}$. Suppose that a^{m+k} occurs as a prefix (the other case is completely symmetric). We have then a subsequence $(g_{i_n})_{n\geq 1}$, where $g_{i_n} = a^{m+k}h_n$, $n \geq 1$, $h_n \in A^*$. Consider the sequence

$$g_1, \ldots, g_{i_1-1}, a^m h_1, \ldots, a^m h_n, \ldots.$$

This sequence is bad. In fact $g_i \not\leq g_j$ for $1 \leq i < j \leq i_1 - 1$. For $1 \leq i \leq i_1 - 1$ and $j \geq 1$, one has $g_i \not\leq a^m h_j$, otherwise $g_i \leq a^m h_j \leq a^{m+k}h_j = g_{i_j}$. Moreover, for $1 \leq s < t$ one has $a^m h_s \not\leq a^m h_t$, otherwise $a^{m+k}h_s \leq a^{m+k}h_t$ and $g_{i_s} \leq g_{i_t}$. In conclusion, we have a bad sequence that contradicts the minimality of the bad sequence $(f_n)_{n\geq 1}$ and this is absurd.

Corollary 6.6.4. *Let L be a 3-permutable language which is quasi-periodic or co-quasi-periodic on the letters. Then L is regular.*

Proof. Since L is 3-permutable, for any sequence w_1, w_2, w_3 of words of A^+ there exists a permutation $\sigma \in \mathcal{S}_3 \setminus \{id\}$ such that $w_1w_2w_3 \equiv_L w_{\sigma(1)}w_{\sigma(2)}w_{\sigma(3)}$; moreover if we suppose that L is quasi-periodic on the letters, then one has that for any letter $a \in A$ there exist integers $m \geq 0$ and $k > 0$ such that $a^m \leq_L a^{m+k}$. We can then introduce a corresponding rewriting system π whose productions are just $w_1w_2w_3 \to w_{\sigma(1)}w_{\sigma(2)}w_{\sigma(3)}$, $w_{\sigma(1)}w_{\sigma(2)}w_{\sigma(3)} \to w_1w_2w_3$, and $a^m \to a^{m+k}$. The language L is, obviously, $\Rightarrow_\pi^*$-closed so that by Theorem 6.6.7 it follows that L is regular. If L is co-quasi-periodic on the letters, then L^c is quasi-periodic and 3-permutable so that L^c as well as L is regular.

The following was proved in [114]:

Corollary 6.6.5. *Let S be a finitely generated 3-permutable semigroup whose generators have a finite order. Then S is finite.*

Proof. Let S be a finitely generated semigroup and $\phi : A^+ \to S$ be the canonical epimorphism, where A is a finite alphabet. Since S is 3-permutable then, denoting by ρ the nuclear congruence $\phi\phi^{-1}$, for any sequence of words $u_1, u_2, u_3 \in A^+$, $u_1u_2u_3 \;\rho\; u_{\sigma(1)}u_{\sigma(2)}u_{\sigma(3)}$, where $\sigma \in \mathcal{S}_3 \setminus \{id\}$ depends on the words u_1, u_2, u_3. Moreover, since the generators of S have a finite order, then for any letter $a \in A$ one has $a^m \;\rho\; a^{m+k}$, $m > 0$, $k > 0$, where the pair (m, k) depends on the letter a. Let us introduce the rewriting system π whose productions are just $u_1u_2u_3 \to u_{\sigma(1)}u_{\sigma(2)}u_{\sigma(3)}$, $u_{\sigma(1)}u_{\sigma(2)}u_{\sigma(3)} \to u_1u_2u_3$, and $a^m \to a^{m+k}$. One has, trivially, that $\Rightarrow_\pi^* \subseteq \rho$. By Theorem 6.6.7 $\Rightarrow_\pi^*$ is a wqo; by Propositions 6.1.1 and 6.3.1 one has that the congruence ρ is of a finite index. Hence, S is finite.

The statement of Corollary 6.6.5 cannot be extended to the case of n-permutable semigroups with $n > 3$; this is a consequence of the fact that there exist infinite groups of the class $\mathcal{P}_4$ generated by two elements of order 2 (see [90], Remark 3). In conclusion, we can formulate the following conjecture which is a natural extension of Corollaries 6.6.2 and 6.6.3:

Conjecture 6.6.1. If L is a linear (one-counter) language which is 3-permutable, then L is regular.

We remark that the above statement is false in the case of the property $\mathcal{P}_4$. Indeed the following holds:

Proposition 6.6.2. *The linear language $L = \{a^nb^n \mid n > 0\}$ on the alphabet $A = \{a, b\}$ satisfies the property $\mathcal{P}_4$.*

Proof. Let $x_1, x_2, x_3, x_4 \in A^+$, and suppose first that $x_1x_2x_3x_4$ is a factor of a word of L, that is, there exist $h, h' \in A^*$ and $n > 0$ such that $hx_1x_2x_3x_4h' = a^nb^n$. Now either $a^n = hx_1x_2z$, $z \in A^*$, that implies $x_1x_2 \in a^*$, or $hx_1x_2 = a^nz$, $z \in A^*$. In the latter case $zx_3x_4h' = b^n$, that implies $x_3x_4 \in b^*$. Hence, one has that in any case if $x_1x_2x_3x_4 \in F(L)$, then either $x_1x_2 \in a^*$ or $x_3x_4 \in b^*$. In the first case one has $x_1x_2x_3x_4 \equiv_L x_2x_1x_3x_4$ and in the second case $x_1x_2x_3x_4 \equiv_L x_1x_2x_4x_3$. Let us now suppose that $x_1x_2x_3x_4 \notin F(L)$. We show that there exists a permutation $\sigma \in S_4 \setminus \{id\}$ such that $x_{\sigma(1)}x_{\sigma(2)}x_{\sigma(3)}x_{\sigma(4)} \notin F(L)$. This would imply, trivially, that $x_1x_2x_3x_4 \equiv_L x_{\sigma(1)}x_{\sigma(2)}x_{\sigma(3)}x_{\sigma(4)}$. Since $x_1x_2x_3x_4 \notin F(L)$ if and only if $x_1x_2x_3x_4 \in A^*baA^*$, it follows that there exists i, $1 \le i \le 3$, such that $x_ix_{i+1} \in A^*baA^*$; hence there exists certainly a nontrivial permutation σ of S_4 such that $x_{\sigma(1)}x_{\sigma(2)}x_{\sigma(3)}x_{\sigma(4)}$ has ba as a factor so that $x_{\sigma(1)}x_{\sigma(2)}x_{\sigma(3)}x_{\sigma(4)} \notin F(L)$.

6.7 Copying systems

Copying systems and languages generated by them were introduced in [60] by Ehrenfeucht and Rozenberg. In their paper it is proved that when the alphabet has cardinality at least three such languages are not, in general, regular (see Theorem 6.7.3). In this section we prove that in the case of a binary alphabet, the languages generated by copying systems are actually all regular [13].

In the following A will denote a binary alphabet $A = \{a, b\}$. Let (A, π) be the rewriting system with $\pi = \{(x, xx) \mid x \in A^*\}$. The derivation relation $\Rightarrow_\pi^*$ is called *copying relation.* We shall consider a restricted copying relation denoted as $\Rightarrow_{\pi'}^*$ where

$$\pi' = \{(a, aa), (b, bb), (ab, abab), (ba, baba)\}.$$

Observe that $\Rightarrow_{\pi'}^* \subseteq \Rightarrow_\pi^*$, thus if we succeed in proving that $\Rightarrow_{\pi'}^*$ is a w.q.o., then, by Proposition 6.1.1, also $\Rightarrow_\pi^*$ turns out to be a w.q.o. We begin with some preliminary lemmas.

Lemma 6.7.1. *Let $\lambda, \mu \in A^*$ such that $aba\lambda \Rightarrow_{\pi'}^* aba\mu$. Then, $ba\lambda \Rightarrow_{\pi'}^* ba\mu$.*

Proof. By induction on the length of the derivation.

Consider first a derivation of length 1, that is, $aba\lambda \Rightarrow_{\pi'} aba\mu$. If the derivation does not use the leftmost symbol a (i.e., $ba\lambda \Rightarrow_{\pi'} ba\mu$), then the assertion follows trivially. Otherwise, the derivation must be of the following type:

$$(ab)a\lambda \Rightarrow_{\pi'} (abab)a\lambda = aba\mu$$

with $\mu = ba\lambda$. On the other hand, $(ba)\lambda \Rightarrow_{\pi'} (baba)\lambda = ba\mu$. The base of the induction is thus proved.

Let us consider derivations of length $n > 1$ of the form

$$aba\lambda \Rightarrow_{\pi'} v_1 \Rightarrow_{\pi'} v_2 \Rightarrow_{\pi'} \cdots \Rightarrow_{\pi'} v_{n-1} \Rightarrow_{\pi'} aba\mu. \tag{6.14}$$

We distinguish the following cases:
Case 1. The first derivation is obtained by applying $\Rightarrow_{\pi'}$ to the word $a\lambda$. In this case we can write

$$aba\lambda \Rightarrow_{\pi'} aba\lambda' = v_1,$$

where $a\lambda \Rightarrow_{\pi'} a\lambda'$ and $aba\lambda' \Rightarrow^*_{\pi'} aba\mu$. Since the latter derivation has length $n-1$, by the induction hypothesis, $ba\lambda' \Rightarrow^*_{\pi'} ba\mu$. On the other hand, $a\lambda \Rightarrow_{\pi'} a\lambda'$ implies $ba\lambda \Rightarrow_{\pi'} ba\lambda'$, thus $ba\lambda \Rightarrow_{\pi'} ba\mu$.
Case 2. The first derivation is obtained by applying $\Rightarrow_{\pi'}$ to the word aba. Let us consider the following subcases:
a). The first derivation is $a(ba)\lambda \Rightarrow_{\pi'} a(baba)\lambda = v_1$. In such a case, v_1 yields $aba\mu$ in $n-1$ steps and, according to the inductive hypothesis, one has $baba\lambda \Rightarrow^*_{\pi'} ba\mu$. Moreover, $ba\lambda \Rightarrow_{\pi'} baba\lambda$, thus $ba\lambda \Rightarrow^*_{\pi'} ba\mu$.
b). The first derivation is $(ab)a\lambda \Rightarrow_{\pi'} (abab)a\lambda = v_1 = a(baba)\lambda$. This subcase is equivalent to subcase (a).
c). The first derivation is $(a)ba\lambda \Rightarrow_{\pi'} (aa)ba\lambda = v_1$. This case cannot occur since v_1 cannot yield $aba\mu$ (a word starting with aa can only yield words starting with aa).
d). The first derivation is $ab(a)\lambda \Rightarrow_{\pi'} ab(aa)\lambda = v_1$. This subcase is included in case 1.
e). The first derivation is $a(b)a\lambda \Rightarrow_{\pi'} a(bb)a\lambda = v_1$. We observe that in the derivation $v_1 = abba\lambda \Rightarrow^*_{\pi'} aba\mu$, there is at least one direct derivation of the first two symbols ab into $abab$. Indeed, if this is not the case, it is easy to see that we could not remove the prefix abb in order to obtain $aba\mu$. Then the derivation (6.14) may be rewritten as

$$aba\lambda \Rightarrow_{\pi'} abba\lambda \Rightarrow^*_{\pi'} abb\lambda' \Rightarrow_{\pi'} ababb\lambda' \Rightarrow^*_{\pi'} aba\mu.$$

The previous derivation may be replaced by the following one

$$aba\lambda \Rightarrow_{\pi'} ababa\lambda \Rightarrow_{\pi'} ababba\lambda \Rightarrow^*_{\pi'} ababb\lambda' \Rightarrow^*_{\pi'} aba\mu.$$

We observe that this derivation has length n and may be dealt with as in subcase (b).

Corollary 6.7.1. *Let $\lambda, \mu \in A^*$ such that $aba\lambda \Rightarrow^*_{\pi'} aba\mu$. Then, for all $k > 0$, $ab^k a\lambda \Rightarrow^*_{\pi'} ab^k a\mu$.*

Proof. According to Lemma 6.7.1, $aba\lambda \Rightarrow^*_{\pi'} aba\mu$ implies $ba\lambda \Rightarrow^*_{\pi'} ba\mu$. Since $\Rightarrow^*_{\pi'}$ is monotone, the result follows.

Corollary 6.7.2. *Let $\lambda, \mu \in A^*$ such that $abab\lambda \Rightarrow^*_{\pi'} abab\mu$. Then, for all $k > 0$, $aba^k b\lambda \Rightarrow^*_{\pi'} aba^k b\mu$.*

Proof. According to Lemma 6.7.1, $aba(b\lambda) \Rightarrow^*_{\pi'} aba(b\mu)$ implies $ba(b\lambda) \Rightarrow^*_{\pi'} ba(b\mu)$. By applying a second time Lemma 6.7.1 (and exchanging the role of a's and b's), $ab\lambda \Rightarrow^*_{\pi'} ab\mu$ follows. Since $\Rightarrow^*_{\pi'}$ is monotone, the result follows.

Clearly for symmetry reasons Lemma 6.7.1, Corollary 6.7.1, and Corollary 6.7.2 hold when λ and μ appear as leftmost words.

Theorem 6.7.1. *The derivation relation $\Rightarrow^*_{\pi'}$ is a well quasi-order.*

Proof. Suppose by contradiction that $\Rightarrow^*_{\pi'}$ is not a wqo. From Proposition 6.1.2 there exists a bad sequence $C = x_1, x_2, \ldots, x_n, \ldots$ which is minimal with respect to the alphabetic order. Clearly C includes infinitely many words starting with the same letter, say a. We distinguish the following cases:

Case 1. C includes infinitely many words starting with the same prefix aa. Denote with $aay_1, aay_2, \ldots, aay_n, \ldots$ such an infinite subsequence and let j be the integer such that $aay_1 = x_j$. Consider the following infinite sequence $D = z_1, z_2, \ldots, z_n, \ldots$ where $z_i = x_i$ for $i = 1, 2, \ldots, j-1$ and $z_i = ay_{i-j+1}$ for $i \geq j$.

The sequence D is bad; indeed, if s and t are such that $s < t$ and $s, t \in \{1, \ldots, j-1\}$, then $z_s \not\Rightarrow^*_{\pi'} z_t$ by assumption since $z_s = x_s$ and $z_t = x_t$. If $s \in \{1, \ldots, j-1\}$ and $t \geq j$, then $z_s = x_s$ and $z_t = ay_k$ with $k = t-j+1$. From $z_s \Rightarrow^*_{\pi'} z_t$, it follows $x_s \Rightarrow^*_{\pi'} ay_k \Rightarrow_{\pi'} aay_k$. This contradicts the assumption that C is a bad sequence. Finally, if $t > s \geq j$, then $z_s = ay_h$ and $z_t = ay_k$ with $h < k$. If $z_s \Rightarrow^*_{\pi'} z_t$, due to left regularity, $aay_h \Rightarrow^*_{\pi'} aay_k$ and this contradicts the assumption that C is a bad sequence.

Case 2. C includes infinitely many words starting with ab. If C includes an infinite number of words belonging to ab^*, then it also includes infinitely many words terminating with bb. This case is similar to the previous one. Let us then suppose that C includes an infinite subsequence $ab^{k_1}ay_1$, $ab^{k_2}ay_2$, $\ldots$, $ab^{k_n}ay_n, \ldots$, with $k_i \geq 1$. One has to consider the following subcases:

a). C includes an infinite subsequence of elements of the form $ab^{k_i}ay_i$ with $k_i \geq 2$ for all $i > 0$. We can always suppose that $(k_i)_{i \geq 1}$ is a non-decreasing sequence. Let j be the positive integer such that $ab^{k_1}ay_1 = x_j$. Consider the following infinite sequence $D = z_1, z_2, \ldots, z_n, \ldots$, where $z_i = x_i$ for $i = 1, 2, \ldots, j-1$ and $z_i = abay_{i-j+1}$ for $i \geq j$. The sequence D is bad; indeed, if s and t are such that $s < t$ and $s, t \in \{1, \ldots, j-1\}$, then $z_s \not\Rightarrow^*_{\pi'} z_t$ by assumption since $z_s = x_s$ and $z_t = x_t$. If $s \in \{1, \ldots, j-1\}$ and $t \geq j$, then $z_s = x_s$ and $z_t = abay_p$ with $p = t - j + 1$. From $z_s \Rightarrow^*_{\pi'} z_t$, it follows $x_s \Rightarrow^*_{\pi'} abay_p \Rightarrow^*_{\pi'} ab^{k_p}ay_p$. This contradicts the assumption that C is a bad sequence. Finally, if $t > s \geq j$, then $z_s = abay_p$ and $z_t = abay_q$ with $p < q$. If $z_s \Rightarrow^*_{\pi'} z_t$, due to Corollary 6.7.1, $ab^{k_p}ay_p \Rightarrow^*_{\pi'} ab^{k_p}ay_q$; since $k_p \leq k_q$ and $ab^{k_p}ay_q \Rightarrow^* ab^{k_q}ay_q$, then $ab^{k_p}ay_p \Rightarrow^*_{\pi'} ab^{k_q}ay_q$ contradicting the assumption.

b). C includes an infinite subsequence of elements of the form $abay_i$, with $i > 0$. Now we have to consider the following subcases:

i). C includes an infinite sequence of elements of the kind $ababy_i$. Denote with $ababy_1$, $ababy_2, \ldots$, $ababy_n, \ldots$ such an infinite subsequence and let j be the integer such that $ababy_1 = x_j$. Consider the following infinite sequence $D = z_1, z_2, \ldots, z_n, \ldots$, where $z_i = x_i$ for $i = 1, 2, \ldots, j-1$ and $z_i = aby_{i-j+1}$ for $i \geq j$. It is easy to verify that D is a bad sequence. Indeed, if s and t are such that $s < t$ and $s, t \in \{1, \ldots, j-1\}$, then $z_s \not\Rightarrow^*_{\pi'} z_t$ by assumption since $z_s = x_s$ and $z_t = x_t$. If $s \in \{1, \ldots, j-1\}$ and $t \geq j$, then $z_s = x_s$ and $z_t = aby_k$ with $k = t - j + 1$. From $z_s \Rightarrow^*_{\pi'} z_t$, it follows $x_s \Rightarrow^*_{\pi'} aby_k \Rightarrow_{\pi'} ababy_k$. This contradicts the assumption that C is a bad sequence. Finally, if $t > s \geq j$, then $z_s = aby_h$ and $z_t = aby_k$ with $h < k$. If $z_s \Rightarrow^*_{\pi'} z_t$, due to left regularity, $ababy_h \Rightarrow^*_{\pi'} ababy_h$ and this contradicts the assumption that C is a bad sequence.

ii). C includes an infinite sequence of elements of the kind $abaay_i$. If C contains infinitely many elements belonging to $abaa^*$, then it also contains infinite elements which terminate with aa and this case is similar to case 1. Thus we can suppose that C includes an infinite subsequence of elements of the form $aba^{k_i}by_i$ with $k_i \geq 2$ for all $i > 0$. We can always suppose that the $(k_i)_{i\geq 1}$ is a non-decreasing sequence. Let j be the integer such that $aba^{k_i}by_1 = x_j$. Consider the following infinite sequence $D = z_1, z_2, \ldots, z_n, \ldots$ where $z_i = x_i$ for $i = 1, 2, \ldots, j-1$ and $z_i = ababy_{i-j+1}$ for $i \geq j$. It is easy to verify that D is a bad sequence. Indeed, if s and t are such that $s < t$ and $s, t \in \{1, \ldots, j-1\}$, then $z_s \not\Rightarrow^*_{\pi'} z_t$ by assumption since $z_s = x_s$ and $z_t = x_t$. If $s \in \{1, \ldots, j-1\}$ and $t \geq j$, then $z_s = x_s$ and $z_t = ababy_p$ with $p = t - j + 1$. From $z_s \Rightarrow^*_{\pi'} z_t$, one derives $x_s \Rightarrow^*_{\pi'} ababy_p \Rightarrow^*_{\pi'} aba^{k_p}by_p$. This contradicts the assumption that C is a bad sequence. Finally, if $t > s \geq j$, then $z_s = ababy_p$ and $z_t = ababy_q$ with $p < q$. If $z_s \Rightarrow^*_{\pi'} z_t$, due to Corollary 6.7.2, $aba^{k_p}by_p \Rightarrow^*_{\pi'} aba^{k_p}by_q$; since $k_p \leq k_q$ and $aba^{k_p}by_q \Rightarrow^*_{\pi'} aba^{k_q}by_q$, then

$$aba^{k_p}by_p \Rightarrow^*_{\pi'} aba^{k_q}by_q$$

which contradicts the assumption.

In all the cases we have considered one derives the existence of a bad sequence D which contradicts the minimality of C.

Let us remark that one can easily prove that the rewriting system π' is, in fact, equivalent to π. Moreover, π' is the smallest set of rules among those which are equivalent to π.

Theorem 6.7.2. *The derivation relation $\Rightarrow^*_\pi$ is a well quasi-order.*

Proof. By the preceding theorem $\Rightarrow^*_{\pi'}$ is a wqo. Since $\Rightarrow^*_{\pi'} \subseteq \Rightarrow^*_\pi$, then from Proposition 6.1.1 the result follows.

Corollary 6.7.3. *Let $L \subseteq A^*$ be a language which is closed with respect to $\Rightarrow^*_\pi$. Then L is a regular language.*

Proof. The statement is a consequence of Theorem 6.7.2 and Theorem 6.3.1.

Let us now consider a free monoid B^* and the copying relation $\Rightarrow_\pi^*$ in B^*. For any $w \in B^*$ we consider the set $L_{w,\pi}$ defined as

$$L_{w,\pi} = \{u \in B^* \mid w \Rightarrow_\pi^* u\}.$$

Now we prove that if in a word w occur at least three distinct letters, then the language $L_{w,\pi}$ is not regular [60]. In order to do this we need some preliminary results.

Lemma 6.7.2. *Let $w \in B^*$. For any $u \in (\mathrm{alph}(w))^*$ there exists $z \in (\mathrm{alph}(w))^*$ such that $w \Rightarrow_\pi^* wuz$.*

Proof. By induction on $|u|$. If $u = \epsilon$, then the result trivially holds. Suppose now that $u = ax$, with $a \in (\mathrm{alph}(w))$ and $x \in (\mathrm{alph}(w))^*$. Since $a \in \mathrm{alph}(w)$, one can write $w = w_1aw_2$ and $w_1aw_2 \Rightarrow_\pi w_1aw_2aw_2$, that is, $w \Rightarrow_\pi waw_2$. By the induction hypothesis, there exists $z \in (\mathrm{alph}(w))^*$ such that $wa \Rightarrow_\pi^* waxz$, which implies

$$w \Rightarrow_\pi waw_2 \Rightarrow_\pi^* waxzw_2,$$

that is, $w \Rightarrow_\pi^* wuz'$, with $z' = zw_2$.

The following technical lemma is proved in [60].

Lemma 6.7.3. *Let $w, u, z \in B^*$, where u is a square-free word. Let $y \in B^+$ such that $y \Rightarrow_\pi^* wuz$. Then $|y| \geq |u|/2^{|wz|}$.*

Proof. By induction on $|wz|$. If $|wz| = 0$, as u is square free, then $y = u$ and the statement trivially follows. Suppose now that $|wz| > 0$. Now either $y = wuz$, in such a case the statement trivially holds, or one can write

$$y \Rightarrow_\pi^* y' \Rightarrow_\pi wuz,$$

with $y' \in B^+$. The statement will be achieved if we prove that $y' = w'u'z'$, where u' is a square free word, $|u'| \geq |u|/2$ and $|w'z'| < |wz|$. In fact by the induction hypothesis one has

$$|y| \geq |u'|/2^{|w'z'|} \geq |u|/2^{|w'z'|+1} \geq |u|/2^{|wz|}.$$

Now we prove that y' can be factorized as above. We have that $y' \Rightarrow_\pi wuz$, that is, $y' = \lambda x\mu$ and $wuz = \lambda xx\mu$, with $\lambda, \mu \in B^*$ and $x \in B^+$. Thus the word y' is obtained from the word wuz deleting one of the two consecutive occurrences of x in the word wuz. Since u is square free, then one of the two occurrences of x is either overlapping with the first occurrence of w in wuz or overlapping with the last occurrence of z in wuz or both. In any case one can easily show that y' has a factorization as required.

Theorem 6.7.3. *Let $w \in B^*$ be a word such that* $\mathrm{card}(\mathrm{alph}(w)) \geq 3$. *Then $L_{w,\pi}$ is not regular.*

Proof. Set $X = L_{w,\pi}$ and suppose, by contradiction, that X is regular. Then there exists an integer k such that for any word $y \in B^*$, one has $y \equiv_X y'$ for some $y' \in B^*$, with $|y'| \leq k$. Since card(alph(w)) ≥ 3, by Proposition 2.6.2 there exists a square free word $u \in (\text{alph}(w))^*$ such that $|u| > |w|2^{|w|+k}$. By Lemma 6.7.2 there exists $z \in (\text{alph}(w))^*$ such that $w \Rightarrow_\pi^* wuz$, i.e., $wuz \in X$. Since $z \equiv_X z'$ for some $z' \in B^*$, with $|z'| \leq k$, one has $wuz' \in X$, that is, $w \Rightarrow_\pi^* wuz'$. By Lemma 6.7.3, one has $|u| \leq |w|2^{|wz'|} \leq |w|2^{|w|+k}$ which is a contradiction.

Proposition 6.7.1. *Let B be a finite alphabet and $w \in B^*$. Then $L_{w,\pi}$ is regular if and only if w contains at most two distinct letters.*

Proof. By Theorem 6.7.3, if w is a word containing at least three distinct letters, then $L_{w,\pi}$ is not a regular language. Hence, if $L_{w,\pi}$ is regular, then card(alph(w)) ≤ 2. Conversely, suppose that $d = \text{card}(\text{alph}(w)) \leq 2$. If $d = 0$, then $w = \epsilon$ and $L_{w,\pi} = \{\epsilon\}$ is regular. If $d = 1$, then $w \in a^*$ with $a \in B$ and $L_{w,\pi} = a^{|w|}a^*$ is regular. If $d = 2$, since $L_{w,\pi}$ is closed with respect to $\Rightarrow_\pi^*$, then from Corollary 6.7.3 the result follows.

References

1. S. I. Adjan, *The Burnside Problem and Identities in Groups.* Springer-Verlag, Berlin, 1978
2. S. I. Adjan and I. G. Lysionok, The method of classification of periodic words and the Burnside problem. *Contemporary Mathematics* **131**, 13–28 (1992)
3. J. Almeida, *Finite Semigroups and Universal Algebra.* World Scientific, Singapore, 1995
4. A. V. Anisimov, Group languages. *Kibernetika* **4**, 18–24 (1971)
5. G. Baron and F. Urbanek, Factorial languages with quadratically upper-bounded growth functions and nonlinearly upper-bounded subword complexities. *Information Processing Letters* **32**, 267–269 (1989)
6. D. B. Bean, A. E. Ehrenfeucht and G. McNulty, Avoidable patterns in strings of symbols. *Pacific J. Math.* **85**, 261–294 (1979)
7. J. Berstel, *Transductions and Context-Free Languages.* Teubner, Stuttgart, 1979
8. J. Berstel and D. Perrin, *Theory of Codes.* Academic Press, New York, 1985
9. J. C. Birget and J. Rhodes, Almost finite expansions of arbitrary semigroups. *J. Pure Appl. Algebra* **32**, 239–287 (1984)
10. R. D. Blyth, Rewriting products of group elements. Ph.D. Thesis, 1987, University of Illinois at Urbana-Champaign, see also Rewriting products of group elements, I. *J. Algebra* **116**, 506–521 (1988)
11. R. D. Blyth and A. H. Rhemtulla, Rewritable products in FC-by-finite groups. *Canad. J. Math.* **41**, 369–384 (1989)
12. L. Boasson, Un critère de rationalité des langages algébriques. In: M. Nivat (ed.), *Automata, Languages and Programming*, pp. 359–365, North-Holland, Amsterdam, 1973
13. D. P. Bovet and S. Varricchio, On the regularity of languages on a binary alphabet generated by copying systems. *Information Processing Letters* **44**, 119–123 (1992)
14. S. Brlek, Enumeration of factors in the Thue–Morse word. *Discrete Applied Mathematics* **24**, 83–96 (1989)
15. T. C. Brown, A semigroup union of disjoint locally finite subsemigroups which is not locally finite. *Pacific J. of Math.* **22**, 11–14 (1967)
16. T. C. Brown, An interesting combinatorial method in the theory of locally finite semigroups. *Pacific J. of Math.* **36**, 285–289 (1971)
17. T. C. Brown, Cancellation in semigroups in which $x^2 = x^3$. *Semigroup Forum* **41**, 49–53 (1990)
18. J. Brzozowski, K. Culik II and A. Gabriellan, Classification of noncounting events. *J. Comput. System Sci.* **5**, 41–53 (1971)
19. J. Brzozowski, Open problems about regular languages, In: R. V. Book (ed.), *Formal Language Theory, Perspectives and Open Problems*, pp. 23–45, Academic Press, London, 1980

20. A. Carpi and A. de Luca, Square-free words in partially commutative free monoids. *Information Processing Letters* **22**, 125–131 (1986)
21. P. Cartier and D. Foata, Problèmes combinatoires et réarrangements. *Lecture Notes in Mathematics*, Vol. 85, Springer-Verlag, Berlin, 1969
22. A. H. Clifford and G. B. Preston, *The Algebraic Theory of Semigroups*. American Mathematical Society, Providence, RI, Vol. 1, 1961, Vol. 2, 1967
23. P. M. Cohn, *Universal Algebra*. Harper and Row, New York, 1965
24. M. Coudrain and M.P. Schützenberger, Une condition de finitude des monoïdes finiment engendrés. *C.R. Acad. Sc. Paris, Sér. A*, **262**, 1149–1151 (1966)
25. E. M. Coven and G. Hedlund, Sequences with minimal block growth. *Mathematical Systems Theory* **7**, 138–153 (1973)
26. M. Curzio, P. Longobardi and M. Maj, Su di un problema combinatorio in teoria dei gruppi. *Atti Acc. Lincei Rend. fis. VIII* **74**, 136–142 (1983)
27. M. Curzio, P. Longobardi, M. Maj and D. J. S. Robinson, A permutational property of groups. *Arch. math.* **44**, 385–389 (1985)
28. F. D'Alessandro, A note on the multiplicative group of a division ring. *Internat. J. Algebra Comput.* **7**, 51–53 (1997)
29. F. M. Dekking, Strongly non-repetitive sequences and progression-free sets. *J. Combin. Theory, Ser. A*, **27**, 181–185 (1979)
30. V. Diekert and Y. Métivier, Commutations and traces. In: G. Rozenberg and A. Salomaa (ed.s), *Handbook on Formal Languages*, Vol. 3, pp. 457–534, Springer-Verlag, Berlin, 1997
31. A. de Luca, A note on the Burnside problem for semigroups. *Semigroup Forum* **31**, 251–254 (1985)
32. A. de Luca, A division property of the Fibonacci word. *Information Processing Letters* **54**, 307–312 (1995)
33. A. de Luca and F. Mignosi, Some combinatorial properties of Sturmian words. *Theoretical Computer Science* **136**, 361–385 (1994)
34. A. de Luca and A. Restivo, A finiteness condition for finitely generated semigroups. *Semigroup Forum* **28**, 123–134 (1984)
35. A. de Luca, A. Restivo and S. Salemi, On the centers of a languages. *Theoretical Computer Science* **24**, 21–34 (1983)
36. A. de Luca and S. Varricchio, On the factors of the Thue–Morse word on three symbols. *Information Processing Letters* **27**, 281–285 (1988)
37. A. de Luca and S. Varricchio, Some combinatorial properties of the Thue–Morse sequence and a problem in semigroups. *Theoretical Computer Science* **63**, 333–348 (1989)
38. A. de Luca and S. Varricchio, Factorial languages whose growth function is quadratically upper-bounded. *Information Processing Letters* **30**, 283–288 (1989)
39. A. de Luca and S. Varricchio, Some combinatorial properties of factorial languages. In: R. Capocelli (ed.), *Sequences, Combinatorics, Compression, Security, and Transmission*, pp. 258–266, Springer-Verlag, New York, 1990
40. A. de Luca and S. Varricchio, On noncounting regular classes. *Lecture Notes in Computer Science*, Vol. 443, pp. 74–87, Springer-Verlag, Berlin, 1990
41. A. de Luca and S. Varricchio, A note on ω-permutable semigroups. *Semigroup Forum* **40**, 153–157 (1990)
42. A. de Luca and S. Varricchio, A combinatorial theorem on p-power-free words and an application to semigroups. *RAIRO, Inform. Théor. Appl.* **24**, 205–228 (1990)
43. A. de Luca and S. Varricchio, A finiteness condition for semigroups generalizing a theorem of Hotzel. *J. Algebra* **136**, 60–72 (1991)

44. A. de Luca and S. Varricchio, Finiteness and iteration conditions for semigroups. *Theoretical Computer Science* **87**, 315–327 (1991)
45. A. de Luca and S. Varricchio, Combinatorial properties of uniformly recurrent words and an application to semigroups. *Internat. J. Algebra Comput.* **1**, 227–245 (1991)
46. A. de Luca and S. Varricchio, On noncounting regular classes. *Theoretical Computer Science* **100**, 67–102 (1992)
47. A. de Luca and S. Varricchio, On finitely recognizable semigroups. *Acta Informatica* **29**, 483–498 (1992)
48. A. de Luca and S. Varricchio, Some regularity conditions based on well quasi-orders. *Lecture Notes in Computer Science*, Vol. 583, pp. 356–371, Springer-Verlag, Berlin, 1992
49. A. de Luca and S. Varricchio, A new unavoidable regularity in free monoids. In: R. Capocelli, A. De Santis, and U. Vaccaro (ed.s), *Sequences II, Methods in Communication, Security, and Computer Science*, pp. 447–456, Springer-Verlag, New York, 1993
50. A. de Luca and S. Varricchio, On a conjecture of Brown. *Semigroup Forum* **46**, 116–119 (1993)
51. A. de Luca and S. Varricchio, A finiteness condition for semigroups. In: C. Bonzini, A. Cherubini, and C. Tibiletti (ed.s), *Semigroups, Algebraic Theory and Applications to Formal Languages and Codes*, pp. 42–50, World Scientific, Singapore, 1993
52. A. de Luca and S. Varricchio, Well quasi-orders and regular languages. *Acta Informatica* **31**, 539–557 (1994)
53. A. de Luca and S. Varricchio, A finiteness condition for semigroups generalizing a theorem of Coudrain and Schützenberger. *Advances in Mathematics* **108**, 91–103 (1994)
54. A. de Luca and S. Varricchio, Regularity and finiteness conditions. In: G. Rozenberg and A. Salomaa (ed.s), *Handbook on Formal Languages*, Vol. 1, pp. 747–810, Springer-Verlag, Berlin, 1997
55. F. Di Cerbo, *Sul Problema di Burnside per i semigruppi.* Tesi Università di Napoli, 1985
56. A. P. do Lago, On the Burnside semigroups $x^n = x^{n+m}$. *Lecture Notes in Computer Science*, Vol. 583, pp. 329–355, Springer-Verlag, Berlin, 1992
57. A. P. do Lago, On the Burnside semigroups $x^n = x^{n+m}$. *Internat. J. Algebra Comput.* **6**, 179–227 (1996)
58. A. P. do Lago, Maximal groups in free Burnside semigroups. *Lecture Notes in Computer Science*, Vol. 1380, pp. 65–75, Springer-Verlag, Berlin, 1998
59. A. Ehrenfeucht, R. Parikh and G. Rozenberg, Pumping lemmas for regular sets. *SIAM J. Comput.* **10**, 536–541 (1981)
60. A. Ehrenfeucht and G. Rozenberg, On regularity of languages generated by copying systems. *Discrete Applied Mathematics* **8**, 313–317 (1984)
61. A. Ehrenfeucht and G. Rozenberg, Strong iterative pairs and the regularity of context-free languages. *RAIRO, Inform. Théor. Appl.* **19**, 43–56 (1985)
62. S. Eilenberg, *Automata, Languages, and Machines.* Academic Press, New York, vol. A, 1974
63. T. Evans, Some connections between residual finiteness, finite embeddability and the word problem. *J. London Math. Soc.* **1**, 399–403 (1969)
64. N. J. Fine and H. S. Wilf, Uniqueness theorem for periodic functions. *Proc. Am. Math. Soc.* **16**, 109–114 (1965)
65. H. Furstenberg, *Recurrence in Ergodic Theory and Combinatorial Number Theory.* Princeton University Press, Princeton, 1981

66. M. Garzon and Y. Zalcstein, On permutation properties in groups and semigroups. *Semigroup Forum* **35**, 337–351 (1987)
67. M. Garzon and Y. Zalcstein, Linear semigroups with permutation properties. *Semigroup Forum* **35**, 369–371 (1987)
68. S. Ginsburg and E. H. Spanier, Bounded Algol-like languages. *Trans. Amer. Math. Soc.* **113**, 333–368 (1964)
69. E. S. Golod, On nil-algebras and finitely approximable p-groups. *Izv. Akad. Nauk SSSR, Ser.Mat.* **28**, 273–276 (1964)
70. R. Graham, B. L. Rothschild and J. H. Spencer, *Ramsey Theory*. J. Wiley, New York, 2-nd edition, 1990
71. J. A. Green and D. Rees, On semigroups in which $x^r = x$. *Proc. Cambridge Philos. Soc.* **48**, 35–40 (1952)
72. R. I. Grigorchuk, Burnside's problem on periodic groups. *Funktsional'nyi Analiz i Ego Prilozheniya* **14**, 53–54 (1980)
73. M. Gromov, Groups of polynomial growth and expanding maps. *Publ. Math. IHES* **53**, 53–73 (1981)
74. V. Guba, The word problem for the relatively free semigroups satisfying $T^m = T^{m+n}$ with $m \geq 3$. *Internat. J. Algebra Comput.* **3**, 335–348 (1993)
75. N. Gupta and S. Sidki, On the Burnside problem for periodic groups. *Math. Z.* **182**, 385–388 (1983)
76. M. Hall, *The Theory of Groups*. Chelsea, New York, 1976
77. K. Hashiguchi, Notes on finitely generated semigroups and pumping conditions for regular languages. *Theoretical Computer Science* **46**, 53–66 (1986)
78. D. Haussler, Another generalization of Higman's well quasi order result on Σ^*. *Discrete Mathematics* **57**, 237–243 (1985)
79. F. C. Hennie, One-tape off line Turing machine computations. *Information and Control* **8**, 553–578 (1965)
80. G. H. Higman, Ordering by divisibility in abstract algebras. *Proc. London Math. Soc.* **3**, 326–336 (1952)
81. E. Hotzel, On finiteness conditions in semigroups. *J. Algebra* **60**, 352–370 (1979)
82. S. V. Ivanov, On the Burnside problem on periodic groups. *Bull. Amer. Math. Soc.* **27**, 257–260 (1992)
83. G. Jacob, La finitude de représentations linéaires des semi-groupes est décidable. *J. Algebra* **52**, 437–459 (1978)
84. N. Jacobson, *Structure of Rings*. American Mathematical Society, Providence, RI, 1964
85. M. Jantzen, *Confluent String Rewriting*. EATCS Monographs on Theoretical Computer Science, Vol. 14, Springer-Verlag, Berlin, 1988
86. J. Justin, Groupes et semi-groupes à croissance linéaire. *C.R. Acad. Sc. Paris, Sér. A*, **273**, 212–214 (1971)
87. J. Justin, Généralisation du théorème de van der Waerden sur les semigroupes répétitifs. *J. Combin. Theory, Ser. A*, **12**, 357–367 (1972)
88. J. Justin, Propriétés combinatoires de partitions finies du demi-groupe libre. In: *Séminaire d'Informatique Théorique Année 1981–82*, pp. 55–66, LITP, Univ. Paris VII
89. J. Justin and G. Pirillo, On a natural extension of Jacob's ranks. *J. Combin. Theory, Ser. A*, **43**, 205–218 (1986)
90. J. Justin and G. Pirillo, A finiteness condition for semigroups generated by a finite set of elements of finite order, *PU.M.A., Ser. A*, **1**, 45–48 (1990)
91. J. Justin and G. Pirillo, Shirshov's theorem and ω-permutability of semi-groups. *Advances in Mathematics* **87**, 151–159 (1991)
92. J. Justin and G. Pirillo, On a combinatorial property of Sturmian words. *Theoretical Computer Science* **154**, 387–394 (1996)

93. I. Kaplansky, *Fields and Rings*. The University of Chicago Press, Chicago, 2nd ed., 1972
94. M. Kargapolov and Iou. Merzliakov, *Eléments de la théorie des groupes*. Editions MIR-Moscou, 1985
95. J. Karhumaki, On cube-free ω-words generated by binary morphisms. *Discrete Applied Mathematics* **5**, 279–297 (1983)
96. O. G. Kharlampovich and M. V. Sapir, Algorithmic Problems in Varieties. *Internat. J. Algebra Comput.* **5**, 379–602 (1995)
97. A. Y. Khinchin, *Three Pearls of Number Theory*. Graylock Press, Baltimore, 1952
98. Y. Kobayashi, Enumeration of irreductible binary words. *Discrete Applied Mathematics* **20**, 221–232 (1988)
99. J. Kortelainen, Every commutative quasi-rational language is regular. *RAIRO, Inform. Théor. Appl.* **20**, 319–337 (1986)
100. J. Kruskal, The theory of well-quasi-ordering: a frequently discovered concept. *J. Combin. Theory, Ser. A*, **13**, 297–305 (1972)
101. G. Lallement, *Semigroups and Combinatorial Applications*. J. Wiley, New York, 1979
102. M. Lanzillotta, *La proprietà di iterazione nella teoria dei linguaggi formali*. Tesi Università di Roma "La Sapienza", 1990
103. M. Latteux and G. Rozenberg, Commutative one-counter languages are regular. *J. Comput. and System Sci.* **29**, 54–57 (1984)
104. M. Lothaire, *Combinatorics on Words*. Encyclopedia of Mathematics and its Applications, Vol. 17, Addison-Wesley, Reading, MA, 1983. Reprinted in the *Cambridge Mathematical Library*, Cambridge University Press, Cambridge, 1997
105. M. Lothaire, *Algebraic Combinatorics on Words*. Cambridge University Press, Cambridge, to appear
106. R. C. Lyndon and P. E. Schupp, *Combinatorial Group Theory*. Springer-Verlag, Berlin, 1977
107. R. C. Lyndon and M. P. Schützenberger, The equation $a^m = b^n c^p$ in a free group. *Michigan Math. J.* **9**, 289–298 (1962)
108. A. Machì, Growth functions and permutation property of groups. *Preprint LITP 89-93*, Univ. Paris VII, 1989
109. W. Magnus, A. Karrass and D. Solitar, *Combinatorial Group Theory*. Dover Publ., New York, 1976
110. A. Mandel and I. Simon, On finite semigroups of matrices. *Theoretical Computer Science* **5**, 101–112 (1977)
111. J. McCammond, On the solution of the word problem for the semigroups satisfying $T^a = T^{a+b}$ with $a \geq 6$. *Internat. J. Algebra Comput.* **1**, 1–32 (1991)
112. R. McNaughton and S. Papert, *Counter-free Automata*. The M. I. T. Press, Cambrige, MA, 1971
113. R. McNaughton and Y. Zalcstein, The Burnside problem for semigroups of matrices. *J. Algebra* **34**, 292–299 (1975)
114. F. Migliorini and J. Szép, On finitely generated semigroups with permutation property. *PU. M. A., Ser. A*, **1**, 33–37 (1990)
115. F. Mignosi, Infinite words with linear subword complexity. *Theoretical Computer Science* **65**, 221–242 (1989)
116. F. Mignosi, On the number of factors of Sturmian words. *Theoretical Computer Science* **82**, 71–84 (1991)
117. F. Mignosi and A. Restivo, Permutation properties and subword complexity. *Preprint Dipartimento di Matematica*, Università di Palermo, 1990

118. M. Morse, Recurrent geodesics on a surface of negative curvature. *Trans. Amer. Math. Soc.* **22**, 84–100 (1921)
119. M. Morse and G. Hedlund, Symbolic Dynamics II: Sturmian trajectories. *Amer. J. Math.* **62**, 1–42 (1940)
120. M. Morse and G. Hedlund, Unending chess, symbolic dynamics and a problem in semigroups. *Duke Math. J.* **11**, 1–7 (1944)
121. W. D. Munn, Free inverse semigroups. *Proc. London Math. Soc.* **29**, 385–404 (1974)
122. I. Niven, *Irrational Numbers.* Carus Monographs, No. 11, J. Wiley, New York, 1956
123. J. Okniński, *Semigroup Algebras.* M. Dekker, New York, 1990
124. J. Okniński and M. S. Putcha, PI semigroup algebras of linear semigroups. *Proc. Amer. Math. Soc.* **109**, 39–46 (1990)
125. P. Perkins, Bases for equational theories of semigroups. *J. Algebra* **11**, 298–314 (1968)
126. M. Petrich, *Introduction to Semigroups.* Merrill, Columbus, OH, 1973
127. G. Pirillo, On permutation properties for finitely generated semigroups. *Ann. Discrete Math.*, Vol. 37, pp. 375–376, North-Holland, Amsterdam, 1988
128. G. Pirillo, Sur les produits directs de semi-groupes répétitifs. *C.R. Acad. Sc. Paris, Sér. I Math.*, **314**, 645–648 (1992)
129. G. Pirillo and S. Varricchio, On uniformly repetitive semigroups. *Semigroup Forum* **49**, 125–129 (1994)
130. G. Pirillo and S. Varricchio, Some combinatorial properties of infinite words and applications to semigroup theory. *Discrete Mathematics* **153**, 239–251 (1996)
131. C. Procesi, The Burnside problem. *J. Algebra* **4**, 421–425 (1966)
132. C. Procesi, *Rings with Polynomials Identities.* M. Dekker, New York, 1973
133. M. E. Prouhet, Mémoire sur quelques relations entre les puissances des nombres. *C.R. Acad. Sci. Paris, Sér. A,* **33**, 225 (1851)
134. F. P. Ramsey, On a problem of formal logic. *Proc. London Math. Soc.*, 2nd ser., **30**, 264–286 (1930)
135. A. Restivo, Permutation properties and the Fibonacci semigroup. *Semigroup Forum* **38**, 337–345 (1989)
136. A. Restivo and C. Reutenauer, Some applications of a theorem of Shirshov to language theory. *Information and Control* **57**, 205–213 (1983)
137. A. Restivo and C. Reutenauer, On the Burnside problem for semigroups. *J. Algebra* **89**, 102–104 (1984)
138. A. Restivo and C. Reutenauer, On cancellation properties of languages which are supports of rational power series. *J. Comput. System Sci.* **29**, 153–159 (1984)
139. A. Restivo and C. Reutenauer, Rational languages and the Burnside problem. *Theoretical Computer Science* **40**, 13–30 (1985)
140. C. Reutenauer, Sur les semigroupes vérifiant le théorème de Kleene. *RAIRO Inform. Théor. Appl.* **19**, 281–291 (1985)
141. C. Reutenauer, Mots de Lyndon et un théorème de Shirshov. *Ann. Sc. Math. Québec* **10**, 237–245 (1986)
142. L. Rowen, *Polynomial Identities in Ring Theory.* Academic Press, New York, 1980
143. J. Sakarovitch, Easy multiplications I. The realm of Kleene's theorem. *Inform. and Comput.* **74**, 173–197 (1987)
144. I. Schur, Über Gruppen periodischer Substitutionen. *Sitzungsber. Preuss. Akad. Wiss.*, 619–627 (1911)
145. A. I. Shirshov, On certain non associative nil rings and algebraic algebras. *Mat. Sb.* **41**, 381–394 (1957)

146. A. I. Shirshov, On rings with identity relations. *Mat. Sb.* **43**, 277–283 (1957)
147. I. Simon, Notes on noncounting languages of order 2. *Manuscript*, 1970
148. I. Simon, Piecewise testable events. *Lecture Notes in Computer Science*, Vol. 33, pp. 214–222, Springer-Verlag, Berlin, 1975
149. I. Simon, Conditions de finitude pour des semi-groupes. *C.R. Acad. Sci. Paris, Sér. A*, **290**, 1081–1082 (1980)
150. I. Simon, Infinite words and a theorem of Hindman. *Rev. Mat. Apl.* **9**, 97–104 (1988)
151. I. Simon, Properties of factorization forests. *Lecture Notes in Computer Science*, vol. 386, pp. 65–72, Springer-Verlag, Berlin, 1989
152. H. Straubing, The Burnside problem for semigroups of matrices. In: L. J. Cummings (ed.), *Combinatorics on words, Progress and Perspectives*, pp. 279–285, Academic Press, New York, 1983
153. A. Thue, Über unendliche Zeichenreihen. *Norske Vidensk. Selsk. Skrifter. I. Mat.-Nat. K1., Christiania* Nr 7, 1–22 (1906)
154. A. Thue, Über die gegenseitige Lage gleicher Teile gewisser Zeichenreihen. *Norske Vidensk. Selsk. Skrifter. I. Mat.-Nat. K1., Christiania* Nr 10, 1–67 (1912)
155. B. L. van der Waerden, Beweis einer Baudet'schen Vermutung. *Nieuw Arch. Wisk.* **15**, 212–216 (1927)
156. B. L. van der Waerden, Wie der Beweis der Vermutung von Baudet gefunden wurde. *Abhandlungen des Mathematischen Seminars der Hanseatischen Universität Hamburg*, 6–15 (1965), also published as: [How the proof of Baudet's conjecture was found. *Studies in Pure Mathematics*, pp. 251–260, Academic Press, New York, 1971]
157. S. Varricchio, A finiteness condition for finitely generated semigroups. *Semigroup Forum* **38**, 331–335 (1989)
158. S. Varricchio, Factorizations of free monoids and unavoidable regularities. *Theoretical Computer Science* **73**, 81–89 (1990)
159. S. Varricchio, A pumping condition for regular sets. *SIAM J. Comput.* **26**, 764–771 (1997)
160. A. I. Zimin, Blocking sets of terms. *Matem. Sbornik* **119**, 363–375 (1982)

Index

Monographs in Theoretical Computer Science · An EATCS Series

C. Calude
Information and Randomness
An Algorithmic Perspective

K. Jensen
Coloured Petri Nets
***Basic Concepts,* Analysis Methods and Practical Use, Vol. 1**
2nd ed.

K. Jensen
Coloured Petri Nets
Basic Concepts, *Analysis Methods* and Practical Use, Vol. 2

K. Jensen
Coloured Petri Nets
Basic Concepts, Analysis Methods and *Practical Use,* Vol. 3

A. Nait Abdallah
The Logic of Partial Information

Z. Fülöp, H. Vogler
Syntax-Directed Semantics
Formal Models Based on Tree Transducers

A. de Luca, S. Varricchio
Finiteness and Regularity in Semigroups and Formal Languages

Texts in Theoretical Computer Science · An EATCS Series

J. L. Balcázar, J. Díaz, J. Gabarró
Structural Complexity I
2nd ed. (see also overleaf, Vol. 22)

M. Garzon
Models of Massive Parallelism
Analysis of Cellular Automata and Neural Networks

J. Hromkovič
Communication Complexity and Parallel Computing

A. Leitsch
The Resolution Calculus

G. Păun, G. Rozenberg, A. Salomaa
DNA Computing
New Computing Paradigms

A. Salomaa
Public-Key Cryptography
2nd ed.

K. Sikkel
Parsing Schemata
A Framework for Specification and Analysis of Parsing Algorithms

Former volumes appeared as EATCS Monographs on Theoretical Computer Science

Vol. 5: W. Kuich, A. Salomaa
Semirings, Automata, Languages

Vol. 6: H. Ehrig, B. Mahr
Fundamentals of Algebraic Specification 1
Equations and Initial Semantics

Vol. 7: F. Gécseg
Products of Automata

Vol. 8: F. Kröger
Temporal Logic of Programs

Vol. 9: K. Weihrauch
Computability

Vol. 10: H. Edelsbrunner
Algorithms in Combinatorial Geometry

Vol. 12: J. Berstel, C. Reutenauer
Rational Series and Their Languages

Vol. 13: E. Best, C. Fernández C.
Nonsequential Processes
A Petri Net View

Vol. 14: M. Jantzen
Confluent String Rewriting

Vol. 15: S. Sippu, E. Soisalon-Soininen
Parsing Theory
Volume I: Languages and Parsing

Vol. 16: P. Padawitz
Computing in Horn Clause Theories

Vol. 17: J. Paredaens, P. DeBra, M. Gyssens, D. Van Gucht
The Structure of the Relational Database Model

Vol. 18: J. Dassow, G. Páun
Regulated Rewriting in Formal Language Theory

Vol. 19: M. Tofte
Compiler Generators
What they can do, what they might do, and what they will probably never do

Vol. 20: S. Sippu, E. Soisalon-Soininen
Parsing Theory
Volume II: LR(k) and LL(k) Parsing

Vol. 21: H. Ehrig, B. Mahr
Fundamentals of Algebraic Specification 2
Module Specifications and Constraints

Vol. 22: J. L. Balcázar, J. Díaz, J. Gabarró
Structural Complexity II

Vol. 24: T. Gergely, L. Úry
First-Order Programming Theories

R. Janicki, P. E. Lauer
Specification and Analysis of Concurrent Systems
The COSY Approach

O. Watanabe (Ed.)
Kolmogorov Complexity and Computational Complexity

G. Schmidt, Th. Ströhlein
Relations and Graphs
Discrete Mathematics for Computer Scientists

S. L. Bloom, Z. Ésik
Iteration Theories
The Equational Logic of Iterative Processes

MIX
Papier aus verantwortungsvollen Quellen
Paper from responsible sources
FSC® C105338

If you have any concerns about our products,
you can contact us on
ProductSafety@springernature.com

In case Publisher is established outside the EU,
the EU authorized representative is:
Springer Nature Customer Service Center GmbH
Europaplatz 3, 69115 Heidelberg, Germany

Printed by Libri Plureos GmbH
in Hamburg, Germany